5th International Workshop on Nuclear Data Evaluation for Reactor Applications (WONDER-2018)

EPJ Web of Conferences Volume 211 (2019)

Aiz-en-Provence, France
8 - 12 October 2018

ISBN: 978-1-5108-8830-2

Printed from e-media with permission by:

Curran Associates, Inc.
57 Morehouse Lane
Red Hook, NY 12571

Some format issues inherent in the e-media version may also appear in this print version.

For additional information, please contact EDP Sciences – Web of Conferences
at the address below.

EDP Sciences – Web of Conferences
17, Avenue du Hoggar
Parc d'Activité de Courtabœuf
BP 112
F-91944 Les Ulis Cedex A
France

Phone: +33 (0) 1 69 18 75 75
Fax: +33 (0) 1 69 28 84 91

contact-edps@webofconferences.org

Additional copies of this publication are available from:

Curran Associates, Inc.
57 Morehouse Lane
Red Hook, NY 12571 USA
Phone: 845-758-0400
Fax: 845-758-2633
Email: curran@proceedings.com
Web: www.proceedings.com

TABLE OF CONTENTS

edp sciences

Web of Conferences

Statement of Peer review

In submitting conference proceedings to *Web of Conferences*, the editors of the proceedings certify to the Publisher that

1. They adhere to its **Policy on Publishing Integrity** in order to safeguard good scientific practice in publishing.
2. All articles have been subjected to peer review administered by the proceedings editors.
3. Reviews have been conducted by expert referees, who have been requested to provide unbiased and constructive comments aimed, whenever possible, at improving the work.
4. Proceedings editors have taken all reasonable steps to ensure the quality of the materials they publish and their decision to accept or reject a paper for publication has been based only on the merits of the work and the relevance to the journal.

Title, date and place of the conference

WONDER - 2018
8th - 12th of October, 2018
Aix en Provence (FRANCE)

Proceedings editor(s):

Dr Olivier SEROT
Dr Abdelhazize CHEBBOUBI

Date and editor's signature

Preface

WONDER-2018 is the fifth International Workshop on Nuclear Data Evaluation for Reactor Applications (the first four workshops were organised respectively in 2006, 2009, 2012 and 2015). It was held on October 8th to 12th, 2018 at the Hotel du Roi René in Aix-en-Provence (France).

About 60 experts in the field of nuclear data were participating at this workshop. The main objective was to review the current modelling and evaluation methods of nuclear data for reactor applications (GEN. II, III, IV) and to discuss possible areas of improvements. Presentations were organised in sessions on: Integral and microscopic measurements, Evaluation of nuclear data (theory, models, codes), Decay data and delayed neutrons, Thermal diffusion laws, Data processing and benchmarking, Nuclear fission (prompt particles emission, fission yields ...), Uncertainties and covariance matrices (methodology and impact on reactor calculations).

We thank the French Alternative Energies and Atomic Energy Commission (CEA), the Nuclear Energy Agency (NEA) and La Communauté du Pays d'Aix (CPA) for their active participation in the organisation of this workshop. Lastly, we would like to warmly thank Mrs Claire Boucher who assumed the secretariat of the workshop.

Taking into account the undeniable success of WONDER-2018, the 6th edition could be born in 2021!

January 2019,
The organizing committee

EPJ Web of Conferences **211**, 01001 (2019) https://doi.org/10.1051/epjconf/201921101001
WONDER-2018

Summation Calculations for Reactor Antineutrino Spectra, Decay Heat and Delayed Neutron Fractions Involving New TAGS Data and Evaluated Databases

M. Estienne[1,*], *M.* Fallot[1], *L.* Giot[1], *V.* Guadilla-Gomez[1], *L.* Le Meur[1], *A.* Porta[1], *A.* Algora[2,13], *J. -L.* Taìn[2], *J. A.* Briz[1], *J.* Agramunt[2], *J.* Äystö[4], *S.* Cormon[1], *A.* Cucoanes[1], *T.* Eronen[7], *L. M.* Fraile[15], *E.* Ganogliu[17], *W.* Gelletly[2,3], *D.* Gorelov[7], *J.* Hakala[7], *A.* Jokinen[7], *M. D.* Jordan[2], *A.* Kankainen[7], *V. S.* Kolhinen[7], *J.* Koponen[7], *M.* Lebois[18], *T.* Martinez[6], *M.* Monserrate[2], *A.* Montaner-Pizá[2], *I.* Moore[7], *E.* Nácher[19], *S. E. A.* Orrigo[2], *H.* Penttilä[7], *I.* Pohjalainen[7], *J.* Reinikainen[7], *M.* Reponen[7,10], *S.* Rinta-Antila[7], *K.* Rytkönen, *B.* Rubio[2], *T.* Shiba[1], *V.* Sonnenschein[7], *A. A.* Sonzogni[11], *E.* Valencia[2], *V.* Vedia[15], *A.* Voss[7], *C.* Weber[7,12], *J. N.* Wilson[18], and *A. -A.* Zakari-Issoufou[1]

[1]Subatech, CNRS/IN2P3, 4 Rue Alfred Kastler, 44307 Nantes, France
[2]IFIC (CSIC-Univ. Valencia), Valencia, Spain
[3]Department of Physics, University of Surrey, Guildford, GUZ7XH, UK
[4]Helsinki Institute of Physics, FI-00014 University of Helsinki, Finland
[5]Universitat Politécnica de Catalunya (UPC), 08034, Barcelona, Spain
[6]CIEMAT, Madrid, Spain
[7]Department of Physics, P.O. Box 35, FI-40014 University of Jyväskylä, Finland
[8]Argonne National Laboratory, Argonne, USA
[9]National Physical Laboratory, Teddington, Middlesex, TW11 0LW, UK
[10]RIKEN, 2-1 Hirosawa, Wako, Saitama 351-0198, Japan
[11]National Nuclear Data Center, Brookhaven National Laboratory, Upton NY 11973-5000, USA
[12]Faculty of Physics, Ludwig-Maximilians University Munich, Am Coulombwall 1, D-85748 Garching, Germany
[13]MTA ATOMKI, Hungary
[14]Petersburg Nuclear Physics Institute, RU-188300 Gatchina, Russia
[15]Universidad Complutense, Grupo de Física Nuclear, CEI Moncloa, E-28040, Madrid, Spain
[16]United Kingdom Atomic Energy Authority, Culham Science Centre, Abingdon, OX14 3DB, UK
[17]Department of Physics, Istanbul University, 34134, Istanbul, Turkey
[18]Institut de Physique Nucléaire d'Orsay, CNRS/IN2P3, F-91406 Orsay, France
[19]Instituto de Estructura de la Materia, CSIC, E-28006, Madrid, Spain

Abstract. Three observables of interest for present and future reactors depend on the β decay properties of the fission products: antineutrinos from reactors, the reactor decay heat and delayed neutron emission. In these proceedings, we present new results from summation calculations of the first two quantities quoted above, performed with evolved independent yields coupled with fission product decay data, from various nuclear data bases or models. New TAGS results from the latest experiment of the TAGS collaboration at the JYFL facility of Jyväskylä will be displayed as well as their impact on the antineutrino spectra and the decay heat associated to fission pulses of the main actinides.

*e-mail: magali.estienne@subatech.in2p3.fr

1 Introduction

The study of β decay is useful to better constrain different domains of physics achievable close to reactors. Its implications for present and future reactors concern the physics of antineutrinos and their applications, the decay heat (DH) understanding and control and the access to the delayed neutron fractions. They are essential respectively for fundamental physics (neutrino oscillation, mass hierarchy and reactor anomaly), reactor monitoring and non proliferation, reactor safety and economy as concerns DH and the operation and control of the chain reactions in reactors. Indeed, in a PWR the thermal power is mainly induced by the fission process undergone by the four isotopes ^{235}U, ^{239}Pu, ^{241}Pu and ^{238}U. Fission products (FP) are released in such process, at the origin of the β decays as they are neutron rich nuclei. In such radioactive processes, a neutron from the parent nucleus converts into a proton in the daughter nucleus and emits an electron and an electron antineutrino. Gamma-rays are also emitted in the de-excitation of the produced daughter nucleus. In the specific case of the production of the daughter in an excited state above the neutron energy threshold, a neutron can be emitted before the de-excitation of a second daughter nucleus. This is known as a β-n decay associated to the emission of a delayed neutron. Whether the γ and β particles, the β delayed neutrons or the neutrinos, their independent measurement allows to assess β decay physics properties and applications. In these proceedings, we will essentially focus on the physics that can be addressed via the γ channel.

In the next section, experimental issues associated to γ measurement will be introduced before to present the Total Absorption Gamma-ray Spectroscopy (TAGS) technique used to overcome experimental biases. Two of the 3 introduced β-decay applications will be commented below: antineutrinos from reactors in Sect. 3 and decay heat in Sect. 4. Both aspects will be discussed in the context of the summation calculations that have been developed by the Nantes group and that will be introduced and commented in Sect. 3.2 and 4.2. The summation method is also a useful tool to give predictions for delayed neutron fractions but these aspects are not presented in the present paper. They have been the focus of a Coordinated Research Program of the IAEA attended by our group dedicated to the establishment of a new β-delayed neutron nuclear database (DB) which outcome is the object of a publication in preparation [1]. A general conclusion will be given in Sect. 5.

2 Pandemonium Effect and Total Absorption Gamma-ray Spectroscopy (TAGS)

2.1 Pandemonium Bias in Nuclear Databases

A precise measurement of the β decay probability, the β intensity I_β, is essential for the determination of the β strength which characterizes the structure properties of a nucleus. It is achievable with a clean determination of the energy levels of the daughter nucleus via the γ-cascade measurement. The experimental knowledge of the endpoints, the Q-value and the Branching Ratios (BR) of a given FP giving access to its β/anti-ν energy spectra, it becomes possible to calculate a complete β/anti-ν spectra of the 4 main fissible nuclei present in the reactor core of a PWR. The access to the mean γ and β energies of a given FP also brings a stone to add to the building of the DH determination and enriches as-well nuclear databases.

Before the 1990s, high resolution γ-ray spectroscopy through the use of germanium detectors was the conventional detection technique used for γ-ray measurement. It offers excellent energy resolution but an efficiency which strongly decreases at high energy and usually limited acceptance. As direct consequences, weak γ-ray cascades are likely difficult to detect and the decay scheme of the parent nucleus may be incomplete. There is a danger of overlooking the existence of β-feeding into the high energy nuclear levels of daughter nuclei especially with decay schemes with large Q-values. This phenomenon is called "Pandemonium effect" highlighted by Hardy in 1977 [2]. The β decay of the ^{105}Mo is a good illustration [3]. A comparison between the calculated antineutrino spectra obtained using the results of a measurement performed with high resolution technique (JEFF 3.1.1) and the TAS one (see Sect. 2.2) which corrects for that effect can be made. A $\simeq$50% discrepancy is observed in the mean energy of the distributions to be compared with the few percent uncertainty quoted in JEFF 3.1.1 DB. This example highlights the strong potential bias present in nuclear DB for nuclei suffering for Pandemonium effect not only concerning the mean energy values, level energies, etc. but also concerning their associated uncertainties. It might have subsequent consequences while using nuclear DB in β decay applications that need careful treatment.

2.2 The Total Absorption γ-ray Spectroscopy

2.2.1 TAS calorimeter and experimental campaigns

A solution proposed to solve the above-mentioned issue was to use a calorimeter containing big crystals ideally covering 4π : the Total Absorption γ-ray Spectroscopy (TAGS). The detection philosophy with respect to HPGe is totally different. Instead of detecting the γ-rays individually, a TAS is used to absorb the full γ-ray energy released by the γ cascades in the β decay process. It offers an excellent efficiency but a lower resolution than HPGE (see below). Apart from the detector inefficiency that has to be corrected for, this technique gives directly access to the energy level in the daughter nucleus. The first TAS developed in the 1970s were too small detectors to be enough efficient. The development of a TAGS method efficient and systematic appeared in the 1990s [4] and has been improved since then. TAS are now segmented allowing "multiplicity" measurements and the analysis methodology has been noticeably improved [5].

The European TAS collaboration has performed so far two experimental campaigns of data taking with two detectors built by the Valencia and Surrey groups at the IGISOL facility in Jyväskylä [7]. The first used in 2009, made of 12 BaF_2 crystals offering a detection efficiency of γ-ray cascade larger than 80%, was coupled with a silicon detector for β coincidences to reduce the background contribution. Seven nuclei of interest for both antineutrino and decay heat applications were measured during this campaign [6]. The second one in 2014, DTAS, a 18-fold segmented $NaI(Tl)$ spectrometer with rectangular crystals of 15x15x25cm^3, has also an excellent efficiency of 80-90% to detect a single γ-ray associated with an individual crystal energy resolution of 7-8%. It is coupled with a plastic detector for β coincidences. This experiment was very successful as 23 nuclei (from the Nantes/Valencia's proposals) of interest for antineutrino and DH[1] [8] were measured. Among the 15 nuclei analyzed so far some preliminary results for the ^{99}Y, ^{138}I, ^{142}Cs [9], ^{100,100m}Nb and ^{102,102m}Nb [10] nuclei are presented below.

2.2.2 Analysis: determination of the β intensity I_β

We show in the top panel of Fig. 1 preliminary results of the β-gated energy spectrum of gammas for the ^{99}Y nucleus, a priority 1 contributor to the spectrum [8]. In the figure, the locations of the Q-value and the neutron emission threshold are indicated by vertical lines. The $\beta - \gamma$ coincidences guarantee a spectrum free from environmental background, however, all other sources of contamination have to be carefully studied in the analysis. The green

[1]The Nantes group is taking part to the TAGS consultant meetings of the IAEA dedicated to the elaboration of the priority list of Pandemonium nuclei that need to be measured.

line represents the contribution of the pile-up signals to the data extracted from a MC study, the blue one the contamination from the daughter ^{99}Zr nucleus determined solving the Bateman equations and the purple line the contamination from β-delayed neutrons. The red line is the result of the reconstruction of the TAGS data with the feeding found with the Bayes method [11]. It is already in very good agreement with the data shown in black. The resulting preliminary I_β distribution (without the β-n contribution) is shown in the bottom panel of Fig. 1 where the feeding is compared with the ENSDF database containing experimental results from high-resolution measurements. The discrepancy between the two feedings highlights how the Pandemonium problem biases high-resolution data and consequently nuclear databases. The shaded areas reflect the envelop of systematic errors obtained in varying the input ingredients entering in the data analysis.

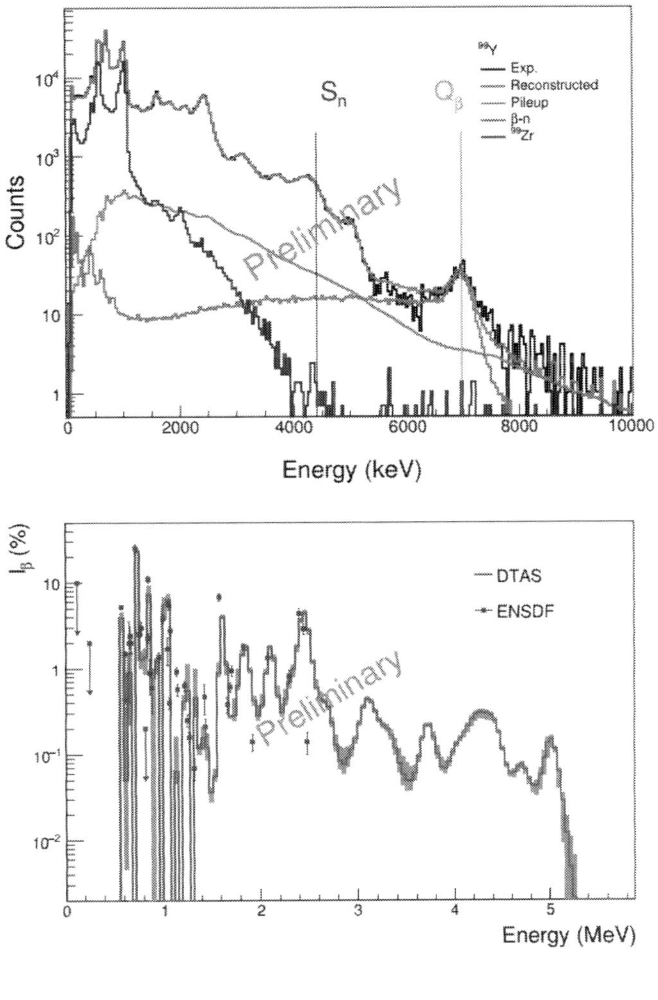

Figure 1. Top: Experimental TAS spectrum of ^{99}Y showing also contaminants (see text) and the reconstruction obtained after the application of the Bayes method (red line). Bottom: Beta intensity of the beta decay of ^{99}Y (see text).

3 Reactor Antineutrinos

3.1 Physics motivations

β decays of FP are responsible for the huge flux of antineutrinos emitted close to reactors. Taking into account that ~6 antineutrinos are emitted per fission[2], a PWR of $1GWe$ releases a flux of the order of 10^{21}/s antineutrinos. With such a neutrino factory, two domains of study hitting for the first applied physics and for the second fundamental physics become attainable. Measuring a flux and a spectral shape should allow to measure the composition of the reactor core. Indeed, it was proposed in the 1970s (and then demonstrated in the 1980s/90s) to use the discrepancy between antineutrino flux and energies from U and Pu isotopes to infer reactor fuel isotopic composition and power. Measuring a flux and a spectral shape would also allow the measurement of an oscillation if it exists. This second domain of study is of the interest of fundamental neutrino physics. Across the years, neutrino experiments have found anomalies in their data, i.e. data which could not be explained by an oscillation scheme with three neutrino flavors. The "reactor anomaly" designates the deficit observed between the detected antineutrino flux by reactor neutrino experiments at less than 100 m from reactors with respect to a new prediction of this detected flux [12]. The latter new prediction has been obtained through a new conversion of the integral β energy spectra [13, 14] measured by Schreckenbach *et al.* [15] into antineutrinos. These changes led to a new normalisation of the converted detected antineutrino flux lying 6% above the detected flux at short distances from reactors, when accounting for out-of-equilibrium effects and a new neutron lifetime value [12]. Moreover, the comparison between the converted spectra and the measured spectra of Daya Bay (DBay), Double Chooz and Reno, apart from the normalisation which confirmed the reactor anomaly, exhibits a large distorsion of the data with respect to the model between 5 and 7 MeV in antineutrino energy called the "bump" [16–18]. Lastly, DBay recently put into question integral β measurement of ^{235}U questioning the degree of trust placed in the conversion method [19].

3.2 Summation Calculations and Recent Experimental Improvements

The summation method was proposed simultaneously to the conversion method as an alternative allowing to study the sub-components of the reactor antineutrino energy spectrum. The antineutrino spectrum associated with one of the four fissioning isotopes in a moderated reactor can be computed as the sum of the contributions of all FP using the full information available per nucleus in the NDB as described in [20]. This method is useful for several reasons. Not only it is the only one that can be adapted to the computation of the antineutrino emission associated to various reactor designs, but also it allows for the computation of antineutrino spectra for which no integral β spectrum has been measured yet taking into account off-equilibrium effects and allowing the use of different energy binnings of interest for reactor neutrino experimental analyses.

As an illustration, we picked-up various results obtained so far. In 2012 [20] we published predictions for the antineutrino spectra for the 4 main isotopes in a reactor core and a direct comparison to the converted spectra highlighting a 5 to 10% deficit in the ^{235}U spectrum in agreement with DBay [19]. We also published the main contributors to a standard PWR antineutrino energy spectrum in bins of energy of interest for the bump study [21] which are of first priority for new TAGS measurements. The impact on the antineutrino energy spectrum of the seven nuclei ^{99}Y, ^{138}I, ^{142}Cs, ^{100,100m}Nb and ^{102,102m}Nb have been studied with the summation method. Their cumulated contributions is illustrated in Fig. 2 for ^{238}U (left),

[2]This is an average value which depends on the fissioning system

^{235}U (middle left), ^{239}Pu (middle right) and ^{241}Pu (right). We observe a maximum negative deviation of $\simeq$4% in an energy range from 3 to 5 MeV and a positive deviation in the region of the bump, deviations which go in the direction of neutrino experiment measurements.

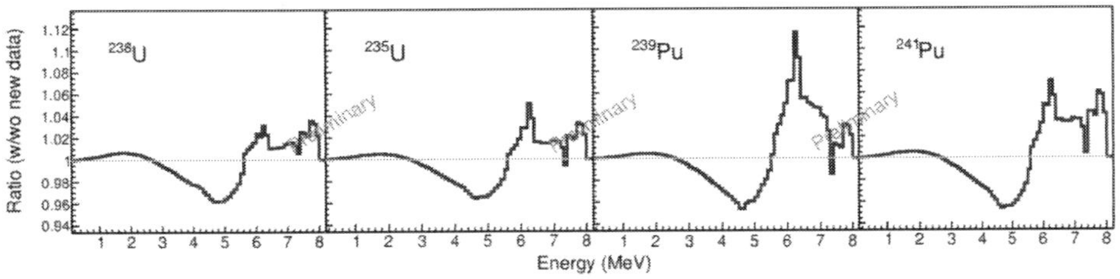

Figure 2. Ratios between 2 spectra built with the dataset described in the text over the same but the TAS data.

4 Decay Heat

4.1 General Motivations for DH Study and Summation Calculations

Following the shut-down of the chain reaction in a reactor, the nuclear fuel continues to release an energy with time called decay heat. It is essentially made up of radioactive (α, β and γ radiations) decays of the FP and actinides. Their released energy as a function of the cooling time distributes very differently. In the first 10s after the reactor stop, the DH is dominated by the delayed fissions, then its main source comes from the FP over 100 years. After this time it is dominated by the actinides decay. On average, it represents a residual power of 6 to 12% of the nominal power of the reactor just after its shut-down. These numbers are huge and motivate the necessity to have a good knowledge of the DH for both the evaluation of the reactor safety and various economic aspects of nuclear power generation. The DH is estimated through the only method able to be predictive for future reactors: the summation method [22]. The decay heat defined as $f(t)$ can be analytically calculated as the sum over the whole FP and actinides inventoried for specific conditions of reactor operation and subsequent cooling period of their mean α, β, γ energies times their total decay constant times their fission yield. In the case of fission pulses, the time ranges up to $\sim 10^6$s and the fission of pure isotopes is considered.

In the 1970s, important discrepancies were observed between summation calculations which used the information stored in nuclear DB taken worldwide and benchmark experiments (see [23] for a review). With the discovery of the Pandemonium effect as potentially responsible for the observed discrepancies, measured TAS data have been used since the 1990s to solve the problem. Figure 2 in [25] illustrates how the long-standing electromagnetic component of the DH puzzle was partially solved in 2010 following the first TAS campaign in Jyväskylä of the Valencian's group. During this experimental campaign, 7 nuclei of first priority for the decay heat interest were measured and analyzed: the ^{101}Nb, ^{105}Mo and the $^{102,104-107}$Tc. Two summation calculations including or not the 7 TAS nuclei were computed using ENDF/B-VII nuclear DB and compared with the Tobias compilation of integral measurements [23]. Five of the 7 nuclei drastically improved the calculation over the whole cooling time covered by the distribution. This important result motivated the necessity to prolongate the experimental effort and is one of the motivations of the second TAS campaign (Sect. 2.2.1). Indeed the γ component of the DH of the most well known nuclides from integral measurements are still not well reproduced by the calculations including different nuclear DB [23].

4.2 Some Preliminary Results from the 2nd TAS Campaign

The mean β and γ energies of the nuclei ^{99}Y, ^{138}I, ^{142}Cs, ^{100,100m}Nb and ^{102,102m}Nb have been computed from the β intensities experimentally determined since the 2nd TAS campaign. They have been compared with ENDF/B-VII.1 and JEFF 3.3 databases. Firstly, they all present the same systematics: a larger $\bar{E}_\gamma$ in TAS compared with the other nuclear DBs and a lower $\bar{E}_\beta$ as expected for Pandemonium nuclei. Secondly, their $\Delta\bar{E}$ with the two other DBs are of the order of few hundreds of keV for both the γ and β components which is on average twice larger than the uncertainties quoted in nuclear DBs. These two observations illustrate the presence of biases in nuclear DBs which need to be improved with such kind of measurements. We started a collaboration with the CEA Cadarache in order to quantify the impact of the bias in the DB uncertainties on the calculation of the total uncertainties associated with the DH.

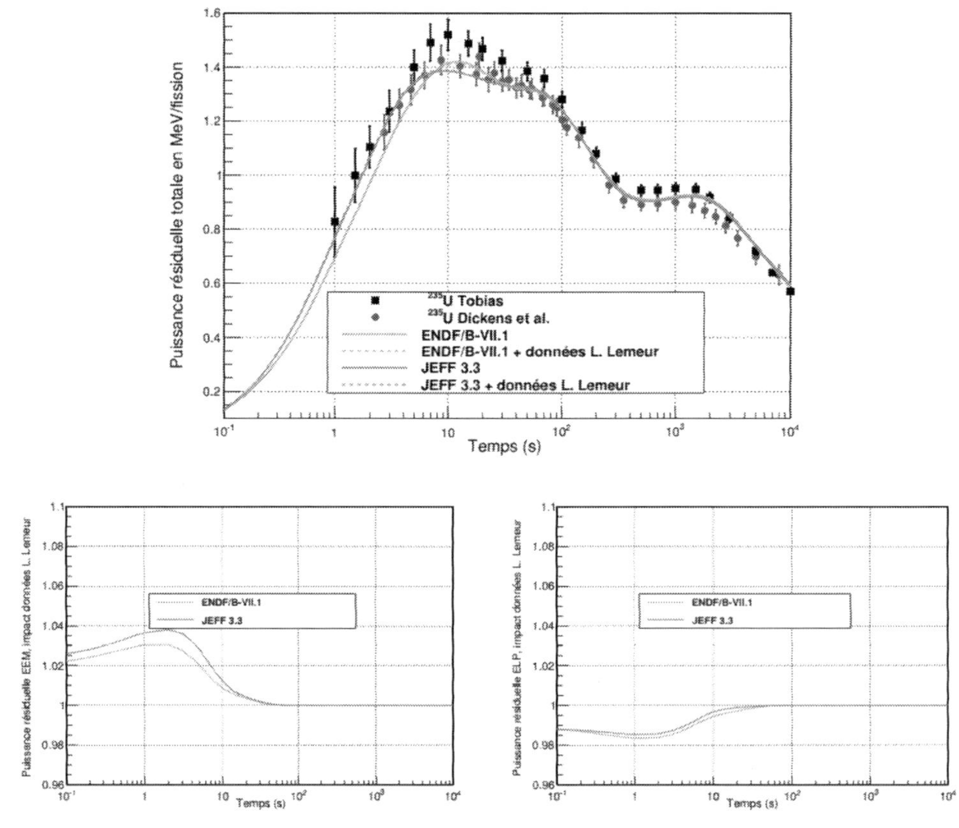

Figure 3. Top: ^{235}U fission pulse total decay heat computed with ENDF/B-VII (green) and JEFF3.3 (blue) and compared to Tobias *et. al* and Dickens *et al* [23]. Bottom: impact of the 3 nuclei ^{99}Y, ^{138}I, ^{142}Cs on the electromagnetic and light particle components of the decay heat respectively, calculated with ENDF/B-VII (green) and JEFF3.3 (blue).

With the knowledge of the new values of $\bar{E}_\gamma$ and $\bar{E}_\beta$ it has then been possible to calculate the impact of all these nuclei on the total DH and on the individual electromagnetic (EEM) and light particle (ELP) components. For that purpose, some summation calculations have been performed using Serpent 2. Serpent is a multi-purpose three-dimensional continuous-energy Monte Carlo particle transport code, developed at the VTT Technical Research Centre of Finland [26]. The decay heat has been computed over 10^4s cooling time. It is illustrated for the case of the three nuclei ^{99}Y, ^{138}I, ^{142}Cs taken together in Fig. 3. The top panel of the figure shows the absolute total DH for ^{235}U while the two other panels depict the relative impact of the 3 nuclei on the EEM and ELP components of the DH respectively. The calculations have been performed with two nuclear databases: ENDF/B-VII.1 (green) and JEFF 3.3 (blue). While the impact of the sum of the 3 nuclei is lower than 1% in both DB on the total DH below $10s$, we can see that it comes from a compensation of the two EEM and ELP components. The left and right panels show that the nuclei have a larger impact on each component taken individually (1-2% on ELP and 3-4% on EEM). Note that the main contributions come from the ^{142}Cs[3] (2%) and the ^{99}Y (1.5%) but this is not shown here. We have similar conclusions concerning the ^{239}Pu. Concerning niobiums, the EEM and ELP components which are also at the level of the percent for both ^{235}U and ^{239}Pu in both DBs compensate for the total DH.

5 Conclusion

In conclusion, we have presented in this paper the motivations for new measurements of β decay properties of fission products, as key ingredients in the estimate of the reactor decay heat, reactor antineutrinos and β-delayed neutron fractions. The Total Absorption Gamma-ray Spectroscopy technique allows to avoid the Pandemonium effect which affects some of the data of evaluated nuclear databases. We have presented new results of a TAGS campaign performed at the IGISOL facility of Jyväskylä, Finland. Their impact on decay heat from fission pulses and on individual antineutrino spectra was assessed and found to be significant.

References

[1] P. Dimitriou *et al.*, submitted to Nuclear Data Sheet, (dec 2018)
[2] J. C. Hardy, *et al.*, Phys. Lett. B **71**, 307 (1977)
[3] M. Estienne, *et al.*, Nucl. Data Sheets **120**, 149 (2014)
[4] R. C. Greenwood et al., Nucl. Instrum. Meth. A **314**, 514 (1992) and references therein
[5] D. Cano-Ott *et al.*, Nucl. Instrum. Methods Phys. Res., Sect. **A430**, 488 (1999); 430, 333 (1999). J. L. Tain and D. Cano-Ott, Nucl. Instrum. Methods Phys. Res., Sect. **A571**, 728 (2007); 571, 719 (2007)
[6] V. Guadilla et al. Characterization and performance of the DTAS detector, Nucl. Instr. Meth. A vol. **910** (2018)
[7] T. Eronen et al., Eur. Phys. J. A **48**, 46 (2012)
[8] P. Dimitriou and A. L. Nichols, IAEA report INDC(NDS)-0676, Feb. 2015, IAEA, Vienna, Austria
[9] L. Le Meur, PhD thesis Nantes University, oct. 2018
[10] V. Guadilla-Gomez, PhD thesis Valencia University, sept. 2017
[11] J.L. Taìn *et al.*, Nucl. Instr. Methods Phys. Res. A **571**, 728 (2012)
[12] G. Mention et al., Phys. Rev. D **83**, 073006 (2011)
[13] Th. Mueller et al., Phys. Rev. C **83**, 054615 (2011)

[3] which was listed as priority 3 in the IAEA list of priority nuclei to measure for DH

[14] P. Huber, Phys. Rev. C **84**, 024617 (2011)
[15] K. Schreckenbach et al., Phys. Lett. **160B**, 325 (1985) and references therein
[16] Y. Abe *et al.*, Phys. Rev. Lett. **108**, 131801 (2012)
[17] F. P. An *et al.*, Phys. Rev. Lett. **108**, 171803 (2012)
[18] J. K. Ahn *et al.*, Phys. Rev. Lett. **108**, 191802 (2012)
[19] F. P. An *et al.* (Daya Bay Collaboration), Phys. Rev. Lett. **118**, 251801 (2017)
[20] M. Fallot *et al.*, Phys. Rev. Lett. 109, 202504 (2012)
[21] A.-A. Zakari-Issoufou *et al.*, Phys. Rev. Lett. **115**, 102503 (2015)
[22] T. Yoshida et al., OECD/NEA Working Party for Interna- tional Evaluation Co-operation, **1425**, 25 (2007). Nuclear Science NEA/WPEC-25 (2007)
[23] M. Fleming and J.-C. Sublet, report 2015 Fispact II, CCFE-R(15) 28 and references therein
[24] K. Takahashi, M. Yamada, T. Kondoh, At. Data Nucl. Data Tables **12 (1)**, 101-142 (1973). T. Yoshida, R. Nakasima, J. Nucl. Sci. Technol. **18 (6)**, 393-407 (1981)
[25] A. Algora, et al., Phys. Rev. Lett. **105**, 202501 (2010)
[26] J. Leppänen *et al.*, Ann. Nucl. Energy **82**, 142-150 (2015)

Energy Dependence of Delayed-Neutron Data

Daniela FOLIGNO[1,*] and *Pierre* LECONTE[1,**]

[1]CEA, DEN, DER/SPRC/LEPh Cadarache, F-13108 Saint Paul-Lez-Durance, France

Abstract. Delayed neutron data is essential in inherent reactor safety and in reactor control. The different sources and results of delayed neutron data in international databases will be discussed. In addition to direct measurement, these data can be obtained by calculations based on fission yields (FY) and radioactive decay data (RDD). Unfortunately, FY are only given at three energies in the European library JEFF. This works proposes to use GEF-FY energy dependence to extrapolate JEFF-FY at other energies. The preliminary results highlight that GEF energy dependence is not always suitable to reproduce experimental values and that more measurements are needed at high energies.

1 Introduction

A DN (*Delayed Neutron*) is a neutron which is not created by the fission event itself but appears during the de-excitation of fission products. For example, ^{137}I is an unstable neutron-rich FP (*Fission Product*) [1] and it undergoes β^--decay, leading to ^{137}Xe. Depending on the excitation energy of the Xenon, and on the separation energy of the last neutron, the Xenon might emit a neutron or a gamma ray, both of which will be called *delayed* because they appear some time after the fission event. In nuclear physics Iodine is considered the DN precursor. In reactor applications it is extremely important to have a good estimate of the reactivity, because it tells how far the system is from criticality. The Nordheim equation (Eq. 1) allows to compute the reactivity (ρ) from a measurable quantity, the reactor period (T)[1, and references therein].

$$\rho = \frac{\Lambda}{T} + \beta_{eff} \sum_{i=1}^{G} \frac{a_i}{1 + \lambda_i\, T} \approx \beta_{eff} \frac{\overline{T_{1/2}}}{T} \tag{1}$$

$$\beta_{eff} = \frac{\sum\limits_{k} \sum\limits_{i=1}^{G} \int\limits_{0}^{\infty} a_{k,i}\, \nu_{d,k}(E)\, \Sigma_{f,k}(E)\, \Phi(E)\, \mathrm{d}E \int\limits_{0}^{\infty} \chi_{d,k,i}(E')\, \Phi^{+}(E')\, \mathrm{d}E'}{\sum\limits_{k} \int\limits_{0}^{\infty} a_{k,i}\, \nu_{t,k}(E)\, \Sigma_{f,k}(E)\, \Phi(E)\, \mathrm{d}E \int\limits_{0}^{\infty} \chi_{t,k}(E')\, \Phi^{+}(E')\, \mathrm{d}E'} \tag{2}$$

$$\overline{T_{1/2}} = \sum_{i=1}^{G} \frac{a_i}{\lambda_i} \tag{3}$$

*e-mail: daniela.foligno@cea.fr
**e-mail: pierre.leconte@cea.fr
[1]It is considered here that the prompt neutron emission had already taken place

In Eq. 1 Λ is the prompt neutron generation time, β_{eff} the effective delayed neutron fraction, a_i and λ_i the kinetic parameters in the group approximation and $\overline{T_{1/2}}$ the mean precursors' half-life. All except Λ are delayed neutrons parameters. β_{eff} represents the fractions of neutrons that caused fission and that were born delayed. In Eq. 2 [6], Φ and Φ^+ are the direct and the adjoint neutron flux, $\Sigma_{f,k}(E)$ the macroscopic fission cross section of the fissioning system k, $\chi_{t,k}$ the prompt neutron spectrum of k, $\chi_{d,k,i}$ the DN spectrum of k per DN group i, $\nu_{t,k}$ and $\nu_{d,k}$ the prompt and delayed neutron yields, respectively. The $\overline{T_{1/2}}$ represents the mean time after which half of the DN have been emitted and can be computed if the kinetic parameters are known. Equations 2 and 3 show how to compute β_{eff} and $\overline{T_{1/2}}$ and are derived from [2]. It is evident the need of mastering the energy dependence of those parameters, especially in fast reactors where the incident neutron energy ranges from some keV to some MeV.

In section 2 the state of the art on delayed neutron data is recalled, together with the history of the libraries updates. Section 3 the methods used to compute the different quantities of interest are introduced. Furthermore, the technique employed to apply the GEF energy dependence to JEFF-3.1.1 FY is illustrated. The results are discussed in section 4 and the conclusions and perspectives in section 5.

2 State of the art

The average delayed neutron yield as a function of the incident neutron energy can be found in the evaluated libraries (MF=1, MT=455).

The origin of the recommended data is not always well defined. According to the references reported in the header of the file, the yields contained in the version 2.2 of the European Library JEFF, come from a semi-empirical calculation performed by Eric Fort using Lendel's model [3]. In the following version, JEFF-3.0, the yields have the reference *England89* [4], the same of the American Library ENDF/B-VI. Overlapping the two sets of data confirmed the coherence of the references. From JEFF-3.1 to JEFF-3.1.2, the reference stayed unchanged, even though a comparison of the curves (see Fig. 1a) shows a different energy dependence. The last two versions, JEFF-3.2 and JEFF-3.3 recursively refer to the previous versions, thus not revealing the origin of the yields reported from JEFF-3.1 on. Following researches led to the conclusion that the mysterious set of data comes from TENDL2009 and that the reference in the header of the file is wrong and misleading.

As far as the American library is concerned, the last version (ENDF/B-VIII.0) takes the ν_d from the previous ENDF/B-VII.1, which in turn took it from ENDF/B-VI, "except between 1.0-5 eV and 50 keV where they were modified to agree with JENDL-3.3 values, following a suggestion from C. Lubitz [Lu06]" [5, ENDF/B-VII.1 library, MT=451]. Finally, ENDF/B-VI seems to have taken the yields from the already mentioned *England89* [4].

Another way to obtain the average delayed neutron yield as a function of the incident neutron energy is to take the fission yields at different energies and to use them, together with a radioactive decay data set, in a summation calculation performed by the author (see the method in Section 3). Figure 1 shows the ν_d as given by JEFF-3.3 (1a) and ENDF/B-VII.1 (1b), but also what it is obtained when using JEFF-3.1.1 FY and JEFF-3.1.1 or ENDF/B-VIII.0 RDD (1c). Since the FY are only given at three energies (thermal, fast and high), only three points can be drawn, and nothing can be inferred on the energy dependence.

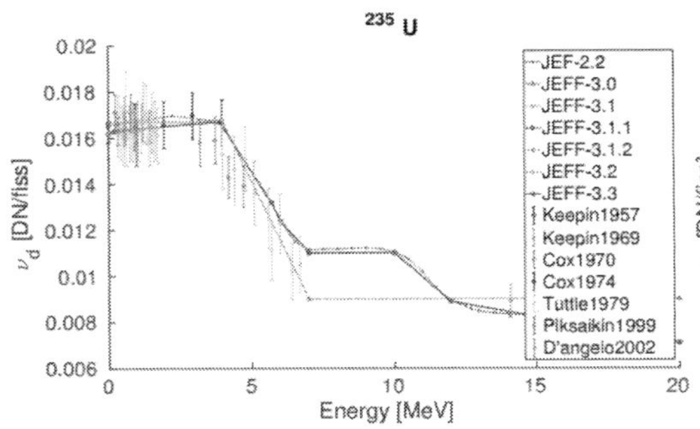

(a) JEFF library

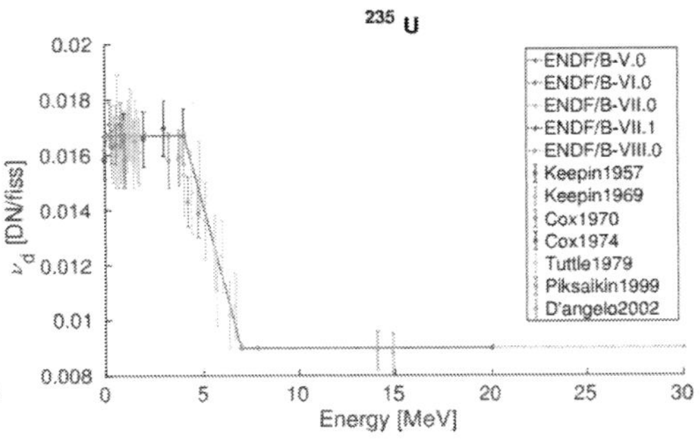

(b) ENDF/B library

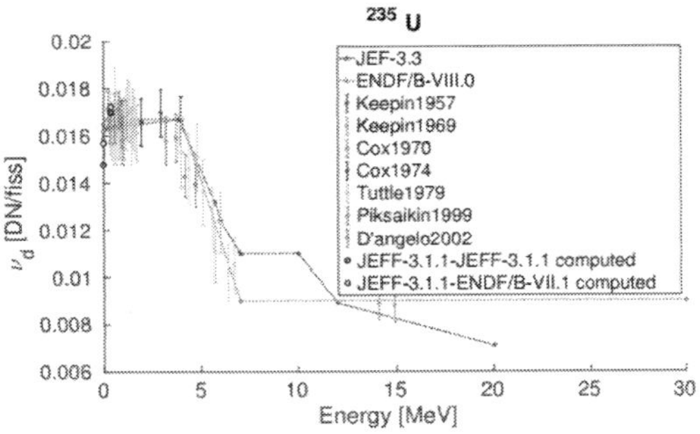

(c) Summation calculation

Figure 1: Library comparison of the average delayed neutron yield as a function of energy for ^{235}U. Fig. 1a contains all the versions of JEFF library from the 2.2 to the 3.3. Fig. 1b contains all the versions of JEFF library from the V.0 to the VIII.0. Fig. 1c contains JEFF-3.1.1, ENDF/B-VII.1 and the points obtained with a summation calculation. All figures show experimental data for comparison.

3 Method

One way of estimating integral delayed-neutron quantities is to perform a summation calculation, in which the microscopic contribution of each delayed-neutron precursor is taken into account and summed up. Equation 4 shows that the contribution of a precursor to the average DN yield is the product of its cumulative fission yield CY, its DN emission probability P_n and the number of neutrons emitted per decay x.

$$\nu_d = \sum_{j}^{N} CY_j \, P_{n,j} \, x_j \tag{4}$$

In the same fashion, Eq. 5 is used to compute the DN activity. In this case, instead of cumulative fission yields one has to take into account the concentration of each precursor at each time t. The system of Bateman differential equations has to be solved in a consistent way, in order to take into account both the precursors created by direct fission and the ones created by the decay of other fission products.

$$\frac{n_d(t)}{n_d(t_\infty)} - \sum_{i=1}^{G}\left(1 - \sum_{\substack{j=1\\ j\neq i}}^{G} a_j\right)\left(1 - e^{-\lambda_i t_{irr}}\right) e^{-\lambda_i (t-t_{irr})} = 0 \tag{5}$$

Equation 5 shows one way to compute the kinetic parameters from the delayed neutron activity curve. The method consists at fitting the difference of two models. The first term is the normalized emission rate computed by summation calculation while the second term is the sum of exponentials we want to fit the curve with. G is the number of DN groups and a_i and λ_i are the abundance and decay constant of the DN group i. Note that the fact of computing a_i as $1 - \sum_{\substack{j=1\\ j\neq i}}^{G} a_j$ is done to normalize to 1 the sum of the abundances without introducing any normalization factor, which would have had the effect of increasing both the uncertainty and the correlations among abundances. With the DN activity (measured or calculated), the kinetic parameters and the mean precursors' half-life are easy to estimate (see Eq. 5 and Eq. 3).

GEF [7] is a Monte Carlo code able to provide FY at almost any neutron energy. When used in a calculation, the resulting ν_d are pretty far from the recommended values. This work aims at testing the energy dependence of GEF fission yields, which will be used to normalize the data of the JEFF library. First of all, one has to choose a reference energy (E_{ref}) per fissioning system, at which JEFF fission yields will be taken ($FY_{JEFF}(E_{ref})$). In this work the reference energy is 0.025 meV for ^{235}U and ^{239}Pu and 400 keV for ^{238}U. For each point in energy, the ratio of GEF fission yields at E and at E_{ref} is computed. The last step consists at applying the GEF energy dependence - the ratio $\frac{FY_{GEF}(E)}{FY_{GEF}(E_{ref})}$ - to the reference $FY_{JEFF}(E_{ref})$ (see Eq. 6).

$$FY_{JEFF}(E) = \frac{FY_{GEF}(E)}{FY_{GEF}(E_{ref})}\, FY_{JEFF}(E_{ref}) \tag{6}$$

The new sets of JEFF fission yields at different energies are then coupled with ENDF/B-VIII.0 radioactive decay data to perform a summation calculaton through which the quantities of interest can be estimated (Eq. 4 and 5).

4 Results

4.1 Average delayed-neutron yield

The results of the calculation and the comparison with experimental data are shown in Fig. 2, where 2a, 2b and 2c, refer to ^{235}U, ^{238}U and ^{239}Pu, respectively. The RDD is taken from ENDF/B-VIII.0. In the three plots the blue line corresponds to the summation calculation performed by using GEF-6.1 FY, while the orange one corresponds to JEFF-3.1.1 FY, corrected with GEF-6.1 energy dependence.

As far as the ^{238}U is concerned, the shape of the orange curve, with a well-defined step-wise behavior, is promising. Furthermore the agreement with the experimental data, especially below 10 MeV, is reassuring. Applying GEF energy dependence to JEFF fission yields for ^{235}U and ^{239}Pu, an unexpected bump appears around 16-17 MeV. Moreover, the expected flat behavior below 4 MeV as well as the experimental slope for ^{235}U are not reproduced by the calculation. Unfortunately, in the intermediate range (7 to 13 MeV) a lack of measurements prevents the author draw any conclusion.

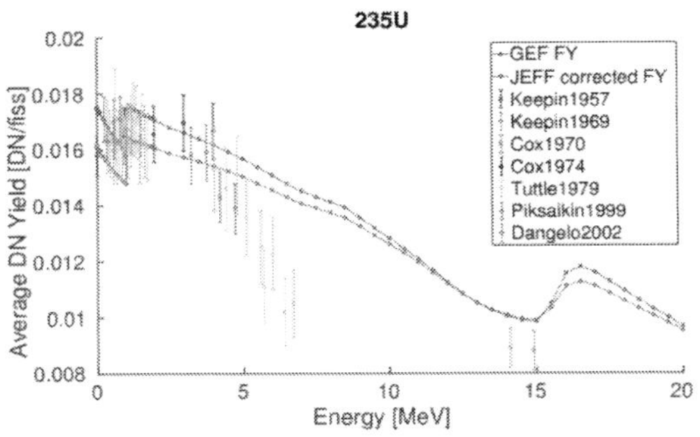

(a) 235U

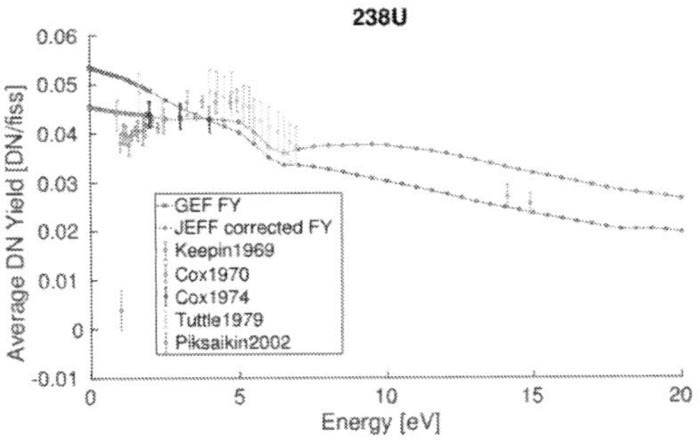

(b) 238U

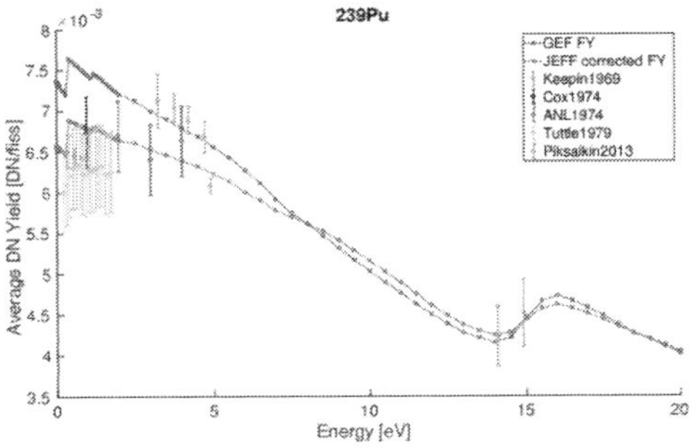

(c) 239Pu

Figure 2: Energy dependence of average delayed neutron yield computed with GEF-6.1 FY and with JEFF-3.1.1 FY modified with GEF-6.1 energy dependence. Experimental values are reported for comparison

4.2 Kinetic parameters and mean precursors' half-life

A set of DN abundances is always associated to a set of decay constants and they cannot be considered separately. One way to evaluate a set of kinetic parameters (a_i, λ_i) is to compute the derived quantity $\overline{T_{1/2}}$ (see Eq. 3). For the sake of clarity, instead of comparing the abundances one by one, only the effects on the $\overline{T_{1/2}}$ are shown and commented (see Fig.3). It seems that for ^{235}U, the JEFF FY modified with GEF-6.1 better reproduce the experimental data than the original ones. The same is true for ^{238}U, except at 14 MeV, where the experiments are spread and far from each other. Finally, for ^{239}Pu, the trend is pretty good but a lack of measurements beyond 5 MeV prevents us to comment on the quality of the results. Unfortunately, measurements in those energy ranges are very difficult to perform, due to the difficulty of finding a good monoenergetic neutron sources in this energy domain.

5 Conclusions and Perspectives

Delayed neutron data is essential in inherent reactor safety and in reactor control. The international databases rarely agree with each other, especially when DN kinetic parameters are concerned.

The ν_d, the average number of delayed neutrons emitted per fission, appears in the equation of the beta effective, which is the effective fraction of neutrons causing fission

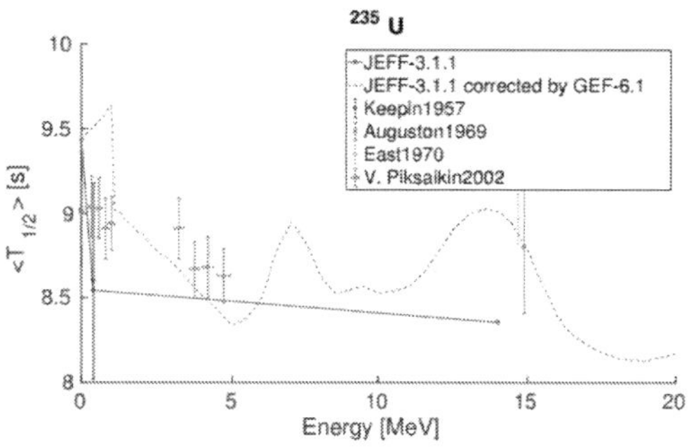

(a) 235U

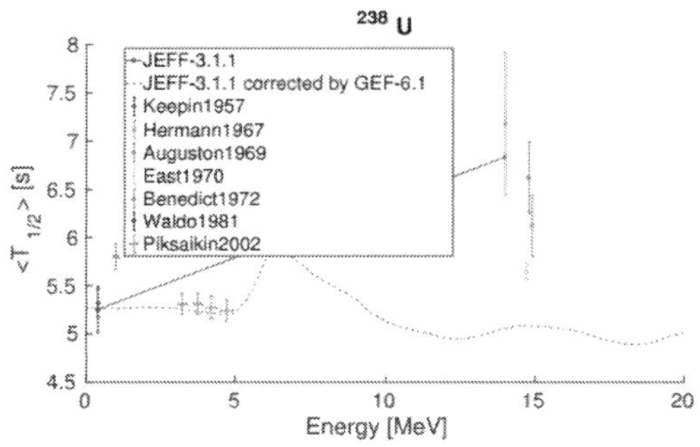

(b) 238U

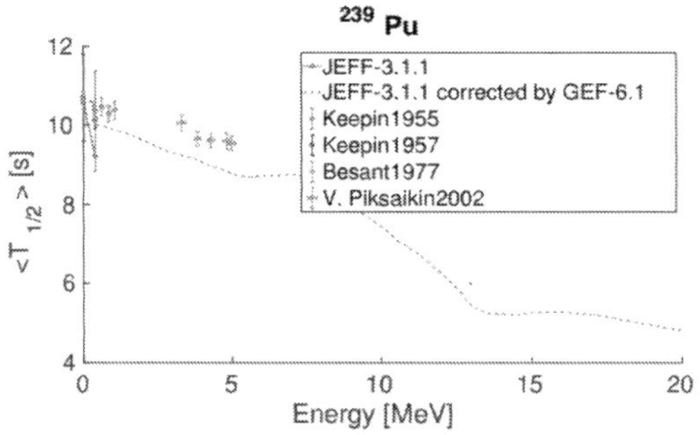

(c) 239Pu

Figure 3: Energy dependence of the mean precursors' half-life computed with the original JEFF-3.1.1 FY and with the with JEFF-3.1.1 FY modified with GEF-6.1 energy dependence. Experimental values are reported for comparison

that are born delayed. The origin of the ν_d as a function of incident neutron energy in the international libraries is not always well reported. Comparing the evaluated values with experimental data (Fig. 2) it is clear that up to 7 MeV and at 14 MeV there is good agreement, while in the intermediate range (7 to 13 MeV) a lack of measurements prevents the author draw any conclusion. Applying GEF energy dependence to JEFF fission yields, the shape of the resulting ν_d presents inexplicable bumps at high energies. Furthermore, the flat behavior below 5 MeV is not reproduced.

The the mean precursors' half-life can be estimated with the same method (Fig. 3). It seems that for ^{235}U, the JEFF modified FY better reproduce the experimental data than the original ones. The same is true for ^{238}U, except at 14 MeV, where the experiments are incoherent. Finally, for ^{239}Pu, the agreement is acceptable until 5 MeV, beyond which a lack of measurements prevents us to infer anything on the quality of the results.

To conclude, the method presented in Section 3 is another way to evaluate the quality of fission yields. GEF-6.1 energy dependence has been tested and for the moment it is not always able to reproduce experimental data. However, the method of applying the energy dependence of a specific set of FY to another set looks promising, especially for estimating quantities of interest at energies where measurements are missing. For example, since JEFF-3.1.1 gives evaluated FY at maximum 3 energies while GEF-6.1 gives calculated FY up to 20

MeV, the mentioned method could provide JEFF FY at almost all energies. Of course, for the method to be exploitable, GEF-6.1 should be able to reproduce JEFF-3.1.1 fission yields at energies different from the one taken as reference. For the moment this is not the case, but an improvement of GEF-6.1 energy dependence of FY could, in the future, find an application for this method.

References

[1] Energy Procedia **71**, 2015, 205-212

[2] Progress in Nuclear Energy **41**, 2002, 317-359

[3] Progress in Nuclear Energy **41**, 2002, 391-412

[4] Los Alamos National Laboratory reports: LA-11151-MS(88), LA-11534T(89), LAUR-88-4118(89)

[5] ENDF/B-VII.1 library, MT=451. The reference Lu06 appeared to be a personal communication in August 2006

[6] V. Zammit-Averlant. Validation Integrale des Estimations du Parametre Beta Effectif pour les Reacteurs MOX et Incineriteurs. PhD thesis, Université d'Aix-Marseille, 1998.

[7] Nuclear Data Sheets **131**, 2016, 170-221

Testing Compound Nucleus Hypothesis Across the ^{17}O Nucleus Excitation

Aloys Nizigama*, *Pierre* Tamagno, and *Olivier* Bouland

CEA, DEN, DER, SPRC, Physics Studies Laboratory, Cadarache, F-13108 Saint-paul-lez-Durance, France

Abstract. The excited compound nucleus $^{17}O^*$ has been studied over (n,α) and (α,n) cross sections modelling, respectively for ^{16}O and ^{13}C targets in their ground states. The modelling is fulfilled within the Reich-Moore formalism. We were able to calculate the (α,n) cross section by two separate ways: the direct kinematic standard route and by inversion of the (n,α) cross section using the compound nucleus hypothesis. Resonance parameters of the resolved resonance range (0 to 6 MeV) were borrowed from the CIELO project. In a first stage, the modelling is carried out in the referential of the incident particle (either way neutron or α) requesting conversion of the CIELO neutron-type resonance parameters to the α-type. In a second stage, the implementation is uniquely designed in the center of mass system of the excited compound nucleus. The resonance parameters are thus converted in that unique reference framework. The present investigation shows the consistency of the kinetic transformation that relies on the compound nucleus hypothesis.

1 Introduction

The study of reactions that refer to the excited compound nucleus $^{17}O^*$ is of strong concern for reactors physics nuclear data evaluators. ^{16}O is the most predominant isotope of oxygen in nature and contributes to ^{4}He gaz production observed in reactor through $^{16}O(n,\alpha)^{13}C$ conversion. On this point of view, the ^{16}O constitutes an important safety issue in the operation of reactors. Furthermore, the oxygen is overriding in the oxide fuels and plays a role in neutron thermalization in water. One can also emphasize the interest of $^{17}O^*$ excited compound nucleus as neutron source, especially meaningful in the *s*-process (astrophysics area), through the $^{13}C(\alpha,n)^{16}O$ reaction. Taking into account nuclear structure aspect, the $^{17}O^*$ compound nucleus is characterized by weak level density and subsequent wide resolved resonance range (RRR) that goes from 0 to 6 MeV. Two reaction types are fully open for this region: the neutron and α reactions. Due to its light nucleus structure feature, the $^{17}O^*$ compound nucleus makes it a nice candidate for models testing in the RRR.

Existing nuclear data discrepancies between measured and evaluated data according to the $^{16}O(n,\alpha)^{13}C$ and reciprocal $^{13}C(\alpha,n)^{16}O$ reactions, have been the subject of strong

* Corresponding author, e-mail : aloys.nizigama@cea.fr

discussions within CIELO [1, 2] project. In 2014, the CIELO team summarized the situation [1] pointing out 30% of discrepancies between lower (IRMM 2008 [3], Harrisopulos 2005 [4]) and higher (original Bair and Haas [5]) cross section trends. These discrepancies were also recorded between major evaluated data libraries. The old ENDF/B-VI evaluation was scaled on the higher cross section magnitude while the lower trend was adopted in the ENDF/VII.0 (unchanged in ENDF/B-VII.1) and JEFF-3.1 according the energy range below 6 MeV. Thereafter, that choice was completely re-examined in the last ENDF/B library. The CIELO team reported in 2018 [2] a raise up of about 40% to be applied to the ENDF/B-VII.1 (n,α) cross section. The modifications were included in the ENDF/B-VIII.0 file. However, the CIELO project recommended new experiments to corroborate these new changes.

Present work takes place in larger demarche that includes redesign of the technique of cross section evaluation meaning the determination of nuclear parameters using a unified framework approach. This paper presents the work on progress using a newly implemented computer branch [TOol for Reactions Analysis (TORA)] of the CONRAD code [6] whose purpose is to reach the above-cited objective. The TORA module has capability of achieving cross sections calculations not only in the laboratory framework attached to the incident particle (either charged or not) but also within the excited compound nucleus center of mass framework taking reference to the fermi energy of the latter. Based on this aspect, the TORA module acts as a complementary tool to the CONRAD code when analyzing data within the center of mass of the excited compound nucleus framework. The tool is tested across dedicated modelling of the $^{16}O(n,\alpha)^{13}C$ and $^{13}C(\alpha,n)^{16}O$ cross sections in both projectile laboratory respective frameworks and center of mass of the excited compound nucleus (defining the unified approach).

2 Theoretical background

This work is fulfilled in the Reich-Moore formalism [7]. The R matrix, in which the gamma channels are beforehand eliminated, is given by:

$$R^{J\pi}_{cc'} = \sum_\lambda \frac{\gamma^{J\pi}_{\lambda c}\gamma^{J\pi}_{\lambda c'}}{E^{J\pi}_\lambda - E - i\left(\gamma^{J\pi}_{\lambda\gamma tot}\right)^2} \tag{1}$$

The quantities $\gamma^{J\pi}_{\lambda c}$ and $\gamma^{J\pi}_{\lambda c'}$ are the reduced width amplitudes, $E^{J\pi}_\lambda$ is the resonance energy and the parameter $\gamma^{J\pi}_{\lambda\gamma tot}$ is the total gamma channel reduced width amplitude of the λth state in a group of states contributing to the same total spin J and parity π of the compound nucleus. E is the energy of the incident particle. Since these resonance parameters are expressed in the neutron laboratory system according to the ENDF format of the evaluated data libraries, a reference framework transformation is usually needed to calculate reciprocal cross sections (exchanging projectile-ejectile pair roles).

3 Direct kinematics calculations according to the $^{16}O(n,\alpha)^{13}C$ and $^{13}C(\alpha,n)^{16}O$ cross sections

The $^{16}O(n,\alpha)^{13}C$ and $^{13}C(\alpha,n)^{16}O$ cross sections calculations are fulfilled respectively in the neutron- and alpha-incident particle laboratory systems using TORA and validated by comparison to the calculations performed using SAMMY [8] and CONRAD [6] codes. The appropriate expressions are used to calculate neutrons and charged particles penetration factors, shift factors and hard-sphere scattering phase shifts. The first step lies in the neutron to alpha laboratory system transformation of the CIELO resonance parameters to reconstruct the direct kinematics $^{13}C(\alpha,n)^{16}O$ cross section. The neutron laboratory frame is by definition scaled on the neutron separation energy (S_n= 4143890 eV) to be juxtaposed to corresponding alpha separation energy (S_α = 6359067 eV). The reaction Q-value is deduced accordingly. To achieve the framework transformation, formulas used are in the spirit of the SAMMY code [8] and have been implemented in the TORA module. The resonance energy $E_\lambda^{J\pi}$ and the reduced width amplitude $\gamma_\lambda^{J\pi}$ are transformed from the neutron laboratory framework to the alpha laboratory framework as below:

$$E_{\lambda(\alpha lab)}^{J\pi} = E_{\lambda(nlab)}^{J\pi} \frac{M_{16_O}}{M_{16_O}+M_n} \frac{M_{13_C}+M_\alpha}{M_{13_C}} - (S_\alpha - S_n) \frac{M_{13_C}+M_\alpha}{M_{13_C}} \quad (2)$$

$$\gamma_{\lambda(\alpha lab)}^{J\pi} = \gamma_{\lambda(nlab)}^{J\pi} \sqrt{\frac{M_{16_O}}{M_{16_O}+M_n}} \sqrt{\frac{M_{13_C}+M_\alpha}{M_{13_C}}} \quad (3)$$

$E_{\lambda(\alpha lab)}^{J\pi}$, $E_{\lambda(nlab)}^{J\pi}$, $\gamma_{\lambda(\alpha lab)}^{J\pi}$, $\gamma_{\lambda n}^{J\pi}$, M_{16_O}, M_{13_C}, M_n and M_{13_C} being respectively the resonance energy in the alpha laboratory framework, the resonance energy in the neutron laboratory framework, the reduced width amplitude in the alpha laboratory framework, the reduced width amplitude in the neutron laboratory framework, the mass of ^{16}O, the mass of ^{13}C, the neutron mass and the alpha particle mass. The subscripts αlab and $nlab$ indicate respectively for alpha laboratory framework and neutron laboratory framework.

The boundary limit (B) is assigned to the shift factor value (S) and channel radii are respectively chosen to a_c= 4.15 fm for the neutron channel and a_c = 6.68 fm for the alpha channel. Figure 1 shows corresponding simulations. The TORA calculations are compared to those fulfilled using the SAMMY code and to the experimental measurements. The (α,n) experimental cross section data are Bair and Haas 1973 [5] data and the forward (n,α) experimental data are Giorginis (IRMM 2008 [3]) data with their fixed EXFOR normalizations. The pseudo (n,α) Bair and Haas 1973 data are obtained by inversion of the forward (α,n) Bair and Haas 1973 data. The data inversion is based on the compound nucleus hypothesis presented in the subsection 4.2. The cross section of the reciprocal reaction is obtained as follow [9]:

$$\sigma_{(n,\alpha)}(E_n)\lambda_\alpha^2 = \sigma_{(\alpha,n)}(E_\alpha) \frac{(2i_\alpha+1)(2I_{13_C}+1)}{(2i_n+1)(2I_{16_O}+1)} \lambda_n^2 \quad (4)$$

E_α, i_α, I_{13_C} and λ_α are respectively the energy of the alpha projectile in the alpha laboratory framework, the spin of the projectile, the spin of the target and the reduced wavelength of the alpha projectile. E_n, i_n, I_{16_O} and λ_n are reciprocal quantities in neutron laboratory framework. E_n is obtained by a transformation of E_α from the alpha laboratory framework to the neutron laboratory framework using equation (2).

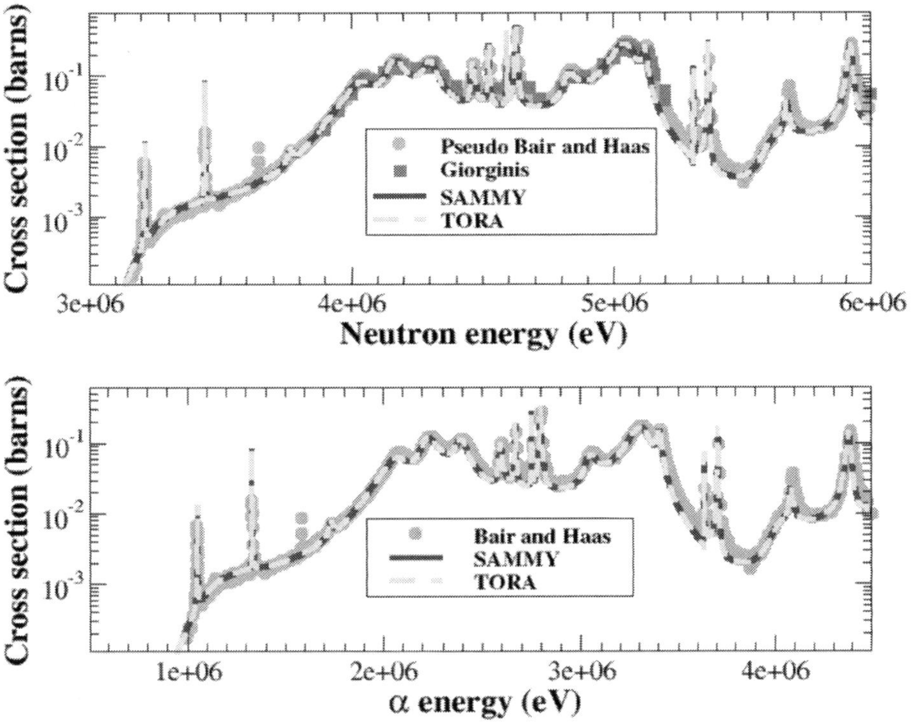

Fig.1. $^{16}O(n,\alpha)^{13}C$ (the top subfigure) and $^{13}C(\alpha,n)^{16}O$ (the bottom subfigure) unbroadened cross sections versus projectile energy. The long dash curves address the results from TORA. They are compared to reference calculations obtained using the SAMMY-8 (solid curve). On the top sub figure, the pseudo-experimental data are obtained by the inversion of the Bair and Haas 1973 (α,n) cross section which are plotted on the bottom subfigure (circular symbols). The forward (n,α) experimental data (square symbols) are Giorginis (IRMM 2008 [3]) data. The modeling is fulfilled using the Reich-Moore parameters kindly released by L.C. Leal [10].

4 Nuclear data computation involving compound nucleus CoM frame

4.1 Various calculated data for $^{17}O^*$ compound nucleus

Modelling of various neutron and α cross sections of reactions related to the $^{17}O^*$ compound nucleus has been fulfilled in the center of mass (CoM) system taking reference to the ground state energy of the $^{17}O^*$ nucleus. The main goal of this approach is to reduce the inconsistencies between data due to the frame reference dependence. All the parameters occurring in the equation (1) are now established in the CoM system. The reduced amplitudes being energy independent physical parameters (see equation 5 below taken from [11]), they become suitable for the purpose of present work suggesting to replace the total gamma width usually occurring in the denominator of the Reich-Moore R-Matrix formulation by its corresponding amplitude as shown in the equation (1).

$$\gamma_{\lambda c} = \left(\frac{\hbar}{2\mathrm{M_c a_c}}\right)^{\frac{1}{2}} \int \varphi_c^* X_{JM}^{\lambda} \mathrm{dS} \qquad (5)$$

φ_c^* is the surface wave function for the channel c and X_{JM}^{λ} is the internal eigenfunction related to the energy E_{JM}^{λ}. The constants a_c and M_c are respectively the channel radius and the reduced mass. The integration is made over the totality of channel surfaces. The modelling of nuclear reactions in the CoM frame of the compound nucleus enables a consistent and simultaneous production of data for nuclear reactions involving the same compound nucleus. Figure 2 shows the results of a simulation of the four reactions: $^{16}O(n,n)^{16}O$, $^{16}O(n,\alpha)^{13}C$, $^{13}C(\alpha,n)^{16}O$ and $^{13}C(\alpha,\alpha)^{13}C$ in the same frame.

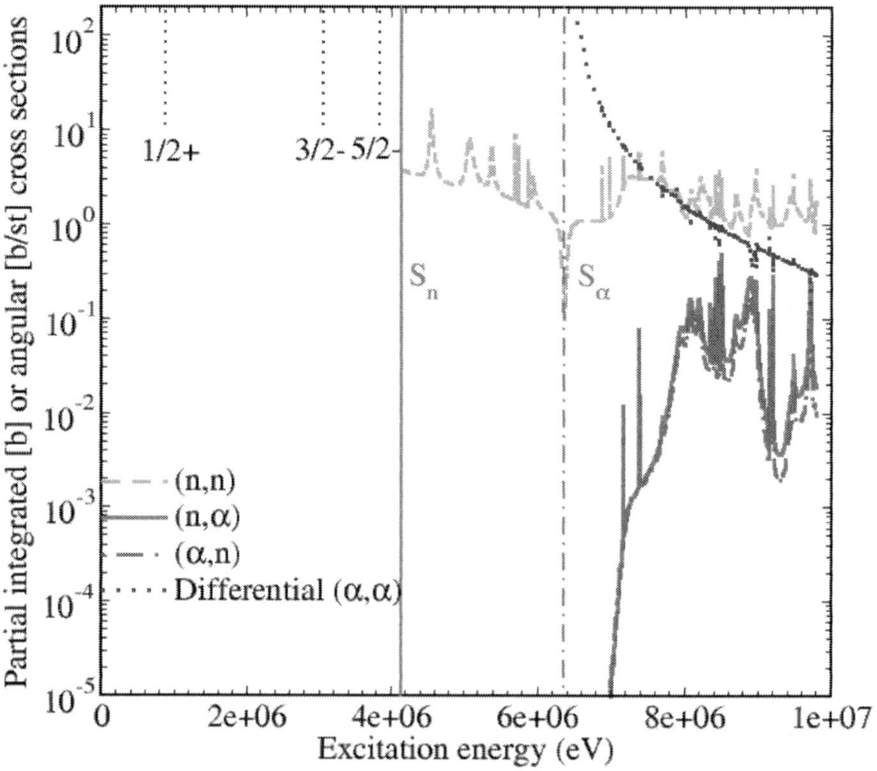

Fig.2. The $^{16}O(n,n)^{16}O$ (long dash curve), $^{16}O(n,\alpha)^{13}C$ (solid curve) ,$^{13}C(\alpha,n)^{16}O$ (long-short dash curve) and $^{13}C(\alpha,\alpha)^{13}C$ (dotted curve) cross sections. The targets are respectively ^{16}O and ^{13}C in their ground states. Vertical solid and dash lines indicate respectively the position of the separation energies of the neutron and α particle on the energy scale. The dotted vertical lines indicate the position of the three bound states of the $^{17}O^*$ nucleus below the separation energy of the neutron. Present differential cross section (α,α) is computed according Blatt and Biedenharn formalism [12] at the CoM angle of 54 degrees (the CoM angle of 54 degrees was chosen in concern of reproducing existing observable based on existing experimental data of Barnes [13] at 54.7 degrees).

In the CoM reference of $^{17}O^*$, the excited states must appear at the same excitation energy for all the simulated reactions since a unique energy scale is used. Furthermore, bound states energies are no longer expressed as a negative value of energy.

The use of the CoM frame gives also the tools to raise inconsistencies in the parameter files that support different energy scales. In the table 1 below, a straightforward comparison between resonance parameters from two separate databases (astrophysics [14] and reactors physics from CIELO project [10]) is done. Inconsistencies are highlighted. The three first total width values in the CIELO base are very large compared to those reported in the astrophysical base, suggesting very low life time assigned to these three levels. One unexpected J^{π} assignment (the second level in the Table 1) can also be noticed. On the contrary, the level energies are in good agreement in spite of the highest level on which the energy is slight lower in the CIELO base.

Table 1. The comparison between resonances parameters for ten major levels from the Astrophysics and the CIELO project database using the unified framework. The bold numbers highlight most important differences between the two databases.

Astrophysics database (reference [14])			**CIELO Reich-Moore parameters (reference [10])**		
J^π	E_λ (eV)	Γ (meV)	J^π	E_λ (eV)	Γ (meV)
1/2+	8.70e+05	**1.27e-03**	1/2+	8.7267e+05	**3.6665e+09**
1/2-	3.05e+06	**2.74**	**3/2-**	3.0552e+06	**2.6721e+07**
5/2-	3.84e+06	**≤ 1.3e+01**	5/2-	3.8434e+06	**4.2045e+04**
3/2-	4.55e+06	4.0e+07	3/2-	4.5522e+06	4.1595e+07
7/2-	6.97e+06	< 1.0e+06	7/2-	6.9726e+06	7.9008e+04
5/2-	7.16e+06	1.38e+06	5/2-	7.1650e+06	1.6502e+06
3/2+	7.202e+06	2.80e+08	3/2+	7.2355e+06	3.0265e+08
5/2+	7.3792e+06	6.4e+05	5/2+	7.3784e+06	4.6646e+05
5/2-	7.3822e+06	9.6e+05	5/2-	7.3812e+06	1.4434e+06
3/2-	**7.559e+06**	5.00e+08	3/2-	**7.4224e+06**	6.7180e+08

4.2 Data reversibility

According to the assumption of Weisskopf and Ewing [15], the cross section of a reaction in which a projectile x collides with a target X to lead to a residual nucleus Y and an ejectile y, can be split as follows:

$$\sigma(x,y) = S_x(E) \times \xi_x(E) \times \eta_y(E_x) \qquad (6)$$

Above hypothesis relies on the fact that given compound nucleus Y* is formed from the couple (x,X) and decays in the couple (y,Y). First product of right side term of Eq. (6) $S_x(E) \times \xi_x(E)$ represents the probability of the formation of compound nucleus. Among those two variables, $S_x(E)$ represents the cross section for reaching the surface of the target nucleus. The incident wave being partially reflected by the potential at the target nucleus surface, $\xi_x(E)$ is the probability denoting the fraction of the incident wave which interacts with the nucleons of the target nucleus. The last variable $\eta_y(E_x)$ quantifies the decaying probability of the compound nucleus giving the couple (y,Y). E is the energy of the projectile and E_x is the excited energy of a state of the compound nucleus created.

The interaction between a particle x and a target nucleus X does not always lead to the formation of a compound nucleus. Other interactions [16] may occur. A competition is possible between the reaction passing through the formation of the compound nucleus and other forms of reaction: surface and volume direct interactions, multiple collisions or even collective nucleon excitation (rotations and vibrations). The evaluation of the $^{16}O(n,\alpha)^{13}C$ cross section has been classically performed using the inverted $^{13}C(\alpha,n)^{16}O$ data from from Bair and Haas [5] and Harissopulos [4]. This inversion of data relies on the compound nucleus hypothesis suggesting that the entire energy brought by the incident wave is right away and equitably shared among all nucleons of the target. The process must last long enough so that the way of the formation of the compound nucleus is "forgotten" by the latter. Subsequent decay probability in given channel is thus independent from the formation channel but both must be connected to the same nuclear state of the compound nucleus. The compound nucleus hypothesis has been illustrated through the computation of the $^{13}C(\alpha,n)^{16}O$ and $^{16}O(n,\alpha)^{13}C$ cross sections. Below are shown results of the comparison between the cross sections obtained by direct kinematics calculation on one side and those obtained by the data inversion on the other side. The inversion method used to obtain the inverse of a reaction cross section $\sigma_{cc'}(E_c^{CoM})$ for which a compound nucleus is formed from a channel c and decays into a channel c' relies on the expression [9]:

$$\sigma_{c'c}(E_{c'}^{CoM})\lambda_c^2 = \sigma_{cc'}(E_c^{CoM})\frac{(2i_c+1)(2I_c+1)}{(2i_{c'}+1)(2I_{c'}+1)}\lambda_{c'}^2 \tag{7}$$

E_c^{CoM}, i_c, I_c and λ_c are respectively the energy of the projectile in the CoM framework, the spin of the projectile, the spin of the target and the reduced wavelength of the projectile in the channel c. $E_{c'}^{CoM}$, $i_{c'}$, $I_{c'}$ and $\lambda_{c'}$ are reciprocal quantities still in CoM system.

A quick eye on the Figure 3 shows that relative differences (down side plots) between the cross sections obtained from either the direct kinematics calculation or the pseudo cross sections obtained by inversion of data are very small. This trend is likely a confirmation of the predominance of a compound nucleus system at low excitation energy.

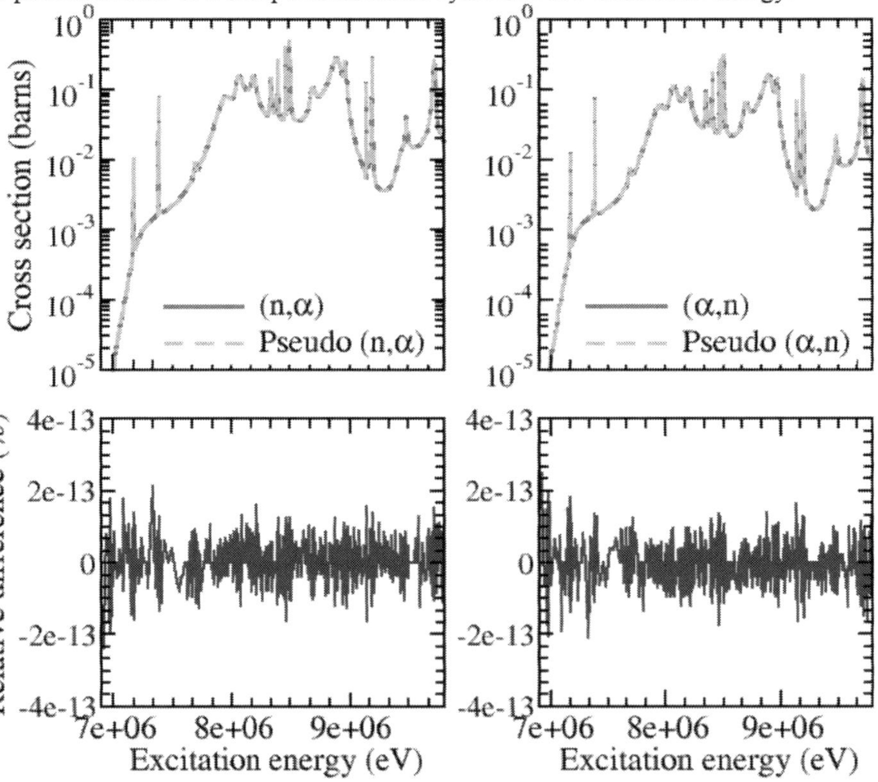

Fig.3. Comparison of the $^{16}O(n,\alpha)^{13}C$ and $^{13}C(\alpha,n)^{16}O$ cross sections obtained using direct kinematics calculation with the data obtained from the inversion of their reciprocal reaction cross sections labelled as pseudo data. The computation is performed using TORA. Relative differences in percentage are computed as the ratio [(Pseudo data – Direct kinematics calculated data)/Pseudo data].

5. Conclusion

A preliminary investigation of the $^{17}O^*$ compound nucleus has been made, using the CIELO Reich-Moore parameters, through a modelling of the $^{16}O(n,\alpha)^{13}C$ and $^{13}C(\alpha,n)^{16}O$ first in the incident particle laboratory frame and secondary in the CoM frame. Both two cross sections have been benchmarked to the inversed cross sections retrieved from the reciprocal reactions. A comparison between nuclear parameters from the Astrophysics and Reactors physics databases highlighted notable differences. A solution to avoid the identified inconsistencies and reach a better accuracy is definitively the use of the CoM of the compound nucleus framework and energy-independent parameters in the nuclear reactions modelling. The approach has been set up by a simulation of the $^{16}O(n,n)^{16}O$, $^{16}O(n,\alpha)^{13}C$, $^{13}C(\alpha,n)^{16}O$ and $^{13}C(\alpha,\alpha)^{13}C$ reactions. A set of parameters obtained from a fit of experimental data within this unified framework, by the mean of a TORA full connexion to the CONRAD code, is foreseen.

6. References

1. M.B. Chadwick, E. Dupont, E. Bauge et al., Nucl. Data Sheets **118**, 1 (2014)
2. M.B. Chadwick, R. Capote, A. Trkov et al., Nucl. Data Sheets **148**, 189 (2018)
3. G. Giorginis et al., EDP Siences, 525 (2008)
4. S. Harissopulos et al., Phys. Rev. C **72**, 062801 (R) (2005)
5. J.K. Bair and F.X. Haas, Phys. Rev. **7**, 1356 (1973)
6. C. De Saint Jean et al., EDP Sciences **1**, 251 (2008)
7. C.W. Reich and M.S. Moore, Phys. Rev. **111**, 929 (1958)
8. M. Larson, Technical Report ORNL/TM- 9179/R8ENDF-364/R2, ORNL (2008)
9. F. H. Fröhner, JEFF Report 18, OECD/NEA (2000)
10. http://www.oecd-nea.org/dbdata/jeff-beta/JEFF33T3/neutrons/files/8-O-16g.jeff33t3
11. A.M. Lane and R.G. Thomas, Rev. Mod. Phys. **30**, 257 (1958)
12. J.M. Blatt and M.C. Biedenharn, Rev. Mod. Phys. **24**, 258 (1952)
13. B. K. Barnes, T. A. Belote, J. R. Risser, Phys. Rev. **140**, B616 (1965)
14. D. R. Tilley et al., Nucl. Phys. A **564**, 1 (1993)
15. V.F. Weisskopf and D.H. Ewing, Phys. Rev. **57**, 472 (1940)
16. V.F. Weisskopf, Nucl. Phys. **29**, 174 (1957)

EPJ Web of Conferences **211**, 02002 (2019) https://doi.org/10.1051/epjconf/201921102002
WONDER-2018

Study of the photon strength functions and level density in the gamma decay of the n + ^{234}U reaction

J. Moreno-Soto[1,*], E. Berthoumieux[1], E. Dupont[1], F. Gunsing[1], O. Serot[2], O. Litaize[2], M. Diakaki[2], A. Chebboubi[2], W. Dridi[3], S. Valenta[33], M. Krtička[33], O. Aberle[4], V. Alcayne[5], J. Andrzejewski[6], L. Audouin[7], V. Bécares[5], V. Babiano-Suarez[8], M. Bacak[4,9,1], M. Barbagallo[4,10], Th. Benedikt[11], S. Bennett[12], J. Billowes[12], D. Bosnar[13], A. Brown[14], M. Busso[10,15,16], M. Caamaño[17], L. Caballero-Ontanaya[8], F. Calviño[18], M. Calviani[4], D. Cano-Ott[5], A. Casanovas[18], F. Cerutti[4], E. Chiaveri[4,12], N. Colonna[10], G. Cortés[18], M. A. Cortés-Giraldo[19], L. Cosentino[20], S. Cristallo[10,15,21], L. A. Damone[10,22], P. J. Davies[12], M. Dietz[23], C. Domingo-Pardo[8], R. Dressler[24], Q. Ducasse[25], I. Durán[17], Z. Eleme[26], B. Fernández-Domínguez[17], A. Ferrari[4], P. Finocchiaro[20], V. Furman[27], K. Göbel[11], A. Gawlik[6], S. Gilardoni[4], I. F. Gonçalves[28], E. González-Romero[5], C. Guerrero[19], S. Heinitz[24], J. Heyse[29], D. G. Jenkins[14], A. Junghans[30], F. Käppeler[31], Y. Kadi[4], A. Kimura[32], I. Knapova[33], M. Kokkoris[34], Y. Kopatch[27], D. Kurtulgil[11], I. Ladarescu[8], C. Lederer-Woods[23], S. J. Lonsdale[23], D. Macina[4], A. Manna[35,36], T. Martínez[5], A. Masi[4], C. Massimi[35,36], P. Mastinu[37], M. Mastromarco[4], E. A. Maugeri[24], A. Mazzone[10,38], E. Mendoza[5], A. Mengoni[39], V. Michalopoulou[4,34], P. M. Milazzo[40], F. Mingrone[4], A. Musumarra[20,41], A. Negret[42], F. Ogállar[43], A. Oprea[42], N. Patronis[26], A. Pavlik[44], J. Perkowski[6], L. Persanti[10,15,21], C. Petrone[42], E. Pirovano[25], I. Porras[43], J. Praena[43], J. M. Quesada[19], D. Ramos-Doval[7], T. Rauscher[45,46], R. Reifarth[11], D. Rochman[24], M. Sabaté-Gilarte[19,4], A. Saxena[47], P. Schillebeeckx[29], D. Schumann[24], A. Sekhar[12], S. Simone[20], A. G. Smith[12], N. V. Sosnin[12], P. Sprung[24], A. Stamatopoulos[34], G. Tagliente[10], J. L. Tain[8], A. Tarifeño-Saldivia[18], L. Tassan-Got[4,34,7], A. Tsinganis[4], J. Ulrich[24], S. Urlass[30,4], G. Vannini[35,36], V. Variale[10], P. Vaz[28], A. Ventura[35], D. Vescovi[10,15], V. Vlachoudis[4], R. Vlastou[34], A. Wallner[48], P. J. Woods[23], T. Wright[12], P. Žugec[13], and the n_TOF Collaboration

[1]CEA Irfu, Université Paris-Saclay, F-91191 Gif-sur-Yvette, France
[2]CEA, DEN, DER, SPRC, Cadarache, F-13108 Saint-Paul-lez-Durance, France
[3]Laboratory on Energy and Matter for Nuclear Sciences Development (LR16CNSTN02), Technopark Sidi Thabet, 2020 Ariana, Tunisia
[4]European Organization for Nuclear Research (CERN), Switzerland
[5]Centro de Investigaciones Energéticas Medioambientales y Tecnológicas (CIEMAT), Spain
[6]University of Lodz, Poland
[7]Institut de Physique Nucléaire, CNRS-IN2P3, Univ. Paris-Sud, Université Paris-Saclay, F-91406 Orsay Cedex, France
[8]Instituto de Física Corpuscular, CSIC - Universidad de Valencia, Spain
[9]Technische Universität Wien, Austria
[10]Istituto Nazionale di Fisica Nucleare, Sezione di Bari, Italy
[11]Goethe University Frankfurt, Germany
[12]University of Manchester, United Kingdom
[13]Department of Physics, Faculty of Science, University of Zagreb, Zagreb, Croatia

*e-mail: javier.moreno-soto@cea.fr

[14]University of York, United Kingdom
[15]Istituto Nazionale di Fisica Nucleare, Sezione di Perugia, Italy
[16]Dipartimento di Fisica e Geologia, Università di Perugia, Italy
[17]University of Santiago de Compostela, Spain
[18]Universitat Politècnica de Catalunya, Spain
[19]Universidad de Sevilla, Spain
[20]INFN Laboratori Nazionali del Sud, Catania, Italy
[21]Istituto Nazionale di Astrofisica - Osservatorio Astronomico di Teramo, Italy
[22]Dipartimento di Fisica, Università degli Studi di Bari, Italy
[23]School of Physics and Astronomy, University of Edinburgh, United Kingdom
[24]Paul Scherrer Institut (PSI), Villingen, Switzerland
[25]Physikalisch-Technische Bundesanstalt (PTB), Bundesallee 100, 38116 Braunschweig, Germany
[26]University of Ioannina, Greece
[27]Joint Institute for Nuclear Research (JINR), Dubna, Russia
[28]Instituto Superior Técnico, Lisbon, Portugal
[29]European Commission, Joint Research Centre, Geel, Retieseweg 111, B-2440 Geel, Belgium
[30]Helmholtz-Zentrum Dresden-Rossendorf, Germany
[31]Karlsruhe Institute of Technology, Campus North, IKP, 76021 Karlsruhe, Germany
[32]Japan Atomic Energy Agency (JAEA), Tokai-mura, Japan
[33]Charles University, Prague, Czech Republic
[34]National Technical University of Athens, Greece
[35]Istituto Nazionale di Fisica Nucleare, Sezione di Bologna, Italy
[36]Dipartimento di Fisica e Astronomia, Università di Bologna, Italy
[37]Istituto Nazionale di Fisica Nucleare, Sezione di Legnaro, Italy
[38]Consiglio Nazionale delle Ricerche, Bari, Italy
[39]Agenzia nazionale per le nuove tecnologie (ENEA), Bologna, Italy
[40]Istituto Nazionale di Fisica Nucleare, Sezione di Trieste, Italy
[41]Dipartimento di Fisica e Astronomia, Università di Catania, Italy
[42]Horia Hulubei National Institute of Physics and Nuclear Engineering, Romania
[43]University of Granada, Spain
[44]University of Vienna, Faculty of Physics, Vienna, Austria
[45]Department of Physics, University of Basel, Switzerland
[46]Centre for Astrophysics Research, University of Hertfordshire, United Kingdom
[47]Bhabha Atomic Research Centre (BARC), India
[48]Australian National University, Canberra, Australia

Abstract. The accurate calculations of neutron-induced reaction cross sections are relevant for many nuclear applications. The photon strength functions and nuclear level densities are essential inputs for such calculations. These quantities for ^{235}U are studied using the measurement of the gamma de-excitation cascades in radiative capture on ^{234}U with the Total Absorption Calorimeter at n_TOF at CERN. This segmented 4π gamma calorimeter is designed to detect gamma rays emitted from the nucleus with high efficiency. This experiment provides information on gamma multiplicity and gamma spectra that can be compared with numerical simulations. The code DICEBOXC is used to simulate the gamma cascades while GEANT4 is used for the simulation of the interaction of these gammas with the TAC materials. Available models and their parameters are being tested using the present data. Some preliminary results of this ongoing study are presented and discussed.

1 Introduction

The aim of this work is to describe the γ-decay of excited nucleus following neutron capture. At low excitation energy, the number of levels per unit energy is rather small and levels can be experimentally resolved. However, as the excitation energy increases the level density is also increasing, so the statistical model is needed to describe the levels and transitions between them – the used quantities are nuclear level density (LD) and photon strength functions (PSFs). Their improved experimental and theoretical description is important for modeling of radiative capture reactions in nuclear astrophysics and nuclear technologies since the neutron capture cross sections above the resolved resonance region are usually calculated using the statistical model of Hauser-Feshbach [1] for which PSFs and LDs are essential inputs.

In this work, the Total Absorption Calorimeter (TAC) at n_TOF facility (CERN) [2–4] was used to measure ^{234}U(n,γ) reaction [5]. The TAC is a 4π detector segmented in 40 BaF_2 crystals with a very high efficiency (almost 100%) to detect the γ rays from the capture cascades. In Fig. 1 (left) one hemisphere of the TAC is shown. The ^{234}U sample is placed in the center and emits γ rays, which are detected by the BaF_2 detectors. Thanks to the segmentation of the detector it is possible to discriminate against the background by putting conditions on the multiplicity and the total deposited energy of events registered by the TAC.

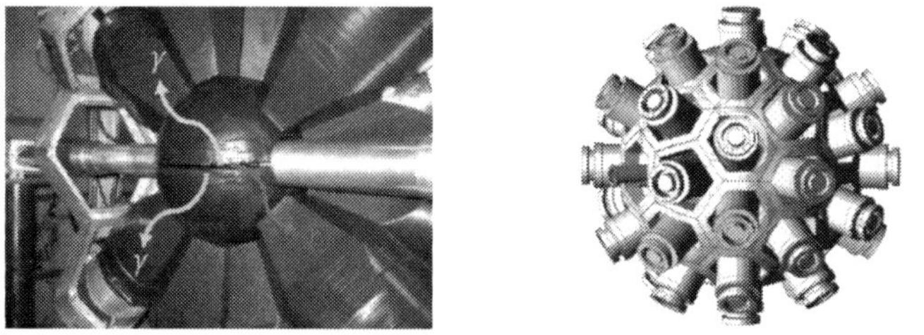

Figure 1. One hemisphere of the TAC consisting of the BaF_2 detectors, the neutron beam tube and the neutron absorber. A cascade event of three γ rays is depicted. If all γ rays are detected the crystal multiplicity is $m_{cr} = 3$ (left). The geometry of the full TAC as implemented in GEANT4 (right).

2 Experimental data

In a radiative capture reaction the compound nucleus decays through a cascade of γ rays. The measured cascade events are reconstructed by taking γ rays detected by the TAC in the BaF_2 detectors in a time coincidence window of 20 ns. A software threshold of 75 keV was set for all BaF_2 detectors to suppress the low energy background. The observables used for the analysis are:

- The neutron energy, E_n, calculated from the measured time of flight.
- The crystal multiplicity, m_{cr}, given by the number of hit crystals in each detected cascade event.
- The total deposited energy or sum energy E_{sum} in the detectors for each cascade event.
- The multi-step cascade spectra for each crystal multiplicity m_{cr}, which are the γ-ray energy spectra for fully detected cascades.

The ^{234}U(n,γ) time-of-flight spectrum is shown in the left panel of Fig. 2. Besides the uranium resonances, a structure due to capture reactions in the Ti canning is observed above a few keV [6].

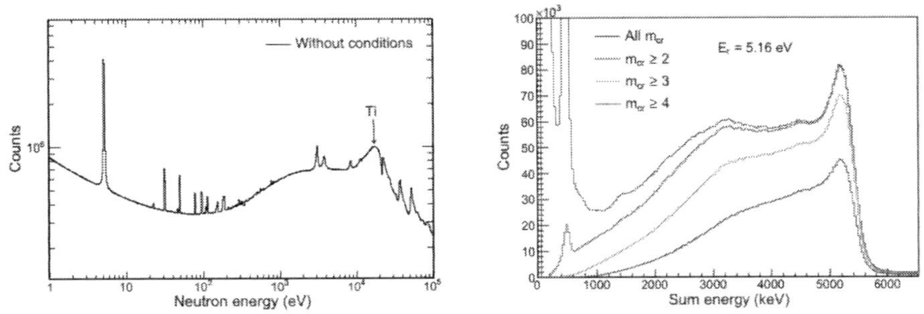

Figure 2. Time-of-flight spectrum converted to neutron energy (left). Total energy deposited for different sets of crystal multiplicities for $^{234}U(n,\gamma)$ in the resonance at 5.16 eV (right).

Fig. 2 (right) shows the sum-energy spectra in the resonance at 5.16 eV, corrected for background, for different multiplicity criteria [7]. All spectra clearly show the sum-peak at 5.3 MeV corresponding to the Q value of the reaction. There are differences between the spectra depending on the considered multiplicities. At low sum-energies (below 1 MeV) the spectrum for all m_{cr} is dominated by the remaining background. However, in the spectra for $m_{cr} \geq 2$ this background is completely absent. For the present study, only cascades with $m_{cr} \geq 2$ are considered to ensure that the background is correctly subtracted, furthermore the multistep cascade spectra are constructed using only the events with sum-energy in the interval $4.8 < E_{sum}$ (MeV) < 6.0.

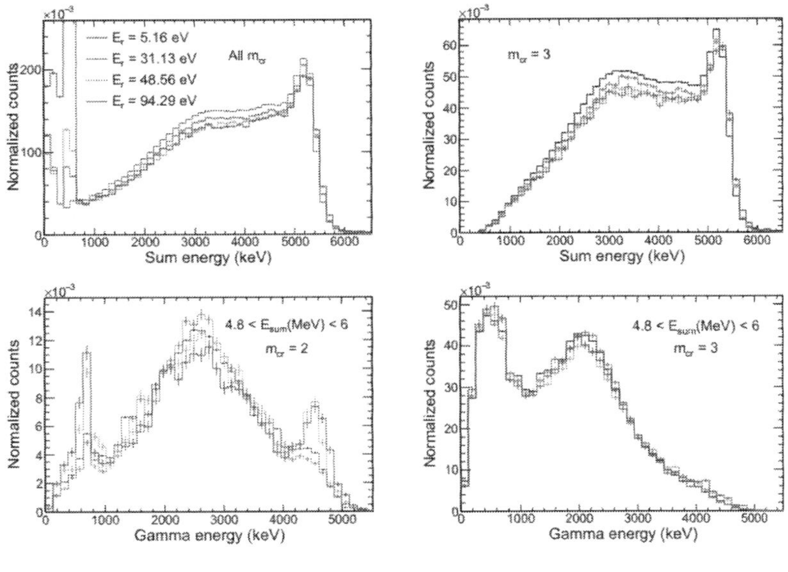

Figure 3. Sum-energy spectra of different resonances for all m_{cr} and $m_{cr} = 3$ (top). Multi-step γ-ray energy spectra of different resonances for $m_{cr} = 2, 3$ (bottom). The statistical uncertainties are small.

As can be seen in Fig. 3 (top), the sum-energy spectra for all m_{cr} of four s-wave resonances show significant differences at low energy only. These differences are due to the fact that the subtraction of the background is approximative – the remaining background induced by the scattered neutrons is most apparent in the resonance at $E_r = 94.29$ eV (blue) due to the larger scattering width. However, these differences do not appear for $m_{cr} \geq 3$ because of the low multiplicity nature of the background. More importantly, the background subtraction gets more accurate with increasing sum-energy, hence all sum-energy spectra have similar behavior for $E_{sum} > 1$ MeV.

The multi-step γ-ray energy spectra, see Fig. 3 (bottom), show similar responses for the different resonances for $m_{cr} \geq 3$. The spectra for $m_{cr} = 2$ show significant differences depending on the considered resonance. These differences can be attributed to the Porter-Thomas fluctuations of primary transition intensities among the resonances, as expected the effects are mostly noticeable at the edges of the $m_{cr} = 2$ multi-step γ-ray energy spectra.

The normalization of all spectra was done by dividing the spectra by the number of counts in the sum-energy spectrum of $m_{cr} \geq 2$ between $E_{sum} = 4.8$ MeV and 6 MeV. The same normalization was applied to the simulations.

3 Simulations

The results presented in this work are based on the comparison of experimental data with statistical model simulations of γ decay. The in-house developed Monte Carlo code DICEBOXC, based on the same algorithm used by F. Bečvář [8] in his code DICEBOX, was used to simulate the gamma cascades while GEANT4 was used for the simulation of the interaction of these gammas with the complete TAC experimental assembly [2].

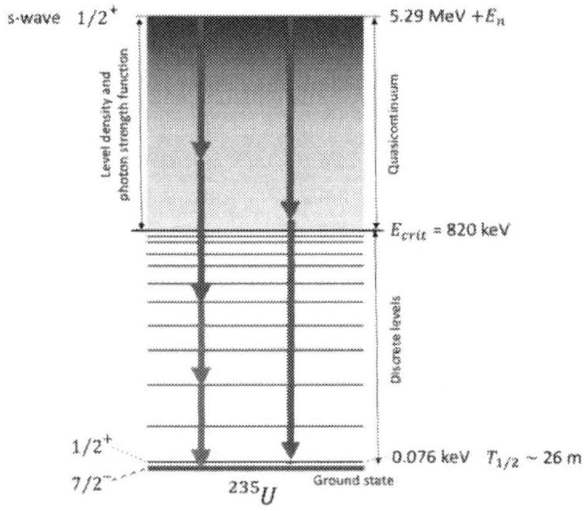

Figure 4. Schema of Monte Carlo cascades generation with DICEBOXC code. Red arrows depict the transitions generated in terms of LD and PSFs, green arows are the transitions among discrete levels taken from spectroscopic data.

DICEBOXC simulates sets of levels and their partial radiation widths known as nuclear realizations [8]. To describe the decay scheme, below a critical energy E_{crit} all the level energies, spins, parities and branching intensities of depopulating transitions are taken from existing experimental data. Above E_{crit}, the level scheme is generated by the code – the levels are obtained by a random discretization of an a priori known LD formula. Further, the PSFs are used to generate probabilities of transitions of type X (electric or magnetic) and multipolarity L. Fig. 4 shows a diagram of the operation of DICEBOXC.

The partial radiation width of an electromagnetic transition from level i to level f, $\Gamma_{i\gamma f}$, is selected from a Porter-Thomas distribution with the mean value $\left\langle \Gamma_{i\gamma f}^{XL} \right\rangle$ defined as

$$\left\langle \Gamma_{i\gamma f}^{XL} \right\rangle = \frac{f^{XL}(E_\gamma) \cdot E_\gamma^{2L+1}}{\rho(E_i, J_i, \pi_i)} \tag{1}$$

where ρ is the LD and $f^{XL}(E_\gamma)$ is the PSF. The γ-ray transition probabilities are corrected for internal conversion using tables from Ref. [9]. The levels and transitions below E_{crit} are taken from ENSDF database [10]. To ensure satisfactory statistics concerning the modelled quantities, 20 nuclear realizations with 10^5 γ-cascades per realization were simulated for each combination of LD and PSFs models.

To simulate the transport and detection of γ-rays, the toolkit GEANT4 is used [11]. The geometry and efficiency of TAC have been accurately modeled following CAD drawings of the engineering design and direct measurements [2, 12]. The modeled geometry is shown in the right panel of Fig. 1. Finally, an amplitude resolution of about 13-17%, depending on the detector, and a threshold of 75 keV for all crystals is assumed.

3.1 Level density models

The LD for given spin and parity is calculated as the product of three factors: the parity distribution $P(E, \pi)$, the spin distribution $R(E, J)$ and the LD $\rho(E)$. In this work, one assumes that both parities are equally probable $P(E, \pi) = 1/2$ at all E, while $R(E, J)$ is

$$R(E, J) = \exp\left(-\frac{J^2}{2\sigma_c^2}\right) - \exp\left(-\frac{(J+1)^2}{2\sigma_c^2}\right) \approx \frac{2J+1}{2\sigma_c^2}\exp\left[-\frac{\left(J+\frac{1}{2}\right)^2}{2\sigma_c^2}\right], \tag{2}$$

where σ_c is the spin cut-off factor. Different forms of spin cut-off factor could be used.

The DICEBOXC code includes various models for the LD $\rho(E)$. The Constant Temperature (CT) [13] model assumes that the number of levels varies according to the constant temperature law and LD is given by

$$\rho(E) = \frac{1}{T}\exp\left(\frac{E-E_0}{T}\right), \tag{3}$$

with parameters E_0 and a nuclear temperature T, which are usually fitted to experimental discrete levels, taken from Ref. [14]. In this work the spin cut-off parameter, which is constant for a given nucleus [15], was used.

The Back-shifted Fermi Gas (BSFG) [16] model assumes the nucleus as a gas of fermions creating pairs and single particle levels are equally spaced and non-degenerated with a LD given by

$$\rho(E) = \frac{\exp\left(2\sqrt{a(E-E_1)}\right)}{12\sqrt{2}\sigma_c a^{1/4}(E-E_1)^{5/4}}, \tag{4}$$

where E_1 is the energy backshift and a is the LD parameter. The energy-dependent spin cut-off factor for the BSFG model was taken from Ref. [15]. The parameters for BSFG model were taken from Ref. [14]. Variations of the BSFG model have been developed, as for example in ref. [17], which incorporates the thermodynamic temperature t. In this case, the spin cut-off is related to a fraction of the moment of inertia of the nucleus that is usually taken between 0.5 and 1. In addition, a BSFG model with energy-dependent LD parameter a and spin cut-off which accounts for the damping of the shell effects was introduced in RIPL-3 [18]. Finally, we used microscopic LD in the form of numerical interpolation tables calculated with the Hartree-Fock-Bogoliubov (HFB) method [19, 20].

3.2 Photon strength function

The statistical decay of compound nuclei from excitation energies above neutron separation energy is dominated by the $E1$ transitions due to the presence of the giant dipole electric resonance (GDER). The shape of the $E1$ PSF for deformed nucleus is usually described by a sum of two standard Lorentzians [18, 21], as a consequence of vibration modes along and perpendicular to the symmetry axis. This description is known as Standard Lorentzian model (SLO):

$$f_{SLO}^{E1}(E_\gamma) = \frac{1}{3(\pi\hbar c)^2} \sum_{i=1}^{2} \frac{\sigma_{G_i} E_\gamma \Gamma_{G_i}^2}{\left(E_\gamma^2 - E_{G_i}^2\right)^2 + E_\gamma^2 \Gamma_{G_i}^2}, \tag{5}$$

where the parameters E_{G_i}, Γ_{G_i} and σ_{G_i} are the energy, width and cross section of the GDER.

Different variations were proposed to better describe the energy region below neutron separation energy. The model by Kadmenskii, Markushev and Furman (KMF) [22] aims only at this energy region while generalised Lorentzian models by Kopecky, Uhl and Chrien (GLO, ELO, EGLO) [23] and other models and calculations attempt to describe the $E1$ PSF in the whole energy region.

The KMF, GLO and ELO models use the damping width $\Gamma_T(E_\gamma, T_f)$ which depends on E_γ and the nuclear temperature T_f in form

$$\Gamma_T(E_\gamma, T_f) = \frac{\Gamma_G}{E_G^2}(E_\gamma^2 + 4\pi^2 T_f^2). \tag{6}$$

Phenomenological modifications of this damping width in which is introduced a k parameter were proposed in the EGLO [23] and the MGLO [24] models. There are other models for $E1$ PSF, we refer the reader to the overview in RIPL-3 [18].

For the decay of levels below the neutron separation energy, $M1$ transitions play an important role. In this work the $M1$ PSF consists of the spin-flip (SF) resonance, which dominates the $M1$ PSF at relatively high energy typically around 7 MeV, and the scissor mode (SC), a concentration of $M1$ strength around 2-3 MeV in deformed nuclei. The SLO model was adopted to describe the $M1$ PSF. For further details see review [25] and references therein.

The electric quadrupole ($E2$) transitions, although playing a minor role with respect to dipole transitions, are also taken into account. The SLO model with a single Lorentzian was used to describe the $E2$ PSF as recommended in [18].

4 Comparison of simulations and measurements

Various combinations of LD and PSF models were checked and compared with the experimental data introduced in Sec. 2. The parameters were taken from RIPL-3 database [18] in which only one SLO term for the $M1$ PSF is recommended, or from original works, (i) the analysis of d- and ^{3}He-induced reactions on actinide targets performed at the Oslo Cyclotron Laboratory (OCL) [26] and (ii) the measurement of multi-step γ-ray energy spectra from resonant neutron capture on uranium samples with DANCE calorimeter [27]. In both a sum of SLO terms was adopted to describe the M1 PSF – one for the SF and two for the SC. The $E2$ transitions were included in the simulations by taking the parameters from [28]. The parameters use for the PSF in the different simulations are collected in Table. 1.

From the $E1$ PSF models introduced in Sec. 3.2 the SLO and KMF models do not reproduce the experimental data in combination with any LD model independently on the chosen parametrisation of the $M1$ PSF. Conversely, the ELO, GLO, EGLO and MGLO $E1$ PSF models, paired with a suitable LD model, allow, by tuning the parameters of the $M1$ PSF and the k parameter, a satisfactory description of the experimental data.

In Fig. 5 we compare experimental data with simulations using the PSFs parameters taken from (i) the RIPL-3 database [18] with GLO for $E1$ and SLO for $M1$ PSF, (ii) the DANCE analysis [27] and (iii) the analysis of d- and ^{3}He-induced reactions at OCL [26]. The standard deviation due to different nuclear realizations is only calculated in the simulation for RIPL-3, for the other simulations the standard deviation shows similar behaviour and is not displayed for a better visualization. The statistical uncertainties are much smaller than the standard deviation. Overall, the introduction of the SC is mandatory for the improvement of the simulation and the increase of the SC strength in the DANCE analysis with respect to OCL improves the description of the experimental data. It may be possible that in order to match experimental data the SC strength has to be further increased and used in conjunction with steeper $E1$ PSF of generalised Lorentzian type.

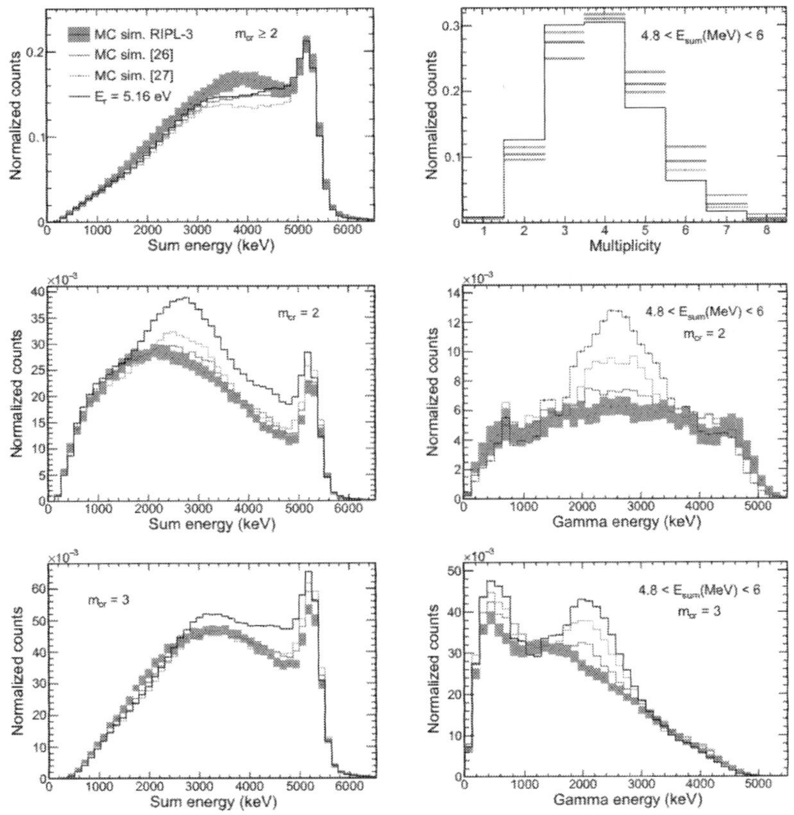

Figure 5. Comparison of experimental data to simulations using the LD and PSF models as recommended in RIPL-3 database [18] (grey filled), and as published in Refs. [26] (red line) and [27] (green line). The left column shows the total deposited energy spectra while on the right the multiplicity distribution and multi-step γ-ray energy spectra are shown. The resonance energy as well as the multiplicity and sum-energy conditions are specified in the figures.

5 Conclusion

The Total Absorption Calorimeter at the n_TOF facility (CERN) was used to measure the γ-ray cascades following the neutron capture in ^{234}U. Simulations of γ decay performed with DICEBOXC for various LD and PSFs combinations were compared with the experimental data. The inadequacy of the SLO and KMF models of $E1$ PSF as well as the necessity of scissors mode contribution to $M1$ PSF was shown. The simulations with model combinations proposed in OCL and DANCE analyses do not reproduce our data. This analysis will continue with a detailed parameter search for analytical models and the use of tabulated PSFs from QRPA calculations [29], as well as the extension to other actinides.

Table 1. Parameters from RIPL-3 [18], OCL [26] and DANCE [27] for the PSFs.

Transition	E_1 (MeV)	Γ_1 (MeV)	σ_1 (mb)	E_2 (MeV)	Γ_2 (MeV)	σ_2 (mb)	E_3 (MeV)	Γ_3 (MeV)	σ_3 (mb)
E1 [18]	11.11	1.12	243.3	13.41	4.98	426	–	–	–
M1 [18]	–	–	–	–	–	–	6.61	4.00	2.35
E2 [28]	10.21	1.18	1.7	–	–	–	–	–	–
E1 [26]	11.40	4.20	572	14.40	4.20	1040	7.30	2.0	15.0
M1 [26]	2.15	0.80	0.45	2.90	0.60	0.40	6.61	4.00	7.00
E2 [28]	10.21	1.18	1.7	–	–	–	–	–	–
E1 [27]	11.28	2.48	325	13.73	4.25	384			
M1 [27]	2.15	0.80	0.60	2.90	0.60	0.53	6.61	4.00	1.50
E2 [28]	10.21	1.18	1.7	–	–	–	–	–	–

References

[1] W. Hauser, H. Feshbach, Phys. Rev. **87**, 366 (1952)
[2] C. Guerrero et al. (n_TOF), Nucl. Instrum. Meth. A **608**, 424 (2009)
[3] E. Chiaveri et al. (n_TOF), Nucl. Data Sheets **119**, 1 (2014)
[4] C. Guerrero et al. (n_TOF), Eur. Phys. J. A **49**, 27 (2013)
[5] W. Dridi, *PhD thesis* (Université d'Evry Val d'Essonne, 2006)
[6] C. Lampoudis et al. (n_TOF), in *Proceedings of the 2007 International Conference on Nuclear Data for Science and Technology* (2008), Vol. 1, pp. 595–598
[7] W. Dridi et al. (n_TOF), in *Conference Proceedings of the American Nuclear Society's Topical Meeting on Reactor Physics* (2006), Vol. C032, pp. 1–10
[8] F. Bečvář, Nucl. Instrum. Meth. A **417**, 434 (1998)
[9] R.S. Hager, E.C. Seltzer, Nucl. Data **A4**, 1 (1968)
[10] E. Browne, J. Tuli, Nucl. Data Sheets **122**, 205 (2014)
[11] S. Agostinelli et al. (GEANT4), Nucl. Instrum. Meth. A **506**, 250 (2003)
[12] C. Guerrero et al. (n_TOF), Nucl. Instrum. Meth. A **671**, 108 (2012)
[13] A. Gilbert, A.G.W. Cameron, Can. J. Phys. **43**, 1446 (1965)
[14] T. von Egidy, D. Bucurescu, Phys. Rev. C **72**, 044311 (2005)
[15] T. von Egidy, H.H. Schmidt, A.N. Behkami, Nucl. Phys. A **481**, 189 (1988)
[16] H.A. Bethe, Phys. Rev. **50**, 332 (1936)
[17] W. Dilg, W. Schantl, H. Vonach, M. Uhl, Nucl. Phys. A **217**, 269 (1973)
[18] R. Capote et al., Nucl. Data Sheets **110**, 3107 (2009)
[19] S. Goriely, S. Hilaire, A.J. Koning, Phys. Rev. C **78**, 064307 (2008)
[20] A.J. Koning, S. Hilaire, S. Goriely, Nucl. Phys. A **810**, 13 (2008)
[21] G.A. Bartholomew, Annual Review of Nuclear Science **11**, 259 (1961)
[22] S.G. Kadmenskii, V.P. Markushev, V.I. Furman, Sov. J. Nucl. Phys. **37**, 165 (1983)
[23] J. Kopecky, M. Uhl, R.E. Chrien, Phys. Rev. C **47**, 312 (1993)
[24] J. Kroll et al., Phys. Rev. C **88**, 034317 (2013)
[25] K. Heyde, P. von Neumann-Cosel, A. Richter, Rev. Mod. Phys. **82**, 2365 (2010)
[26] M. Guttormsen et al., Phys. Rev. C **89**, 014302 (2014)
[27] J.L. Ullmann et al., Phys. Rev. C **96**, 024627 (2017)
[28] *Handbook for Calculations of Nuclear Reaction Data Reference Input Parameter Library*, IAEA-TECDOC-1034 (Vienna, 1998)
[29] S. Goriely et al., Phys. Rev. C **98**, 014327 (2018)

High-resolution evaluation of the U5(n,f) cross section from 3 keV to 30 keV

I. Duran[1], *C. Paradela*,[1,2], *M. Caamaño*[1], *P. Cabanelas* [1], *L. Tassan-Got*[3], *L. Audouin*[3], *on behalf of the CERN - nTOF Collaboration*

1) Universidad de Santiago de Compostela, 15782, Spain
2) EC-JRC-IRMM, Retieseweg 111, B-2440 Geel, Belgium
3) CNRS/IN2P3-Université Paris-Sud- IPN, Orsay,

Abstract. ^{235}U neutron-induced fission cross-section is commonly used as reference for determining other isotope fission cross-section. However, below 150 keV this cross section is only included as Standard at the thermal point and recently its integral value between 7.8 eV and 11 eV [1]. The resolved resonance region, spanning up to 2.25 keV, has been re-evaluated with high resolution in the last ENDF/B-VIII release [2] and a SAMMY resonance analysis was done by L. Leal et al. [3] including the CERN-nTOF experimental work of Paradela el al. [4] up to 10 keV, taken into account the IAEA Reference file.
In this work the ^{235}U(n,f) low-background and high-resolution experimental data obtained at the CERN-nTOF facility is combined with previous high-resolution experimental data, in order to produce a very fine grid dataset with normalisation to the IAEA Reference file. The extremely-high energy calibration required to reproduce the resonance sharp profiles is based on the nTOF DAQ system with a resolution below 0.1% with reference to the 8.78 eV resonance and to the sharp Al(n,g) capture dip at 5.904 keV.
The comparison of the so-evaluated profile with the experimental data and with the evaluated ones will be discussed.

1 Introduction

Accurate data on the fission of heavy nuclei at energies beyond the resolved resonance region (RRR) are of a renewed interest for both fundamental and applied nuclear physics.
Most neutron cross section measurements are made relative to the neutron cross section standards, being so the basis for measurements and evaluations. Very few cross sections can be measured absolutely and in most cases cross sections are measured relative to the standards or to the data files retrieved from the evaluated libraries. An important point is that the accuracy of a cross section is limited by the uncertainty in the reference cross section relative to which it is measured. There is so a need for additional data related to standard isotopes used as reference and the extension in energy range of certain cross sections, as remarked in Ref [1].
Concerning the ^{235}U neutron-induced fission cross-section, below 150 keV this cross section is only included as Standard at the thermal point and recently its integral value between 7.8 eV and 11 eV [1]. The resolved resonance region, spanning up to 2.25 keV, has been re-evaluated in this energy range with high resolution in the last ENDF/B-VIII release [2] and a SAMMY resonance analysis was done by L. Leal et al. [3] after including the work of Paradela el al., presented at WONDER-2015 [4], and taken into account the IAEA Reference file. This result of L. Leal has been included in the last up-dating version

of JEFF (JEFF-3.3). Above 2.25 keV the current ENDF/B-VIII shows a coarse energy resolution, notwithstanding its integral mean values are in fine agreement with the point-wise dataset in the IAEA Reference file. JEFF-3.3 and JENDL-4.0, on the other hand, are giving only a smooth line.

^{235}U(n,f) reaction has been amply measured and a bunch of entries can be retrieved at the EXFOR library, many of them consisting on single points, mean cross-section values in a broad energy interval or low energy resolution cross-section spectra. The aim of this work is to determine the shape of the ^{235}U(n,f) cross section with an improved energy resolution, in the energy region of the Unresolved Resonance Region (URR) up to 30 keV. So that, after inspection of the EXFOR library some selected datasets have been retained arriving to the conclusion that, besides the CERN-nTOF-PPAC results, the dataset of Weston and Todd [5] is the one having the best compromise between statistics and energy resolution, even though its bad energy calibration.

In the RRR, below 2.25 keV, most of the resonances found are of Class I, having so its energy intervals increasingly closer. On the other hand, in the transition from the RRR to the URR, the resonance structures due to the coupling of Class II states to Class I ones become more and more probable, resulting in a too high density of barrier transition states that make it almost impossible to resolve (for a detailed explanation see Ref. [6] and references therein). Above 3 keV, nevertheless, the cross-section spectrum is dominated by the grouping of the observed cross section peaks, pointing to the existence of resonance structures that have been so far not well resolved.

2 Evaluation procedure

When looking for the best energy achievable from the experimental data sets, only a few can be retrieved having both high statistics and an energy resolution high enough, as otherwise is difficult discerning statistical artifacts from real resonances. To overcome these difficulties different datasets must be compared between them, but only after a careful energy re-calibration and a suited re-normalisation. In Fig 1 can be seen some results from

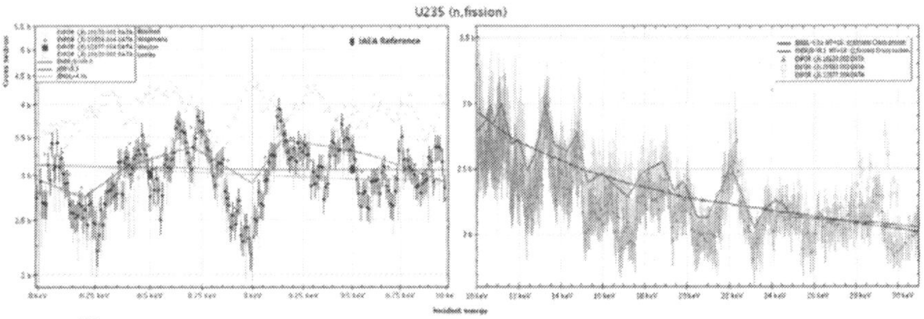

Fig. 1. ^{235}U(n,f) experimental data from EXFOR. The ENDF/B-VIII. JENDL-3.2 and the IAEA evaluations are also included.

EXFOR [5] and [7-12] for two energy ranges. In both panels can be perceived that an evident correlation exists, even though the scatter of the points due to not having a correct calibration. It is worth of mention the coarse miscalibration of the Bowman's dataset [7] shown in the left panel (light blue).

The low-background and high-resolution ^{235}U(n,f) experimental data obtained at the CERN-nTOF facility can not be taken alone because of its limited statistics. We started so by combining different data sets from measurements done with different detector setups but having the same time-to-energy calibration. The CERN-nTOF facility is described in detail in Refs. [13] and [14], and the data analysis is described in Refs [4] and [15 - 17]. We take advantage of the very high stability of both the neutron flux and the DAQ system timing that having a clock frequency of 2 GHz gets an energy resolution well below 0.1%, all along this energy range,

2.1 Data reduction and conditioning

As above mentioned, there is in EXFOR many entries for the $^{235}U(n,f)$ reaction measurements, but only a few having both high statistic and fine energy resolution. So that, this work is based in only those susceptible of passing a suited filter without loss of the required information. Every dataset was managed to have around 2000 bin/decade in order to cope with the extremely-high energy calibration required to reproduce the resonance sharp profiles. Then a triangular Barlett filter was applied in order to reduce the statistical scatter without reducing the binning. The nTOF energy calibration has been taken as reference with reference to the $^{235}U(n,f)$ 8.775(3) eV resonance and to the sharp Al(n,g) capture dip at 5.904(5) keV. In Fig 2 (left) is shown the nTOF data (in blue) matching the ENDF/B-VIII evaluation and in the right plot is the dip produced by the capture and by the

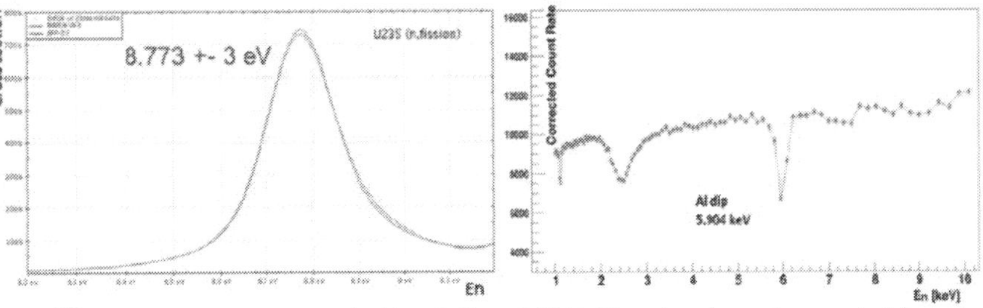

Fig. 2. $^{235}U(n,f)$ peak at 8.773 eV and Al(n,tot) dip at 5.904 keV used as time references in this work.

scattering of neutrons produced in the windows of the neutron-beam vacuum pipe. In this way we know that the accuracy of the energy calibration is below 0.1% in all the energy range. The cross section of the nTOF datasets has been normalised in order to match the IAEA standard value of the cross section integral in the energy interval 7.8 to 11 eV (247.5 b · eV), as recommended in [1]. Then, both the energy calibration and the cross section normalisation of the nTOF datasets were kept untouched along the rest of the procedure.

Once the nTOF dataset was accurately calibrated, different resonance peaks along the energy range were used to re-calibrate and re-normalise the datasets of Weston and Bowman. It is worth to mention that whereas the Weston data were quite fine calibrated (a few percent both in energy and in cross section) the Bowman data required severe shifts (the original energy calibration was found to be non-linear) and the cross section normalisation do not match correctly all along the energy range, as it will be shown later.

The thereafter called USC-Eval file was then built starting by introducing points (energy-cross section) at those resonances having maxima and minima clearly seen in both the PPACs nTOF dataset and the Weston dataset, because these data files are the most reliable. Plotting the USC-Eval together with the experimental profiles, more points were added in an iterative process, in order to reproduce as close as possible both experimental datasets.

At this point, we realised that number of resonance peaks were composed of two or even few peaks not being resolved in one or other data files. This ambiguity required the comparison with other experiments and so the Bowman data [7], and sometimes also the Lemley data [8], were used to cross-check the assignment of the peaks. Moreover, the data

from an experiment performed at CERN-nTOF using a FIC detector has been used too, as it can be taken as an independent experiment. The analysis of this experiment [18] is not yet finished but it shows a very high energy resolution and has been useful to prove the existence of some doubtful peaks. So that, every peak introduced in the USC-Eval file required to be present in at least two different experiments and having some trace in others. It is worth to mention that, after the correct calibration of all the datasets, the grade of coincidence has been surprisingly high, as it will be shown in the next section.
Finally, the whole USC-Eval dataset was normalised in order to fit with the point-wise values recommended by the IAEA Reference file [19]. This result is shown in Table 2, at the end of the text, compared with the recent evaluations.

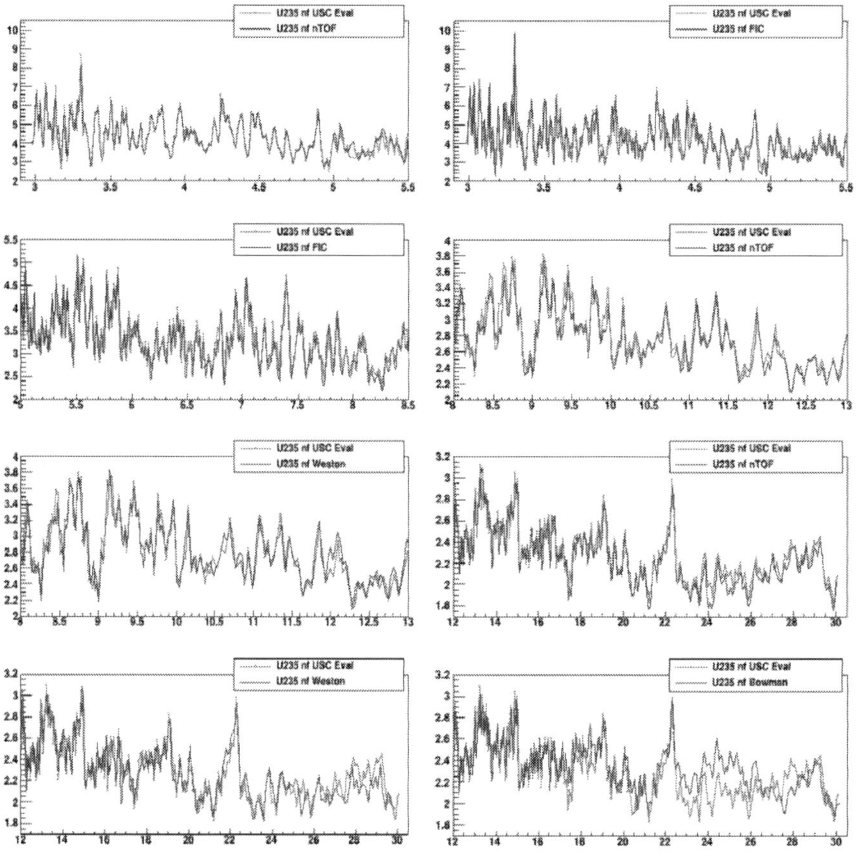

Fig. 3. Comparison of the USC-Eval with experimental datasets at different energy intervals.

3 Results

In this section are shown the results of the ^{235}U(n,f) evaluation based on the CERN-nTOF PPACs experiments, combined with the Weston data and compared to the ones of Bowman, Lemley and Furman. In this way we can said that the shape of the cross section spectrum reproduce well the experimental datasets and, on the other hand, its integral mean values are in a fair agreement with the GMA nodes used by the IAEA for the Reference files. In Fig.3 are shown some plots comparing the result of our evaluation with the experimental data at different energy intervals. In Fig 4 are summarised the statistics of such a comparison for the four experimental datasets: namely: nTOF-PPACs, nTOF-FIC, Weston and Bowman. In the left panel is the histogram of the energy residuals, showing that after the energy re-calibration the dispersion of the energy at the resonance peaks is 0.11% which is compatible with the energy resolution of each individual setup. In the right is the histogram with the residual of the cross-section values at the peaks with a RMS dispersion of 5.2% with maximum values of 20%. As the evaluated values are a compromise between the experimental ones, the differences from experiments to evaluation are lower than the differences between experiments. It must be noted that this differences arose because most resonances have a very narrow peak, depending so its apparent height on the energy binning.

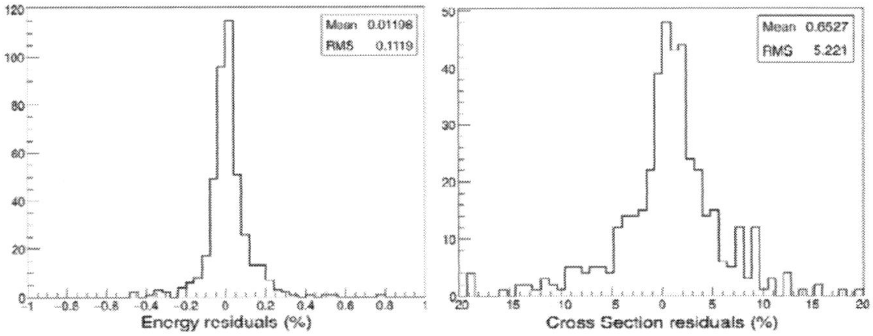

Fig. 4. Statistics of the energy and cross section values of the resonance peaks, taken as the differences between evaluation and experiments in percent.

3.1 Main findings

The high degree of coincidence found between the Paradela data at CERN-nTOF and those of Weston at ORNL is confirmed by other high-resolution datasets. More than 280 peaks have been unambiguously identified, distributed as follows: 70 from 3 to 5 keV; 73 from 5 to 8 keV; 56 from 8 to 13 keV; and 82 from 13 to 30 keV. The position in energy of these points is statistically distributed with a RMS of 1.1 % (Fig. 3-left).

In a recent paper of G.F. Bertsch *et al.* [20] the authors examine the autocorrelation function of the $^{235}U(n,f)$ reaction with a view to quantify the presence of intermediate structures in the cross section. Finally they claim that the apparent fluctuations due to compound nucleus resonances, on the order of the eV energy scale, do not present structure but in the region around 20 keV. They based the study mainly in the Perez data from ORNL [21], discarding among others the Weston data, with the argument that the Perez dataset has a superior energy resolution, what it is false in the range below 30 keV (from the information available in EXFOR). This bad choice is leading us to criticize the Bersch assertion. In Fig. 5 is presented the smoothed USC-Eval which suggests the existence of correlated peaks on the scale of 1 keV, that have the same profile as the 22 keV structure analysed in [20]. That structure shape can be suspected around 9 keV, 15 keV, and 19 keV, for instance. The existence of narrowest structures is also suggested below 9 keV in the plots presented in Fig. 3.

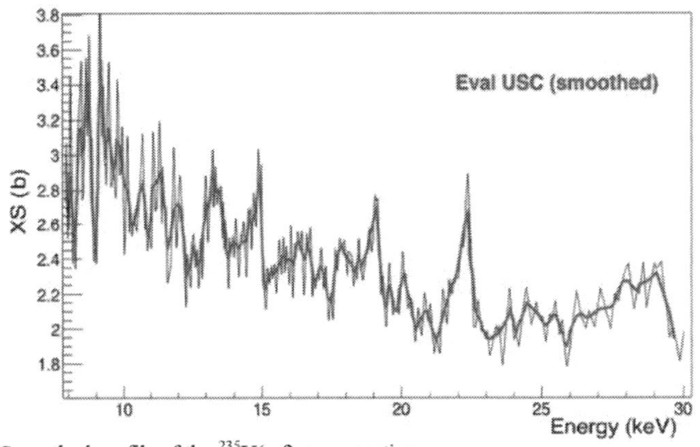

Fig. 5. Smoothed profile of the ^{235}U(n,f) cross section

3.2 Numerical results

The numerical values of the USC-Eval data file are at the end, in Table 1. The uncertainty in energy is 0.1 % and in the cross section it can be estimated to be 2.6 %, assuming that the evaluated values are the mean of the experimental data with an energy binnig of 2000 bin / decade. In Table 2 are the integral mean values compared with last evaluated libraries

4 Conclusions

An evaluation of the ^{235}U(n,f) reaction has been made in the energy range from 3 keV to 30 keV, comparing datasets from different experiments having high statistics and a fine energy resolution. The very high coincidence between those experimental datasets allow us to unambiguously assign the energy and cross section value at the identified peaks with the resolution equivalent to 2000 bin/decade, with an uncertainty in energy below 0.1% and in cross section of 2.6%.

The shape of the cross section suggest the existence of resonance structures paving the way to a further understanding of the compound nucleus fission barriers.

References

[1] A. D. Carlson *et al.*, Nuclear Data Sheets **148** (2018) 143-188
[2] M. B. Chadwick *et al.*, EPJ WoC **146** (2017) 02001
[3] L. Leal *et al.* EPJ WoC **146** (2017) 02021
[4] C. Paradela *et al.*, EPJ WoC **111** (2016) 02003 EXFOR 23294.003
[5] L.M. Weston and J.H. Todd, . Nucl. Sci. and Eng. **88** (1984) 567 EXFOR 12877.004
[6] M. Sin, R. Capote, M.W. Herman and A. Trkov, Nuclear Data Sheets **139** (2017)138
[7] C.D. Bowman *et al.* Nucl.Data React. Conf (Helsinki) **2** (1970) 65 EXFOR 10170.002
[8] P. Lemley *et al.,* Nucl. Sci. and Eng. **43** (1971)281 EXFOR 10120.002
[9] T.A. Mostovaya *et al.,* Conf. Neutron Phys (Kiev) **3** (1980)30 EXFOR 40616.002
[10] J. Blons *et al.,* Nucl. Phys. **69** (1965)545 EXFOR 201483. 002
[11] C. Wagemans *et al.,* Annals of Nucl. Ener. **3**(1976)437
[12] R.B. Perez *et al.,* Nucl. Sci. and Eng. **55** (1974)203 EXFOR 10302.002
[13] C. Guerrero *et al.,* Eur. Phys. J. A**49** (2013)27-42
[14] M. Barbagallo *et al.,* Eur. Phys. J. A**49** (2013)12-24
[15] D. Tarrío *et al.,* Nucl. Instrum. Meth. A **743** (2014)79

[16] C. Paradela *et al.*, Phys. Rev. C **82** (2010)034601
[17] E. Leal-Cidoncha *et al.* EPJ WoC **146** (2017) 04057
[18] nTOF-FIC U5(n,f) dataset. W. Furman, private communication (not yet published)
[19] IAEA NDS Reference file
https://www-nds.iaea.org/standards/references/235U_nf_Reference_xs_data.txt
[20] G.F. Bertsch, D. Brown and E.D. Davis, Phys. Rev. C **98** (2018)014611
[21] R.B. Perez *et al.*, Nucl. Sci. Eng. **55** (1974)203

==

Table 1. USC-Eval [Energy in eV – Cross section in barn]

3000	5.167	3005	6.420	3009	6.802	3013	6.144	3019	5.288
3023	4.650	3030	5.192	3036	6.134	3041	4.765	3046	4.483
3053	4.804	3058	3.884	3065	5.757	3070	7.204	3077	6.199
3084	5.112	3091	4.757	3098	5.027	3102	5.116	3110	4.683
3116	4.029	3119	3.681	3124	4.426	3131	5.244	3135	6.976
3142	5.375	3149	4.262	3156	4.883	3163	4.198	3168	3.201
3172	2.652	3185	4.157	3193	6.153	3201	5.032	3204	3.485
3208	4.038	3211	3.060	3223	4.332	3231	6.026	3238	5.726
3245	6.115	3253	5.203	3260	4.954	3268	5.202	3275	4.673
3283	6.220	3290	5.699	3297	7.358	3301	8.489	3309	7.811
3317	5.043	3321	4.551	3332	4.678	3335	5.133	3344	4.501
3352	4.343	3360	3.561	3367	2.780	3373	2.761	3383	3.510
3391	4.940	3399	5.620	3407	5.633	3416	5.877	3423	5.581
3431	4.407	3438	3.401	3446	3.121	3454	3.626	3462	4.510
3470	4.933	3478	4.587	3486	4.230	3495	5.028	3503	6.374
3511	6.069	3518	4.464	3525	3.758	3535	4.255	3543	5.168
3551	5.124	3559	4.049	3566	4.021	3576	5.467	3582	5.978
3592	5.060	3600	5.330	3608	5.835	3617	4.668	3625	3.717
3631	3.707	3641	4.466	3650	4.749	3658	4.153	3667	3.870
3675	3.772	3684	3.876	3692	4.536	3701	5.046	3709	4.690
3718	3.812	3725	3.438	3735	3.772	3744	4.227	3752	4.469
3761	4.810	3770	5.367	3776	5.494	3787	5.102	3796	4.791
3805	4.816	3813	5.216	3820	5.738	3830	5.545	3839	5.791
3846	6.151	3854	5.060	3861	3.859	3873	3.865	3883	3.905
3892	3.535	3901	3.214	3911	3.276	3920	3.836	3926	4.374
3939	4.424	3949	5.280	3956	5.872	3966	5.815	3973	6.035
3981	5.249	3991	4.096	4000	4.288	4008	4.894	4014	4.375
4021	4.759	4030	4.213	4039	4.716	4041	4.832	4053	4.482
4068	4.788	4077	3.961	4088	4.221	4099	3.926	4103	4.068
4116	4.131	4125	3.871	4130	3.604	4144	3.613	4154	3.722
4162	3.993	4172	3.939	4183	3.923	4191	4.753	4199	5.377
4211	4.900	4220	4.309	4230	4.145	4240	4.824	4250	6.391
4258	6.601	4269	5.551	4279	5.155	4290	5.524	4299	5.388
4307	4.846	4318	4.754	4328	4.488	4338	3.970	4345	3.819
4358	4.797	4370	5.168	4380	5.482	4387	5.569	4398	4.551
4409	3.835	4419	3.378	4429	3.385	4439	4.484	4447	5.641
4460	5.563	4470	4.870	4480	4.911	4489	5.641	4491	5.656
4508	4.981	4513	4.926	4521	4.394	4533	4.334	4543	4.479
4552	4.179	4564	3.444	4570	3.729	4585	4.002	4594	4.732
4603	4.888	4615	4.162	4626	3.779	4637	3.744	4649	3.137
4660	3.267	4669	4.219	4672	4.773	4691	4.361	4701	3.820
4712	3.391	4724	2.891	4736	3.168	4745	3.649	4757	3.627
4769	3.687	4778	4.240	4786	3.924	4802	3.714	4812	3.052
4822	3.291	4834	3.840	4845	3.869	4854	3.820	4868	4.050
4879	4.625	4890	5.480	4896	5.765	4912	4.487	4924	3.067
4935	2.670	4947	2.951	4958	3.171	4965	2.918	4972	2.428

4986	3.508	5000	4.415	5021	3.502	5038	4.040	5046	5.045
5055	4.172	5061	4.517	5079	3.471	5092	3.877	5109	3.309
5127	4.344	5148	3.187	5166	3.502	5178	3.167	5195	3.816
5215	3.278	5235	3.706	5242	3.136	5254	3.502	5265	3.238
5282	4.314	5301	3.542	5323	4.688	5345	3.705	5355	4.009
5366	3.400	5383	4.507	5405	3.410	5421	4.415	5437	3.400
5455	3.532	5470	2.944	5490	4.283	5512	5.228	5526	3.938
5542	4.689	5552	4.375	5562	5.085	5600	3.116	5634	4.649
5664	3.055	5686	3.816	5706	3.025	5723	3.542	5742	3.055
5772	4.796	5796	3.715	5810	4.141	5818	3.705	5838	4.517
5852	4.111	5872	4.892	5902	3.014	5931	3.502	5945	3.133
5975	3.918	5992	3.217	6000	3.512	6025	3.715	6041	3.502
6055	3.827	6066	3.252	6071	3.370	6078	3.268	6093	3.228
6106	2.994	6120	3.258	6135	2.791	6148	3.228	6171	2.649
6191	2.964	6199	2.933	6224	3.674	6242	2.994	6278	3.360
6299	2.830	6336	3.735	6351	3.238	6373	3.745	6385	3.390
6396	3.380	6411	3.998	6424	3.502	6432	3.735	6447	3.228
6464	3.725	6488	2.852	6511	3.248	6520	3.055	6535	3.288
6557	2.750	6581	3.705	6598	3.471	6614	3.502	6632	2.771
6647	3.441	6661	2.700	6675	2.751	6693	2.984	6707	2.771
6728	2.558	6745	3.075	6761	3.258	6773	3.552	6792	3.675
6810	3.938	6825	3.563	6838	2.446	6850	2.548	6866	3.461
6884	2.954	6899	3.471	6915	3.217	6952	4.425	6971	3.319
6986	3.745	7000	3.502	7014	3.197	7041	4.659	7053	4.172
7060	3.918	7076	4.111	7089	3.837	7109	3.055	7125	3.573
7158	2.598	7200	3.647	7225	2.751	7245	2.984	7257	2.832
7270	3.685	7285	3.107	7305	3.258	7324	2.755	7333	2.517
7351	3.593	7369	3.644	7387	4.395	7399	4.841	7415	3.877
7438	2.548	7461	2.923	7470	2.883	7505	3.776	7522	3.461
7540	3.705	7570	2.497	7595	2.842	7610	3.218	7625	2.801
7645	3.461	7661	3.177	7677	3.471	7698	2.923	7712	2.960
7741	2.252	7770	3.157	7786	3.685	7824	2.588	7859	3.918
7875	3.461	7904	2.619	7940	3.146	7970	2.598	8000	2.820
8018	2.870	8032	2.710	8059	3.235	8073	3.013	8087	3.490
8115	3.207	8138	2.513	8169	2.659	8200	2.487	8227	2.669
8260	2.311	8290	2.954	8320	2.823	8356	3.127	8377	2.811
8400	3.278	8429	3.146	8458	3.623	8485	3.532	8503	2.766
8520	2.827	8548	2.750	8570	3.239	8595	3.230	8618	3.718
8655	3.623	8685	3.146	8708	3.090	8735	3.796	8765	3.755
8795	2.943	8817	2.883	8832	3.126	8845	3.146	8875	2.436
8895	2.492	8930	2.334	8960	2.545	8990	2.420	9000	2.298
9010	2.385	9047	3.018	9086	2.953	9110	3.700	9130	3.684
9144	3.859	9170	3.654	9206	3.136	9230	3.126	9257	3.497
9312	2.994	9340	3.009	9365	2.842	9402	3.360	9443	3.644
9466	3.323	9501	3.319	9534	2.831	9561	2.893	9590	2.817
9620	2.944	9650	2.842	9679	2.984	9711	2.710	9757	3.547
9783	3.197	9802	3.329	9852	2.908	9892	3.040	9916	3.027
9951	3.345	9970	3.197	10000	2.517	10026	2.465	10055	2.487
10090	2.740	10120	2.730	10150	3.207	10192	2.548	10228	2.710
10275	2.538	10300	2.771	10378	2.509	10410	2.598	10442	2.522
10474	2.750	10533	2.603	10591	2.841	10646	2.791	10699	3.233
10744	2.952	10764	2.599	10840	2.690	10890	2.455	10941	2.690
10997	2.386	11080	3.274	11115	3.128	11151	2.801	11187	2.832
11224	2.690	11252	2.791	11320	3.146	11345	3.288	11375	2.984
11415	2.559	11460	3.004	11530	2.639	11570	2.720	11643	2.279
11711	2.456	11782	2.416	11854	3.141	11892	2.817	11951	2.450
11990	2.499	12010	2.700	12060	2.781	12092	2.937	12130	2.750
12170	2.548	12220	2.619	12275	2.131	12308	2.064	12355	2.460
12405	2.416	12460	2.213	12510	2.508	12545	2.483	12595	2.559
12620	2.355	12682	2.492	12715	2.386	12744	2.201	12777	2.345
12819	2.512	12882	2.264	13008	2.908	13023	2.802	13075	2.455
13164	2.853	13210	2.898	13240	3.100	13293	2.705	13333	2.994

13407	2.730	13442	2.852	13505	2.799	13540	2.850	13580	2.537
13610	2.791	13650	2.487	13678	2.669	13710	2.601	13780	2.212
13810	2.253	13850	2.548	13892	2.477	13942	2.563	13990	2.431
14044	2.678	14140	2.295	14218	2.588	14250	2.527	14325	2.619
14420	2.275	14495	2.722	14528	2.676	14572	2.741	14619	2.443
14677	2.835	14760	2.639	14795	2.619	14874	3.050	14921	2.806
14975	2.999	15011	2.761	15047	2.259	15081	2.155	15118	2.154
15118	2.154	15193	2.369	15264	2.246	15338	2.330	15381	2.249
15460	2.365	15527	2.162	15645	2.516	15712	2.372	15777	2.400
15822	2.285	15895	2.507	15992	2.224	16065	2.604	16119	2.142
16182	2.527	16220	2.436	16290	2.603	16362	2.497	16441	2.603
16535	2.154	16615	2.608	16672	2.497	16755	2.575	16825	2.152
16863	2.350	16922	2.220	17022	2.355	17066	2.254	17137	2.445
17217	2.183	17295	2.214	17389	1.940	17460	2.123	17560	2.021
17625	2.386	17690	2.455	17780	2.436	17855	2.391	17944	2.471
18056	2.274	18095	2.487	18200	2.426	18250	2.426	18309	2.222
18360	2.380	18405	2.334	18470	2.578	18514	2.375	18562	2.367
18632	2.284	18820	2.588	18930	2.426	19030	2.829	19075	2.741
19111	2.771	19241	2.224	19325	2.341	19412	2.045	19560	2.443
19590	2.385	19690	2.091	19760	2.139	19830	2.099	19870	2.172
19920	2.101	20000	2.395	20055	2.520	20090	2.299	20160	2.335
20220	2.110	20325	2.113	20385	1.942	20406	1.896	20520	1.912
20580	2.102	20715	1.944	20808	2.222	20909	2.071	21007	2.192
21205	1.829	21230	1.928	21325	1.918	21430	2.273	21580	2.055
22612	2.049	22670	2.190	22715	2.199	22800	2.011	22880	2.040
22960	1.979	23040	2.020	23136	1.866	23230	1.939	23330	2.050
23395	2.005	23460	2.070	23580	1.839	23740	2.249	23850	2.293
23980	1.989	24105	2.010	24185	1.879	24375	2.243	24458	2.281
24620	2.064	24790	2.184	24910	2.081	25040	2.116	25150	1.959
25270	2.133	25380	2.090	25499	2.212	25620	2.017	25730	2.113
25858	1.855	25950	1.887	25980	1.989	26130	2.121	26225	2.250
26330	2.106	26420	2.152	26560	2.035	26660	2.142	26850	2.040
27000	2.274	27120	2.137	27250	2.213	27410	2.035	27488	2.028
27590	2.296	27724	2.233	27890	2.416	27990	2.345	28080	2.360
28270	2.142	28390	2.289	28475	2.228	28605	2.396	28730	2.162
28980	2.410	29070	2.375	29240	2.451	29395	2.162	29520	1.979
29660	2.060	29890	1.837	29910	1.847	29955	1.908	30000	2.040

Table 2. Integral mean values compared with last evaluated libraries.
E_n is in keV and the cross section in barn.
In brackets are the ratios of the USC-Eval to the evaluated libraries

E_{ini}	– E_{fin}	EvalUSC	ENDF-VIII	JEFF-3.3	IAEA[Ref 19]
3.00	4.00	4.802	4.804 (1.00)	4.653 (1.03)	4.792 (1.00)
4.00	5.00	4.286	4.294 (1.00)	4.273 (1.00)	4.268 (1.00)
5.00	6.00	3.850	3.842 (1.00)	3.823 (1.01)	3.847 (1.00)
6.00	7.00	3.270	3.368 (0.97)	3.352 (0.98)	3.298 (0.99)
7.00	8.01	3.253	3.229 (1.01)	3.215 (1.01)	3.242 (1.00)
8.00	9.00	3.015	3.062 (0.98)	3.094 (0.97)	3.014 (1.00)
9.00	10.0	3.117	3.219 (0.97)	3.064 (1.02)	3.130 (1.00)
10.0	12.0	2.720	2.928 (0.93)	2.950 (0.92)	
12.0	14.0	2.568	2.715 (0.95)	2.722 (0.94)	
14.0	16.0	2.482	2.591 (0.96)	2.536 (0.98)	
16.0	18.0	2.365	2.370 (1.00)	2.432 (0.97)	
18.0	20.0	2.378	2.440 (0.97)	2.369 (1.00)	
20.0	25.0	2.164	2.234 (0.97)	2.227 (0.97)	
25.0	30.0	2.175	2.153 (1.01)	2.121 (1.03)	

EPJ Web of Conferences **211**, 02004 (2019)
WONDER-2018
https://doi.org/10.1051/epjconf/201921102004

Study of the (n,γf) process on ^{239}Pu

Esther Leal-Cidoncha[1,*], *Gilles* Noguere[1], *Olivier* Bouland[1], and *Olivier* Serot[1]

[1]CEA/DEN/DER/SPRC/LEPh/Cadarache center, F-13108 Saint Paul Lez Durance (France)

Abstract. In the neutron resonance range, fission cross section of ^{239}Pu can be seen as a sum of the immediate (n,f) and the two-step (n,γf) fission reactions. In that case, five channel widths should be considered for a proper evaluation, those are: two opened fission channels for $J^\pi = 0^+$, one opened fission channel for $J^\pi = 1^+$ and two J-dependent for the (n,γf) reaction. The sizeable contribution of the (n,γf) process should have an impact on the determination of the capture and fission widths involved in the Reich-Moore approximation of the R-matrix theory. The present work aims to investigate this impact by using the CONRAD code and the $\Gamma_{\gamma f}$ available from literature. Prompt neutron multiplicity (ν_p) has been also reproduced including the contributions of the (n,γf) process.

1 Introduction

When a low-energy neutron collides with a target nucleus there is a high probability to form a compound nucleus in a given excited state. This compound nucleus may decay in different manner as by neutron or γ-ray emission, or by fission. The probability of each of these reactions to occur is weighted through partial reaction cross sections: scattering, capture or fission in each case. In fission reactions, neutrons may be emitted in different time windows, depending on which they are called *prompt* and *delayed* neutrons. It may happen that one γ-ray is emitted before fission occurs, through a two-step fission reaction, contributing to the total fission cross section. In this case, the available excitation energy to the fission fragment is lower and less prompt neutrons are emitted, leading to an anticorrelation between prompt-neutron emission and γ-ray emitted before fission multiplicities.

The fission process can be described using a potential barrier expressed as a function of the compound nucleus deformation. When this potential barrier has a double-humped shape, it presents two wells populated by energy states through which fission may happen. Two-step fission reactions are related to dipole electric (E1) and magnetic (M1) transitions between states, occurring in the first well of the fission barrier [1].

Following the studies initiated by E. Fort [2], this work aims to provide a description of the prompt-neutron multiplicity for neutron-induced fission reactions on ^{239}Pu by analyzing together a set of experimental data, including in the calculations the partial cross sections of the different channels, and taking into account the immediate and the two-step fission reactions for the total fission cross section.

*e-mail: Esther.LEALCIDONCHA@cea.fr

2 Neutron-induced fission on ^{239}Pu

2.1 ^{239}Pu(n,f) cross section

The ^{239}Pu(n,f) cross section can be described by:

$$\sigma_f(E_n) = \sigma(E_n)_{(n,f)} + \sigma(E_n)_{(n,\gamma f)} \tag{1}$$

where the two-step fission cross section may be expressed as:

$$\sigma(E_n)_{(n,\gamma f)} = \sigma_\gamma(E_n) \cdot \frac{\Gamma_{\gamma f}}{\Gamma_\gamma} \tag{2}$$

The partial fission widths are required to perform the calculations in R-Matrix Theory. Three immediate fission channels have been considered in this work, two for $J^\pi = 0^+$ with partial fission widths $\Gamma_{1f}(0^+)$ and $\Gamma_{2f}(0^+)$ and one for $J^\pi = 1^+$, with partial fission width $\Gamma_f(1^+)$ [3]. Because of its fluctuating feature, only two γ-fission channels have been considered in the two-step fission process, with partial fission widths $\Gamma_{\gamma f}(0^+)$ and $\Gamma_{\gamma f}(1^+)$, and the second partial width for $J^\pi = 0^+$ has been considered negligible. Resonances 0^+ dominate for immediate fission, while resonances 1^+ dominate for two-step fission reactions.

2.2 Prompt neutron multiplicity (ν_p) experimental data

The experimental prompt neutron multiplicity datasets available in EXFOR, measured by Gwin [4, 5] and Fréhaut [6] (Figure 1), have been analyzed.

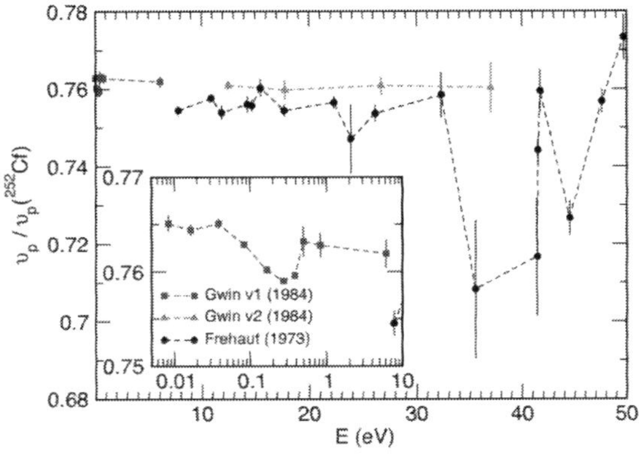

Figure 1. Prompt neutron multiplicity experimental data from Fréhaut [6] and Gwin [4, 5] up to 50 eV normalized to the ν_p of ^{252}Cf.

There are some issues concerning these data that had to be examined. There is an incongruity between the normalization of both datasets. In addition the experimental data provided in the EXFOR library by Fréhaut are given in a pointwise description. In order to treat these data using the CONRAD code, they have been averaged, specifying the edges of each energy interval, see Figure 2. Unfortunately this format is not a good description of the data, because the values given by the points are not centered in the energy interval.

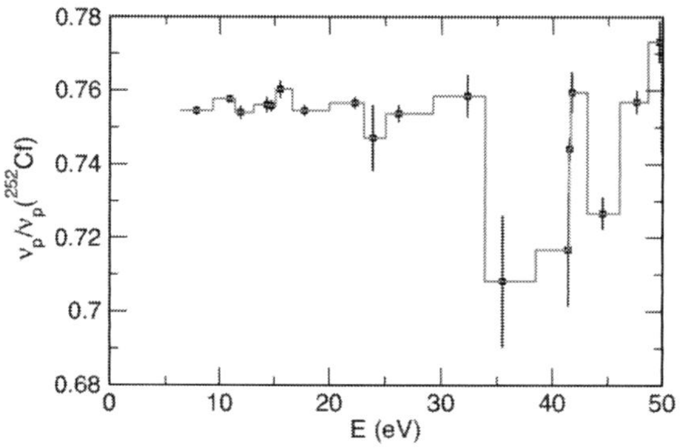

Figure 2. Comparison between the averaged and pointwise descriptions of the data from Fréhaut [6].

2.3 CONRAD resonance analysis of ^{239}Pu data

In the evaluated file of the fission cross section, the $J^\pi = 1^+$ resonances are considered as being produced only through immediate fission, however, they are also produced by two-step reactions, as previously mentioned. In order to make a more accurate evaluation including the (n,γf) channel for $J^\pi = 1^+$, a new resonance analysis has been done using the R-Matrix code CONRAD[1] (COde for Nuclear Reaction Analysis and Data Assimilation) [7]. Total and partial cross sections have been analyzed together with the neutron multiplicities. The values of the partial fission widths for immediate fission have been obtained from the JEFF-3.3 evaluated data library, and the partial fission widths for the two-step fission reactions have been taken from the work of Trochon et al. [8] and Lynn et al. [1], see Table 1. This parameter was calculated by Trochon et al., considering a single-humped fission barrier and taking into account only E1 transitions between states. A double-humped fission barrier was considered instead in the work from Lynn et al., taking into account a low-lying M1 resonance, known as the scissors mode, as well as E1 transitions between states. For a more detailed description of these calculations, consult the references [1, 8]. The averaged radiative width ($<\Gamma_\gamma>$) has been taken from Mughabghab [9], who gives a value of 43 ± 4 meV.

Table 1. Partial widths available in the literature for the two-step fission process.

Reference	$\Gamma_{\gamma f}(J^\pi{=}0^+)$ (meV)	$\Gamma_{\gamma f}(J^\pi{=}1^+)$ (meV)
[8]	7.3	4.2
[1]	1.5	2.29

The experimental data from Fréhaut have been introduced in the CONRAD code in both formats: the pointwise and the averaged, to fit the $\nu_{n\gamma f}$, ν_0, ν_1, and the normalization parameter of the experimental data from Fréhaut. Results of the fit to the prompt neutron multiplicity from Fréhaut using both $\Gamma_{\gamma f}$ parameters from [1, 8] are shown in Figure 3. In both cases the pointwise treathment of the data gives the best result. The use of $\Gamma_{\gamma f}$ from references [1, 8], does not change substantially the prompt neutron multiplicity, as this is compensated by fitting the $<\nu_{n\gamma f}>$ parameter.

The averaged values obtained using both values of $\Gamma_{\gamma f}$ are given in Table 2, similar results are obtained for $<\nu_0>$, $<\nu_1>$ and the normalization of the data from Fréhaut, however the $<\nu_{n\gamma f}>$ obtained in each case is slightly different.

[1]Developed at CEA/Cadarache

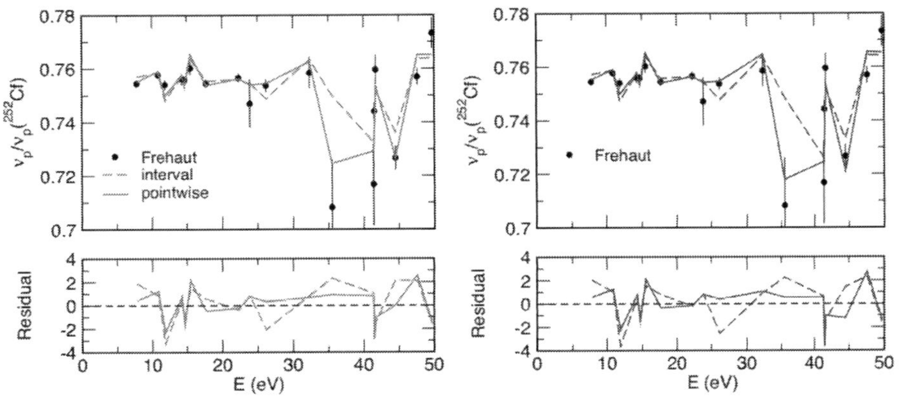

Figure 3. Comparison between the results obtained fitting the averaged and pointwise data from Fréhaut, using the $\Gamma_{\gamma f}$ from the works of Trochon et al. (left panel), and Lynn et al. (right panel).

Table 2. Output parameters obtained from the fit to the ν_p experimental data.

Parameter	Result with $\Gamma_{\gamma f}$ from [8]	Result with $\Gamma_{\gamma f}$ from [1]
$<\nu_{n\gamma f}>$	2.60 ± 0.02	2.41 ± 0.03
$<\nu_0>$	2.897 ± 0.002	2.894 ± 0.002
$<\nu_1>$	2.874 ± 0.002	2.873 ± 0.002
$N_{Frehaut}$	0.9963 ± 0.0007	0.9965 ± 0.0007

The results of the fit to the ν_p data from Gwin are shown in Figure 4. Large error bars are associated to the data in the eV energy region, detracting credibility from the data. Hence, new experimental data are required.

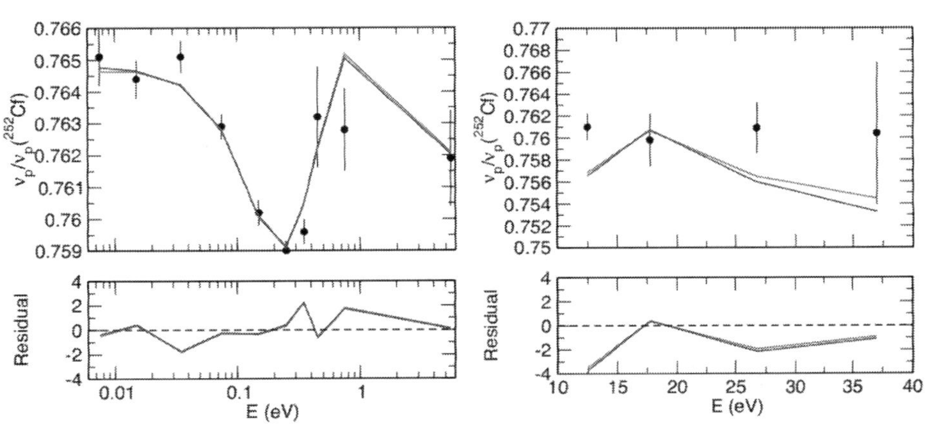

Figure 4. Comparison between the fits to the ν_p experimental data from Gwin using $\Gamma_{\gamma f}$ from Trochon et al. (red line) and Lynn et al. (blue line).

https://doi.org/10.1051/epjconf/201921102004

2.4 Results of ν_p up to 50 eV

The ν_p obtained in this work using both values for the $\Gamma_{\gamma f}$ parameter show that the spin contribution can reproduce the experimental data at low energies, below ~25 eV, above this energy and up to 50 eV, the dips of the ν_p are quite well reproduced by the $\nu_{n\gamma f}$ contribution, see Figure 5.

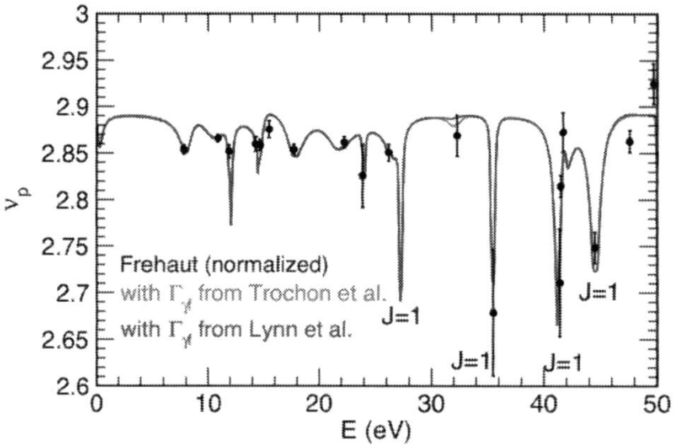

Figure 5. Comparison between the ν_p results obtained in this work and the normalized experimental data from Fréhaut up to 50 eV.

Both results are compared in Table 3 with the normalized data from Fréhaut up to 50 eV.

Table 3. Prompt neutron multiplicity obtained using both values for $\Gamma_{\gamma f}$ compared with the normalized data from Fréhaut up to 50 eV. The ratio from the experimental data to the results are given in parenthesis.

Energy (eV)	Fréhaut	Result with $\Gamma_{\gamma f}$ from [8]	Result with $\Gamma_{\gamma f}$ from [1]
7.8	2.854 ± 0.005	2.852 (1.001)	2.853 (1.000)
10.9	2.866 ± 0.004	2.867 (1.000)	2.867 (1.000)
11.9	2.852 ± 0.007	2.832 (1.007)	2.832 (1.007)
14.3	2.860 ± 0.008	2.862 (0.999)	2.862 (0.999)
14.7	2.859 ± 0.006	2.845 (1.005)	2.845 (1.005)
15.5	2.876 ± 0.009	2.892 (0.994)	2.891 (0.995)
17.7	2.854 ± 0.006	2.847 (1.002)	2.848 (1.002)
22.3	2.862 ± 0.006	2.857 (1.002)	2.857 (1.002)
23.9	2.83 ± 0.03	2.848 (0.994)	2.848 (0.994)
26.2	2.851 ± 0.009	2.850 (1.000)	2.850 (1.000)
32.3	2.87 ± 0.02	2.881 (0.996)	2.889 (0.993)
35.5	2.68 ± 0.07	2.736 (0.980)	2.711 (0.989)
41.4	2.71 ± 0.06	2.754 (0.984)	2.737 (0.990)
41.5	2.82 ± 0.01	2.777 (1.015)	2.768 (1.019)
41.7	2.88 ± 0.02	2.848 (1.011)	2.848 (1.011)
44.5	2.75 ± 0.02	2.744 (1.002)	2.723 (1.010)
47.6	2.87 ± 0.01	2.890 (0.993)	2.892 (0.992)
49.7	2.93 ± 0.02	2.891 (1.013)	2.891 (1.013)

2.5 Discussion above 50 eV

At higher energies, above 50 eV, the data from Fréhaut can not be described by the equations due to their large dispersion (1.2%), see Figure 6. New experimental data with lower statistical uncertainties are then requested in this energy region in order to correctly determine the ν_p.

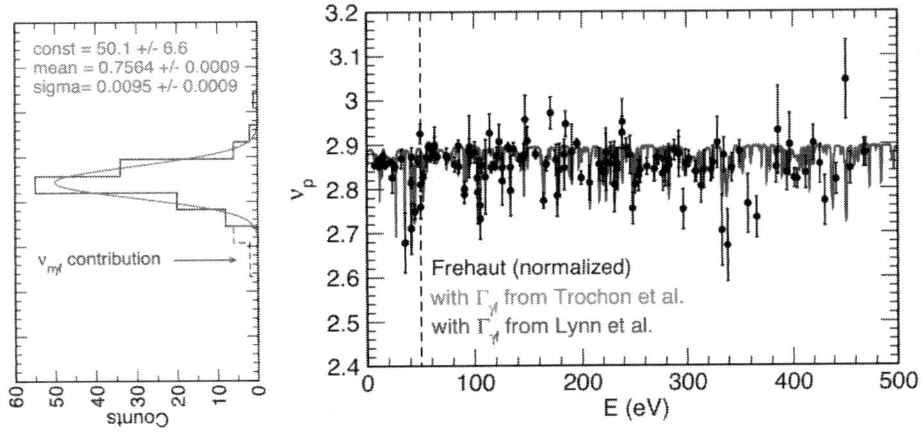

Figure 6. Comparison between the ν_p results obtained up to 500 eV using $\Gamma_{\gamma f}$ from Trochon et al. and Lynn et al., and the normalized data from Fréhaut (right panel). The ν_p histogram of the data from Fréhaut has been fitted to a gaussian distribution (left panel), showing a dispersion of the 1.2%.

3 Conclusions and perspectives

The prompt neutron multiplicity (ν_p) of the ^{239}Pu(n,f) reaction has been reevaluated taking into account both, the "immediate" and the "two-step" contributions to the total fission cross section. The multiple analysis of the total and partial cross sections, and the total and prompt neutron multiplicity has been performed using the CONRAD code developed at CEA/Cadarache. For the prompt neutron multiplicity, two sets of experimental data are provided in the EXFOR database at low energies (below 50 eV), it includes the data measured by Fréhaut and Gwin. Below 50 eV, both datasets are not normalized among themselves and, in addition large error bars are associated to the higher energy data from Gwin making them not reliable at such energies. Only the data from Fréhaut are provided above 50 eV, however the large dispersion (1.2%) does not allow to reproduce them. New experimental data need to be measured in order to perform a more accurate evaluation of the ν_p in a wide energy region.

The ν_p experimental data have been fitted in order to calculate $\nu_{n\gamma f}$, ν_0, ν_1, and the normalization parameter of the measurement from Fréhaut. The values of the resonance parameters for the immediate reactions have been extracted from the JEFF-3.3 evaluation. Two values for the $\Gamma_{\gamma f}$ parameter have been tested, the ones from Trochon et al. and Lynn et al., both giving similar results, although a lower $<\nu_{n\gamma f}>$ has been obtained using the $\Gamma_{\gamma f}$ from [1]. The model that has been tested in this work will be officially released in CONRAD. The good results here obtained aim to apply the same model to analyze the ^{235}U(n,f) reaction, which will be the next step in this work line.

References

[1] J.E. Lynn, P. Talou, and O. Bouland, PRC **97**, 064601 (2018).

[2] E. Fort et al., *Evaluation of ν_p for ^{239}Pu: Impact for Applications of the Fluctuations at Low Energy*, Nucl. Sci. Eng. **99**, 375 (1988).

[3] A. Michaudon, *Advances in Nuclear Physics*, **Vol. 6** (M. Baranger and E. Vogt, eds.), Plenum Press, New York (1973).

[4] R. Gwin, R.R. Spencer, and R.W. Ingle, *Nucl. Sci. E,g.*, **87**, 381 (1984).

[5] R. Gwin, R.R. Spencer, R.W. Ingle, J.H. Todd, and H. Weaver, (a) ORNL/TM-6246 (1978) and (b) ORNL/TM-7148 (1980), Oak Ridge National Laboratory; see also R. Gwin, R.R. Spencer, and R.W. Ingle, *Nucl. Sci. E,g.*, **94**, 365 (1986).

[6] J. Fréhaut, and D. Shackleton, *Proc. 3rd Symp. Physics and Chemistry of Fission*, Rochester, New York, Aug. 13-17 (1973).

[7] P. Archier et al., EPJ Web of Conferences **42**, 02004 (2013).

[8] Trochon, *Proc. 3rd Symp. Physics and Chemistry of Fission*, Rochester, New York, Aug. 13-17 (1973).

[9] S. F. Mughabghab, Atlas of Neutron Resonances, 5th ed. (Elsevier, Amsterdam, 2006).

Trends on Major Actinides from an Integral Data Assimilation

Gerald Rimpault[1,**], *Virginie* Huy[1,2], *Gilles* Noguère[1]

[1]CEA, DEN, DER, SPRC, Cadarache, F-13108 St Paul-Lez-Durance, France
[2]Aix Marseille University, Doctoral school 352

Abstract: Nuclear data evaluations on major actinides can be improved by integral data assimilation. Appropriate integral measurements with reliable experimental techniques have been selected such as ICSBEP, IRPhE and MASURCA critical masses, PROFIL irradiation experiments and the FCA-IX experimental programme (critical masses and spectral indices). Highly reliable analyses are possible with the use of as-built geometries calculated with the TRIPOLI4 Monte Carlo code. The C/E values have been used in an integral data assimilation solving the Bayes equation. The trends on the JEFF3.1.1 ^{235}U capture cross section are quite consistent with recent differential measurements. Assimilation results suggest up to a 2.5% decrease for ^{238}U capture from 3 keV to 60 keV, and a 4-5% decrease for ^{238}U inelastic in the plateau region. For this energy range, uncertainties are respectively reduced from 3-4 to 1-2% and from 6-9% to 2-2.5% for 238U capture and ^{238}U inelastic. Results on ^{239}Pu fission cross sections are included in posterior uncertainties. The increase trends on ^{239}Pu capture cross-section is of around 3% in the [2 keV-100 keV] energy range. For ^{240}Pu capture cross section, the increase is of around 4% in the [3 keV-100 keV] energy range and goes in the same direction as a recent evaluation.

1. INTRODUCTION

Five out the six Generation IV concepts are Fast Reactors with breeding capability (they produce Pu out of depleted Uranium) and with ability to use possibility Pu and MA, that would otherwise be a waste (Pu from PWR-MOX not suitable for PWR).

In France, 2006 June 28 act requires the design of a generation of sodium fast reactors, those being the most mature Generation IV concept. It is the reason for which CEA with its industrial partners started the design of ASTRID[1,2] (Advanced Sodium Technological Reactor for Industrial Demonstration), a 600 MWe sodium-cooled fast reactor concept. Its construction should demonstrate the feasibility of a SFR design with enhanced safety and its ability to achieve a zero breeding gain hence enabling to use the entire Uranium ore and not only ^{235}U.

[*] Corresponding author : gerald.rimpault@cea.fr

Sodium Fast Reactors require the development and validation of scientific calculation tools as a proof of their characteristics used in their safety dossiers. Nuclear data, the input parameters of the neutronic codes, constitute the main source of uncertainty in neutronic calculations[3]. For instance, uncertainty (in pcm) on critical mass for the ASTRID core (using JEFF3.1.1 and its associated COMAC-V1 nuclear data covariances) amounts up to 1558 pcm. Major actinides are contributing significantly to this uncertainty as Table 1 illustrates.

Table 1. Uncertainty breakdown (in pcm) on critical mass for the ASTRID core (using COMAC-V1 nuclear data covariances).

Isotope	Fission	Capture	Elastic	Inelastic	NxN	Nu	Fission Spectrum	Total
^{16}O	0	110	50	0	0	0	0	120
^{23}Na	0	20	10	10	0	0	0	30
^{56}Fe	0	80	30	50	0	0	0	100
^{238}U	439	869	30	649	30	70	60	1168
^{238}Pu	10	20	0	0	0	80	40	100
^{239}Pu	699	190	10	60	10	100	210	759
^{240}Pu	669	300	0	60	10	20	170	619
^{241}Pu	70	90	0	10	0	70	150	200
^{242}Pu	40	50	0	20	0	20	40	80
TOTAL	1069	859	60	649	30	170	320	1558

Existing fertile blankets with depleted U induces great sensitivities to ^{238}U cross-sections, notably inelastic. The use MOx fuel with Pu retrieved from PWR-MOx spent fuel induces high sensitivities not only to ^{239}Pu cross sections but also to other Pu isotopes cross sections (notably ^{240}Pu).

2. STRATEGY USED

Integral Data Assimilation (IDA) can contribute to nuclear data improvement if attention is being given to ways that minimize the possibility of creating compensating errors. This approach has been done in the past and it is important to learn from past integral data assimilation works to develop a strategy that can avoid important compensating errors.

2.1 Previous integral data assimilation works

Since the early seventies, several integral experiments assimilation work for fast reactor applications have been conducted.

Among them, the statistical adjustment method (called BARACCA) was used to produce two CARNAVAL libraries (III and IV)[4] with the objectives of answering needs for PHENIX and SUPER-PHENIX respectively. For the adjustment, for each nuclear data, equiprobable probabilities were associated to an interval of possible values. This choice was done because assuming that the discrepancy in C/E values for integral experiments came mainly from normalization issues when performing microscopic measurements. Because, there were no constraints in the a priori multigroup cross sections other than the boundary of the initial interval, work resulted in compensating errors between ^{238}U capture cross sections and ^{239}Pu fission cross section.

ERALIB1[5] is a library resulting from an integral data adjustment whose experimental database includes 355 integral data (critical masses, spectral indices, buckling etc.) coming from 71 distinct experimental programs. Prior variances-covariances matrices associated to nuclear data were produced from expert judgements. The improvements made on the assimilation process include:

- The use of Gaussian distributions for multigroup nuclear data to perform adjustments based on Bayes Inference using Generalized Least Square Methods.
- The use of statistical indicators (χ^2 criteria) to identify inconsistent integral data, which need to be eliminated from the experimental database.

However, the experimental database included many unreliable measurements and some of covariance were lacking physical meaning (especially for ^{23}Na). This led in particular to discrepant sodium void worth.

2.2 Current Strategy

To avoid deficiencies identified in previous Integral Data Assimilation (IDA) work, a more reliable nuclear data library was used JEFF3.1.1[6] for all isotopes with the exception of ^{23}Na for which JEFF3.2 was used[7]. The associated COMAC V1 covariance[8] were used as they mimic the evaluation process itself.

Only reliable integral experiments have been selected. They are ICSBEP, IRPhE and MASURCA critical masses, PROFIL irradiation experiments and the FCA-IX experimental programme (critical masses and spectral indices). Highly reliable experiment analyses are now possible avoiding method approximation and using as-built geometries as it is possible with the TRIPOLI4 Monte Carlo code[9].

Marginalization technique has been used for light and structural isotopes for which approximations (anisotropy, secondary neutron energy distribution) in the integral data assimilation technique are rather high. A test on whether Prompt Fission Neutron Spectrum (PFNS) needs to be fitted or not (such as in ERALIB1). It appears from this test that PFNS is a major source of uncertainty and requires to be fitted[12].

Attention has been given to ways that minimize compensating errors. Compensating errors between ^{238}U capture and ^{239}Pu fission (such as in CARNAVAL IV) has been eliminated by using first only U-fueled experiments then adding Pu-fueled experiments.

The IDA has been using Bayesian inference[10] with a minimization of a cost function (1) in which M_σ is the covariance matrix associated to nuclear data and M_E the one associated to integral experiments.

$$\chi^2_{GLS} = (\boldsymbol{\sigma}-\boldsymbol{\sigma}_{a\,priori})^T M_\sigma^{-1}(\boldsymbol{\sigma}-\boldsymbol{\sigma}_{a\,priori}) + \left(\boldsymbol{E} - C(\sigma)\right)^T M_E^{-1}\left(\boldsymbol{E} - C(\sigma)\right) \quad (1)$$

Solutions for nuclear data are given by equation (2) where S is the sensitivity matrix calculated with the ERANOS code system[11] while associated uncertainties are given by equation (3):

$$\boldsymbol{\sigma}'-\boldsymbol{\sigma}_{a\,priori} = M_\sigma \,.\, S^T\left(M_E + S.\,M_\sigma \,.\, S^T\right)^{-1}.\left(\boldsymbol{E} - C\left(\sigma_{a\,priori}\right)\right) \quad (2)$$

$$\boldsymbol{M'}_{\boldsymbol{\sigma}} = M_\sigma - M_\sigma \,.\, S^T\left(M_E + S.\,M_\sigma \,.\, S^T\right)^{-1} S.\,M_\sigma \quad (3)$$

3. INTEGRAL DATA ASSIMILATION STUDIES

3.1 Integral experiments

The Uranium configurations chosen for the exercise are MASURCA R2, MASUSCA 1B, FCA IX-1 to 7, GODIVA and FLATTOP U235. The Plutonium configurations chosen are MASURCA 1A', MASURCA ZONA2, ZPPR10A, ZPR6/7, ZPR6/7 high Pu240, BFS 82-2, MASURCA PRE-RACINE 2A & 2B, SNEAK 7A & 7B , JEZEBEL Pu239 & Pu240, FLATTOP Pu239. Also, FCA-IX fission chambers C/Es of Np237, Am241, Am243, Pu238, Pu242, Cm244 have been used as well as PHENIX PROFIL irradiated samples C/E on U238, Pu238, Pu239, Pu240, Pu241, Pu242, Am241, Np237. C/E values display a great dispersion in results. Nuclear data uncertainties (of the order of 1000 to 2000 pcm) are far beyond experimental uncertainties, which justify the use of IDA method.

3.2 Trends for ^{235}U Capture

As seen on Figure 1, the trends were a -27 to -33 % decrease at 1-2 keV (end of RRR) which is in agreement with Danon measurements (2017)[13] and a +8 to +10% increase in the 10-100keV region (URR) which is in agreement with Jandel measurements (2012)[14]. Uncertainties are significantly reduced.

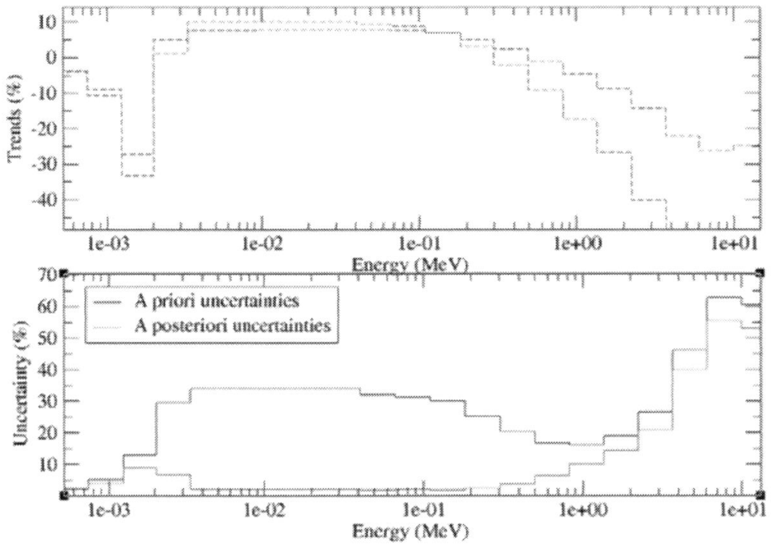

Figure 1: Trends on ^{235}U capture

The consistency of integral data assimilation results on ^{235}U capture cross section with recent differential measurements encourages us to rely on the simultaneous use of PROFIL C/E and various U235 enriched critical masses to reassess capture cross sections.

3.3 Trends for ^{238}U Capture and Inelastic

For ^{238}U inelastic and capture cross sections, a comparison of ^{235}U-^{238}U assimilation results and U-Pu assimilation results allows us to conclude that the trends proposed for these cross sections are not the result of compensating errors with Pu nuclear data. Assimilation results suggest up to a 2.5% decrease for ^{238}U capture from 3 keV to 60 keV, and a 4-5% decrease for ^{238}U inelastic in the plateau region. For these energy range, uncertainties are respectively reduced from 3-4 to 1-2% and from 6-9% to 2-2.5% for ^{238}U capture and ^{238}U inelastic. The simultaneous use of GODIVA and FLATTOP-235U and JEZEBEL 239Pu and FLATTOP-239Pu is relevant to reassess ^{238}U inelastic cross sections. Their critical mass C/E are greatly affected by the reflecting properties of the FLATTOP core depleted Uranium blankets very sensitive to this cross section.

Assimilation results for ^{238}U capture are compared with differential measurements[15] and with the "a priori" JEFF-3.1.1, CIELO and JEFF-3.3 evaluation (Figure 2). Posterior uncertainties for assimilation results are in dotted line.

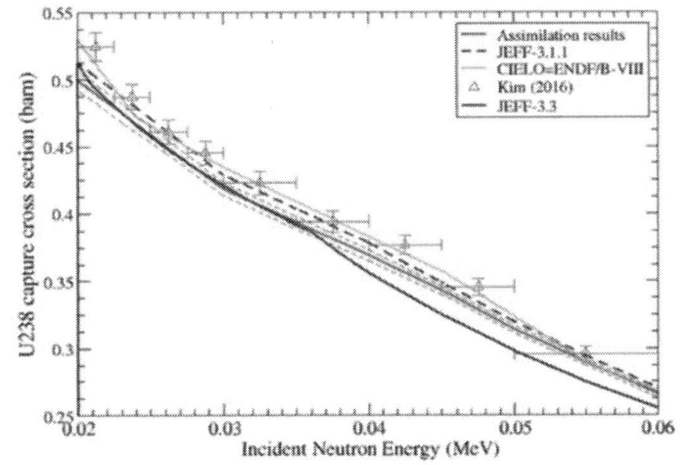

Figure 2: Assimilation Results compared to major evaluations for ^{238}U capture

Because Prompt Fission Neutron Spectrum (PFNS) lacks some differential measurements, variances are large. Since integral experiments available cannot distinguish between ν and χ (PFNS) the possibility of having compensating errors is high. Hence, a parametric study has been conducted with ^{235}U and ^{238}U PFNS fitted or not through assimilation. The magnitude of trends on ^{238}U capture cross sections differ on whether ^{235}U and ^{238}U PFNS is fitted or not through assimilation.

In the plateau region (~1 MeV to 6 MeV), a 5-6% decrease is suggested for ^{238}U inelastic. This decrease trend is in agreement with recent evaluations (see suggested value compared to CIELO and JEFF-3.3 ones on Figure 3; Posterior uncertainties for assimilation results are in dotted line).

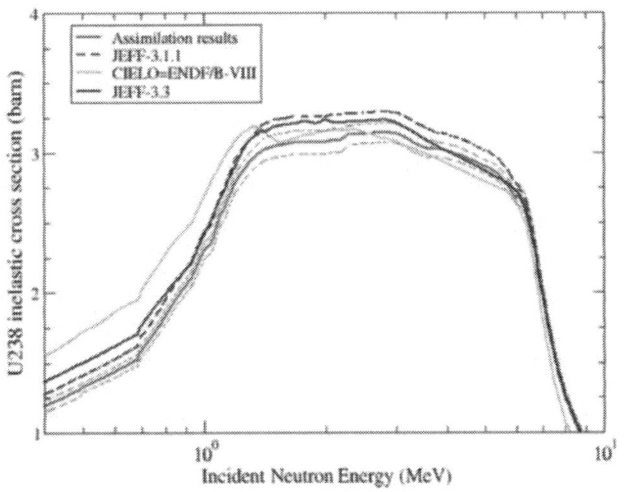

Figure 3: Assimilation Results compared to major evaluations for ^{238}U inelastic

Notably, the simultaneous use of FLATTOP-U235/GODIVA and FLATTOP-Pu239/JEZEBEL enables us to discriminate contributions from U238 and fissile isotopes. Indeed, GODIVA is a bare U235 sphere while FLATTOP-U235 is surrounded by a $U_{depleted}$ blanket. The same occurs for JEZEBEL and FLATTOP-Pu.

3.4 Trends for Pu isotopes

Summary of trends on Pu isotope data, which can have a significant impact for the ASTRID core, are given in Table 2.

Table 2. Trends on Pu isotope cross sections

Cross section	Trends (%)	Energy range	Experiments	Comments
^{239}Pu fission	<1% (+/- 1.7% or less)	0.1 keV- 500 keV	Pu-fueled critical masses	Trends included in posterior unc.
^{239}Pu capture	from -1.4 to -4.8 % (+/- 2-5%)	0.04 keV - 300keV	PROFIL and critical masses	
^{240}Pu fission	**from - 7.5 to -9.6% (+/-8-11%)**	**0.75 keV-100 keV**	**Mostly JEZEBEL- ^{240}Pu**	**Risk of compensating errors with multiplicity and PFNS. Trends are close to posterior uncertainties.**
	from -3.5 to -4.6% (+/- 4-7%)	**1 keV-3.7 MeV**		
^{240}Pu capture	from 2.5 to -4.2 % (+/- 2-4.5%)	0.7 keV - 100 keV	PROFIL	
^{242}Pu fission	**- 10% (+/- 3.5%)**	**500 keV- 4 MeV**	**FCA-IX IS**	**Underestimated prior uncertainties (~2% in the plateau region)**
^{242}Pu capture	from -3 to -9% (+/- 2-5%)	3 keV - 300 keV	Trends are driven by PROFIL (C/E~1.14).	IDA trends have an unreliable energy breakdown due to Covariances
	around -20 % (+/- 12%)	0.4 keV- 3 keV		

Results on ^{239}Pu fission cross sections are included in posterior uncertainties which have been significantly reduced through the IDA process (from 3.5-2% to 2.5-0.5%) . There is no differential measurement able to reach that type of uncertainty. This means that IDA process will be needed in a way or another. The increase trends on ^{239}Pu capture cross-section is of around 3% in the [2 keV-100 keV] energy range. From 4keV up to 30 keV, the ENDF/B-VIII evaluation corresponds to JEFF-3.1.1. For ^{240}Pu capture cross section, the increase is of around 4% in the [3 keV-100 keV] energy range and goes in the same direction as the recent ENDF/B.VIII evaluation though at a much lower level.

4. IMPACT ON THE ASTRID CFV CORE

There is a significant impact of ^{238}U capture and inelastic on the critical mass of the ASTRID core but also ^{239}Pu, ^{240}Pu and ^{242}Pu fission cross-sections. With a prior value for the critical mass of 1.02688 (1.02908 with JEFF-3.1.1 for all isotopes except ^{23}Na for which JEFF3.2 is used: Impact of using ^{23}Na evaluation from JEFF-3.2 instead of JEFF-3.1.1 on the ASTRID core reactivity: -220 pcm.) the IDA conducted led to a posterior value of 1.02255 (-435 pcm). The Integral Data Assimilation reduces uncertainties associated to nuclear data significantly with, in particular, a significant reduction on ^{239}Pu and ^{240}Pu fission cross-sections.

Table 3. Impact of the nuclear data trends (in pcm) on the critical mass of the ASTRID core and associated uncertainty (a priori uncertainties using COMAC-V1 nuclear data covariance are also given).

all values in pcm	Impact on k_{eff}	Posterior associated uncertainty contribution	Prior associated uncertainty contribution
U238_CAPTURE	249	180	869
U238_NU	0	40	70
U238_FISSION_SPECTRUM	12	40	60
U238_ELASTIC	-7	30	30
U238_INELASTIC	342	60	649
Pu238_CAPTURE	6	10	20
Pu238_FISSION	-82	20	10
Pu239_CAPTURE	-96	20	190
Pu239_FISSION	-143	120	699
Pu239_NU	-49	40	100
Pu239_FISSION_SPECTRUM	-37	50	210
Pu240_CAPTURE	-100	40	300
Pu240_FISSION	-323	170	669
Pu240_NU	-2	10	20
Pu240_FISSION_SPECTRUM	-66	90	170
Pu241_CAPTURE	7	70	90
Pu241_FISSION	-8	60	70
Pu241_NU	-6	60	70
Pu241_FISSION_SPECTRUM	-29	130	150
Pu242_CAPTURE	99	30	50
Pu242_FISSION	-197	20	40
Total actinides (pcm)	**-435**	**490**	**1558**

5. CONCLUSION AND PERSPECTIVES

Integral Data Assimilation (IDA) in this work has proved to be efficient in identifying the sources of possible normalization problems in the differential measurements. This achievement is mainly due to the progresses in identifying the different sources of uncertainties whether they are from nuclear data evaluations themselves (through covariance) or from integral experiments (either for their set up or their modelling).

The trends on the JEFF3.1.1 ^{235}U capture cross section are quite consistent with recent differential measurements.

IDA results suggest also a 2.5% decrease for ^{238}U capture from 3 keV to 60 keV, and a 4-5% decrease for ^{238}U inelastic in the plateau region. For this energy range, uncertainties are respectively reduced from 3-4 to 1-2% and from 6-9% to 2-2.5% for ^{238}U capture and ^{238}U inelastic.

The increase trend on ^{239}Pu capture cross section is of around 3% in the [2 keV-100 keV] energy range. For ^{240}Pu capture cross section, the increase is of around 4% in the [3 keV-100 keV] energy range and goes in the same direction as the recent ENDF/B.VIII evaluation though at a much lower level.

As perspectives for future works, there is a need of more differential measurements and more reliable nuclear data covariance.

For instance, IDA has identified the lack of differential measurements, in particular for prompt fission neutron spectrum or ^{238}U inelastic and parametric studies have shown that nuclear data covariance data with better reliability are required (for PFNS and some capture cross sections).

There is also a request to have covariance associated to anisotropy of scattering and distribution of secondary energy neutrons in particular for light and structural isotopes. This will enable to improve the IDA by incorporating all sources of uncertainties. Also adding more integral experiments will help increasing the reliability of the IDA, specifically if these are defined to target a given spectrum or nuclide.

References

1. M. Saez, J.-C. Robin, B. Riou, A. Villedieu, D. Deprest, and G. Prele, "Status of ASTRID nuclear island pre-conceptual design," Fast Reactor and Related Fuel Cycles FR13 Proc. Int. Conf., 2015.
2. F. Varaine et al., "Pre-conceptual design study of ASTRID core," 2012 International Congress on Advances in Nuclear Power Plants (ICAPP '12), Chicago, IL, June 24-28, 2012
3. G. Rimpault, "Nuclear Data Needs for the Assessment of Generation IV Nuclear Energy Systems",", International Workshop on Nuclear Data Needs for Generation IV Nuclear Energy Systems, Antwerpen 5-7 April 2005
4. J. P. Chaudat, "The CARNAVAL 4 adjusted library," Trans Am Nucl. Soc., vol. 27, no. 877, 1977.
5. E. Fort, W. Assal, G. Rimpault, J. Rowlands, P. Smith, and R. Soule, "Improved performances of the fast reactor calculational system ERANOS-ERALIB1 due to improved a priori nuclear data and consideration of additional specific integral data," Ann. Nucl. Energy Oxf., vol. 30, no. 18, pp. 1879–1898, 2003.
6. A. Santamarina, D. Bernard et al., "The JEFF-3.1.1 Nuclear Data Library. Validation results from JEF-2.2 to JEFF-3.1.1," JEFF Report 22, OECD/NEA Data Bank 2009.
7. P. Archier, "Contribution à l'amélioration des données nucléaires neutroniques du sodium pour le calcul des réacteurs de génération IV," Grenoble Alpes, 2011.
8. P. Archier et al., "CONRAD Evaluation Code: Development Status and Perspectives," Nucl. Data Sheets, vol. 118, pp. 488–490, Apr. 2014.
9. E. Brun et al., 'TRIPOLI-4®, CEA, EDF and AREVA reference Monte Carlo code', Ann. Nucl. Energy, vol. 82, pp. 151–160, Aug. 2015.
10. T. Bayes, "An Essay Toward Solving a Problem in the Doctrine of Chances," Philos. Trans. R. Soc. Lond., vol. 53, pp. 370–418, 1763.
11. G. Rimpault et al, "The ERANOS Code and Data System for Fast Reactor Neutronic Analyses, International Conference on the New Frontiers of Nuclear Technology : Reactor Physics, Safety and High-Performance Computing", PHYSOR'02, October 7-10, 2002, Seoul, Korea.
12. L. Berge, "Contribution à la modélisation des spectres de neutrons prompts de fission. Propagation d'incertitudes sur un calcul de fluence cuve," Grenoble Alpes, 2015.
13. Y. Danon et al. Simultaneous Measurement of ^{235}U Fission and Capture Cross Sections From 0.01 eV to 3 keV Using a Gamma Multiplicity Detector. Nuclear Science and Engineering, vol. 187, 291-301, 2017.
14. M. Jandel et al. New Precision Measurements of the ^{235}U(n; γ) Cross Section: Physical Review Letters; 109(20);November 2012:
15. H. I. Kim et al., Neutron capture cross section measurements for ^{238}U in the resonance region at GELINA, Phys:J:A; 52(6) : 170; June 2016:

 https://doi.org/10.1051/epjconf/201921103002

Total Monte Carlo acceleration for the PETALE experimental programme in the CROCUS reactor

Axel Laureau[1,*], *Vincent* Lamirand[1,2], *Dimitri* Rochman[2], and *Andreas* Pautz[1,3]

[1]Laboratory for Reactor Physics and Systems behaviour (LRS), Ecole Polytechnique Fédérale de Lausanne (EPFL), CH-1015 Lausanne, Switzerland
[2]Laboratory for Reactor Physics and Thermal Hydraulics (LRT), Paul Scherrer Institut (PSI), CH-5232 Villigen, Switzerland
[3]Nuclear Energy and Safety Research Division (NES), Paul Scherrer Institut (PSI), CH-5232 Villigen, Switzerland

Abstract. The Bayesian Monte Carlo technics requires individual evaluations of random cross section files based on a Total Monte Carlo propagation. This article discusses the use of a Correlated Sampling acceleration applied to TMC calculations for experiments where a brute force technics is too expensive. An efficient estimation of the reaction rate uncertainties in small dosimeters is obtained, together with the inter-dosimeter correlation associated to the cross section uncertainties.

1 Introduction

The Bayesian Monte Carlo (BMC) is one of the assimilation technics used to provide feedback from critical reactors to the nuclear cross sections themselves. This technics is working together with a Total Monte Calo (TMC) uncertainty propagation able to provide observables of interest (k_{eff}, activity, ...) calculations associated to a set of random cross sections such as displayed in Fig. 4 for the ^{56}Fe. The comparison of measurements and calculations provides a χ^2 value to each random cross section and then a weight is calculated: $\exp -\chi^2/2$.

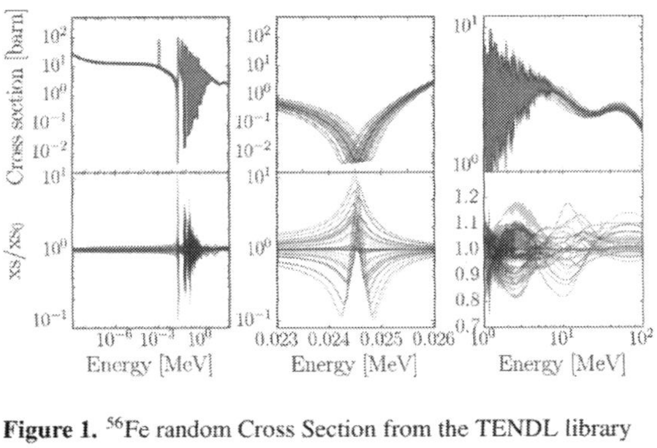

Figure 1. ^{56}Fe random Cross Section from the TENDL library

A very large number of random cross sections can be required in order to apply the BMC assimilation. This approach is already used [1] on fast critical systems with a large prior uncertainty and a short calculation time. If the evaluation of the random cross sections requires a large computation time, this step represents a limitation for the applicability of the method. For the PETALE experimental program at EPFL, the observables are the reaction rates in small dosimeters located inside a metallic reflector of a water reactor that requires Monte Carlo transport calculations to be correctly estimated. The estimation of the dosimeter activities for a single random cross section can require days of calculation.

For this reason, as presented in this article, an acceleration of the TMC uncertainty propagation has been developed, based on a Correlated Sampling (CS) technics as detailed in Sec. 2. Two application cases are used to compare the developed method with reference and legacy approaches (Sec. 3): the HMI-001 benchmark and the PETALE setup itself. Finally, an approach able to quantify the correlation between the dosimeters for the nuclear data uncertainty is proposed in Sec. 4.

2 TMC acceleration using the Correlated Sampling technics

The correlated sampling [2] technics is a numerical method for Monte Carlo particle transport that allows to maintain the representativeness of two systems using a single particle track. For that, at each sampled event, the probability to do this event is calculated for the reference and the modified systems. For example, the system can be 'modified' according to dimensions, concentration, temperature or cross sections, the latter being the one of interest in this study. Then with a "modified weight" associated to the neutron transport, and then by multiplying this weight by the ratio of the two probabilities and finally by correcting all the scores using this weight, all the scores are representative of the two systems. A detailed presentation of this implementation in the Serpent2 [3] code is available in [4] for thermal feedback estimations. For

*e-mail: axel.laureau@epfl.ch

example, for a neutron distance sampling d with a total macroscopic cross section of Σ_t in the reference system and Σ'_t in the modified system, the 'modified weight' multiplication term is: $\frac{\Sigma'_t \exp \Sigma_t d}{\Sigma_t \exp \Sigma'_t d}$. Note that this weight has to be propagated to the next generations of neutrons in order to take into account that the fission source is also modified.

In this study, the calculations are performed using the TENDL library. The first set of cross sections is considered arbitrarily as the "reference system". All the other random cross sections are considered as "modified systems". For each transported neutron, an array of modified weights corresponds to each modified system. An array of arrays of the neutron father's weight is also generated to take into account the fission source modification induced by the modification of the random cross sections.

The sampling distance and the type of interaction are considered here in the modified weight calculation. An unbiased estimation of the score modification due to these phenomena is expected even if a large cross section modification implies a slower convergence. A possible bias can be observed regarding the scattering anisotropy since the associated ratio of probabilities is not taken into account in this first implementation.

Considering an iron material definition with a perturbation of the ^{56}Fe, and assuming that the Fe56.####c element exists in the xsdata file where #### are the "0000, 0001, 0002, ..." random cross section files, the modified Serpent2 input file using the correlated sampling on 256 libraries is given in Fig. 2

```
mat mat_iron autopert 0 256 -7.874
     Fe54.03c     0.05845      Fe56.0000c    0.91754
     Fe57.03c     0.02119      Fe58.03c      0.00282
```

Figure 2. Iron material definition with 256 different ACE files

During the Serpent2 initialisation phase, an optional "autopert" key has been added in the material definition. When declared, all the isotopes defined with ".0000c" are then automatically duplicated for the correlated sampling with the labels ".0000c", ".0001c", ".0002c"...

3 Validation

The overall objective of the current work being the integral experimental program PETALE (see section 3.3) associated to the stainless steel at high neutron energy, the HMI-001 benchmark has first been selected for its high sensitivity to the ^{56}Fe uncertainty. It consists of a highly enriched uranium/iron metal core reflected by a stainless-steel reflector. In a second step, the approach is also validated on the reaction rates of dosimeters located in the PETALE experimental setup.

3.1 Comparison of uncertainty propagation approaches

Three technics are considered here to compute the TMC uncertainty propagation.

The "TMC-Ref" is the classic Total Monte Carlo uncertainty propagation technics. It requires n independent calculations for n random cross section files. This technics provides a reference distribution of the propagated uncertainty.

The "TMC-CS" is the new TMC uncertainty propagation using the correlated sampling technics. Up to 128 random cross section files are propagated in a single calculation. Multiple calculations are then combined altogether using the same "reference system".

Finally the "TMC-sensi" is based on sensitivity calculations. The sensitivity array is estimated using Serpent2, and represents the k_{eff} variation due to a 1 % perturbation of the cross sections in a given energy range. Associated to the actual cross section relative variations between the cross section files Fe56.####c and Fe56.0000c, the uncertainty is propagated for each random cross section file.

3.2 HMI-001 benchmark

3.2.1 Classic TENDL library

The effective multiplication factor distribution according to the random cross sections is presented in Fig. 3 We can

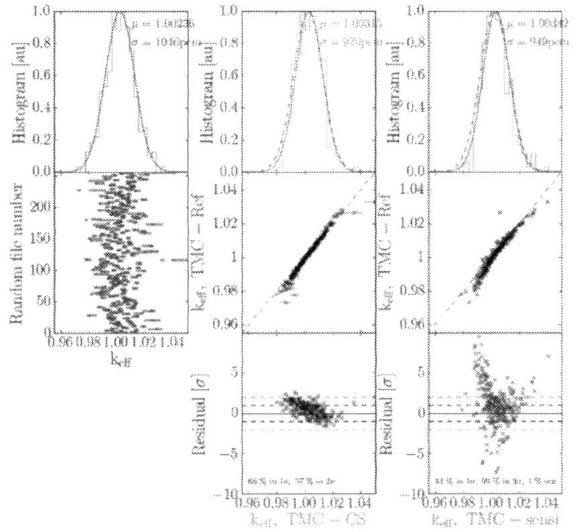

Figure 3. k_{eff} distribution associated to the ^{56}Fe uncertainty. The plots on the first line respectively present the results of the TMC-Ref, TMC-CS and TMC-sensi approaches. The second line corresponds to the k_{eff} values and the agreement between the three approaches for each value. The third line presents the associated residuals.

see that the impact of the ^{56}Fe cross section on the k_{eff} is significant, around 1000 pcm of uncertainty. We also observe that the three uncertainty propagation technics provide a standard deviation with a very good agreement.

Concerning the difference between the k_{eff} calculation for a same random cross section but using different methods (second line Fig 3), the global trend is linear. Using the TMC-CS, all the k_{eff} calculations are compatible with the TMC-Ref results. The residuals are good even if a slightly non-linear effect seems to appear, possibly due to the angular effect that is not taken into account. A non-linearity appears on the sensitivity approach when the reactivity variation (and then the cross section variation) is too large. The global trend is not linear but looks like a coma curve, resulting in larger residuals even if the global agreement is good.

3.2.2 Extended TENDL library

The classic TENDL library performs a sampling of the nuclear parameters and compares the generated cross sections to the EXFOR database in order to obtain random cross sections. In the extended TENDL library used here, the sampling is performed on the nuclear models themselves in order to have a larger cross section dispersion and thus to better check the validity of the methods.

Figure 4 shows the dispersion of the cross sections (left) using different sets of nuclear models, each model being declined with the sampling of the model parameters. We can see that two packs of cross sections are visible at high energy for the inelastic and capture cross sections. Fig. 4-right presents the uncertainty propagation using these cross sections. A larger non-linearity appears with two distinct coma lines, each line being associated to a pack of cross sections with a discontinuity between the two.

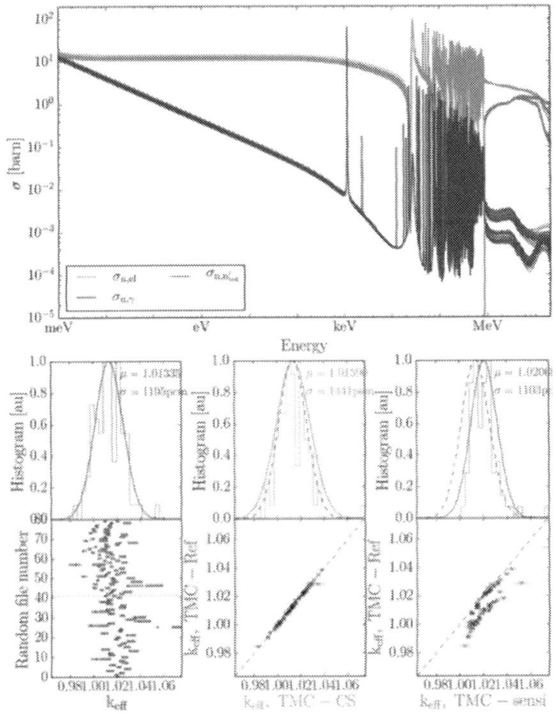

Figure 4. Extended TENDL library cross sections and corresponding uncertainty propagation (see legend of Fig. 3

3.3 Dosimetry in the PETALE setup of the CROCUS reactor

The HMI-001 benchmark illustrates the good capability of the TMC-CS technics to predict reactivity variations associated to the random cross sections. This section aims to validate the method for reaction rate estimation in dosimeters.

Figure 5 presents the PETALE setup in the CROCUS reactor. A full description of the experiment is available

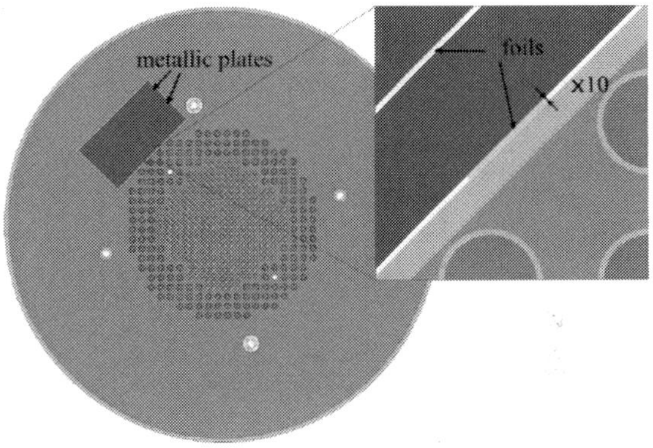

Figure 5. Serpent2 CROCUS geometry with the PETALE experimental setup

in [5] and the description of the experimental program is available in [6]. The PETALE setup is composed of 8 metallic plates (in gray) interleaved with nine dosimeters (in green). In this paper the metallic reflector is composed of pure iron, and the dosimeters are made of indium in this section.

During the calculations, the reaction rate score is declined using different modified weights that correspond to the different sets of random cross sections. The corresponding reaction rates are presented in Fig. 6. We can see that the reaction rates as a function of the dosimeter number (top) have a very different pattern between the capture (left) and the inelastic (right) reactions. This is due to the different characteristic energies of the neutron doing the interaction: at high energy for the threshold inelastic cross section and at thermal energy for the capture reaction. The impact of the random cross sections is visible through the curve dispersions, and the relative variation is presented on the second line. We can see that this dispersion reaches a maximal value of around 15%. A very good agreement is obtained between the TMC-Ref (dashed lines) and the TMC-CS (plain lines) calculations. We also observe that the statistical uncertainty on the reaction rate variation is reduced with the TMC-CS. With the classic TMC-Ref approach, a difference between two independent calculations with their independent statistical uncertainties σ_1 and σ_2 leads to a total uncertainty of $\sqrt{\sigma_1^2+\sigma_2^2}$, while the correlated sampling technics directly provides the estimation of the score difference using the same neutron tracks and then this results in a reduced statistical uncertainty.

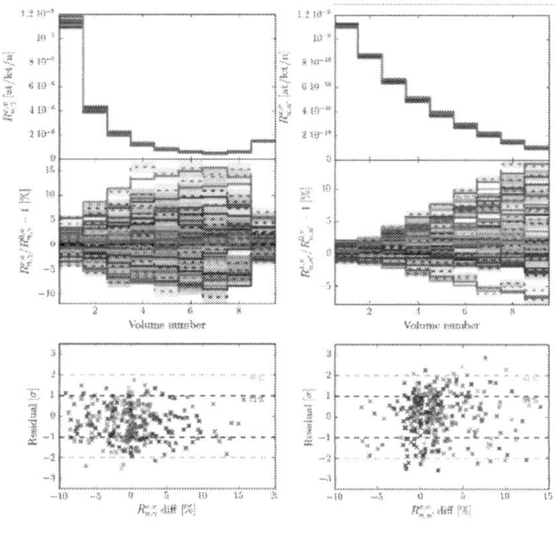

Figure 6. Reaction rate distributions (top) for the capture (left) and inelastic (right) reactions; relative variations with one color per random cross section of the classic TENDL library (second line) with the TMC-CS in plain lines and the TMC-Ref in dashed lines; and residuals (bottom)

4 Dosimeter correlations

The TMC-CS is used to estimate the reaction rate values in dosimeters for multiple random cross sections with a single calculation. Multiple dosimeters can be used, by putting the dosimeters side by side or using distinct calculations. The first solution is considered in this section.

Figure 7 shows the reaction rate dispersion for two threshold reactions: the rhodium and the indium inelastic reactions. One can see that the ordering between the

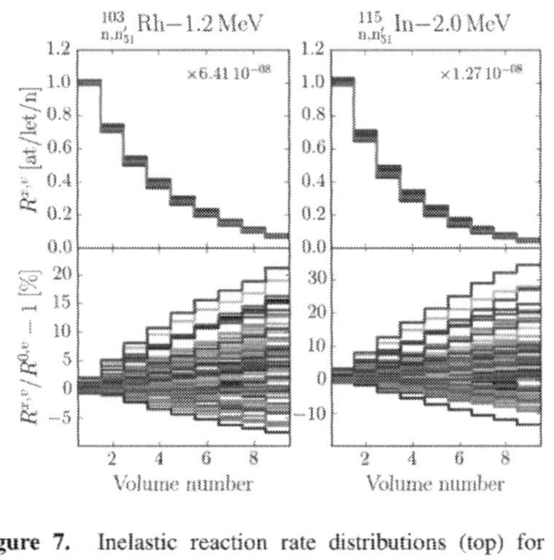

Figure 7. Inelastic reaction rate distributions (top) for the rhodium (left) and indium (right) elements and the relative variation between the random cross sections (bottom)

random cross section curves is very similar. For example, in volume 8, even if the relative variation is larger for the rhodium (20 %) compared to indium (30 %), the ordering is the same: brown curve for the maximal, then the yellow curve, the blue, the brown, the green one... This means that, if we assume that the measurement has no experimental uncertainty and that there is no nuisance parameters, the two dosimeters will provide the same information. For example if we also assume for a simplification purpose that one of the TENDL random cross sections (corresponding for example to #42 associated to the blue color in the figure) is very close to the real one, then the indium measured value will be the blue curve, and the BMC will provide a large weight for all the curves with a similar value (the yellow, blue, brown ones). And then since the ordering is the same with the rhodium dosimeter, the selected curves will be identical. In the opposite situation, a different *uncorrelated* distribution would provide a distinct information and allows to discriminate between the remaining cross section files.

For this reason we can use the correlation matrix of this reaction rate dispersion in order to discriminate between independent dosimeters and quantify the level of correlation. Figure 8 presents this correlation matrix for a set of dosimeters. We can see that as expected the rhodium and

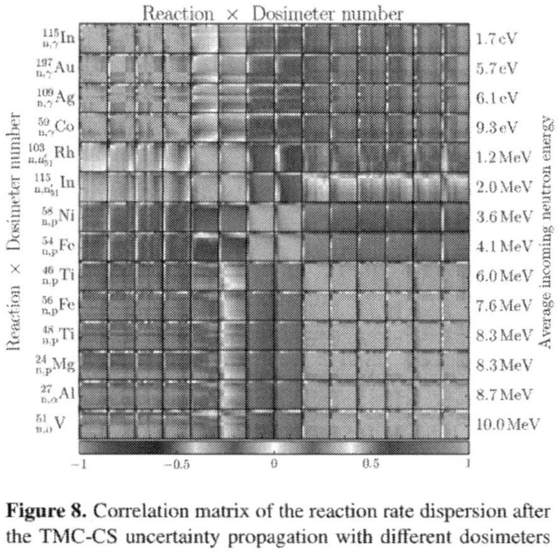

Figure 8. Correlation matrix of the reaction rate dispersion after the TMC-CS uncertainty propagation with different dosimeters ordered by the lethargy average incoming neutron energy. Each sub-matrix in the diagonal corresponds to one element with the correlation of the different positions (nine distinct dosimeters), and the sub-matrix off-diagonal corresponds to the inter-element correlations.

the indium results are very correlated. We also obtain a correlation coefficient for all the different dosimeters. This correlation will be useful for the BMC assimilation in order to avoid assimilating several times the same information. This is a useful element regarding the experimental program as discussed in [6].

5 Conclusions

The correlated sampling technics is used to propagate random cross sections through a single Monte Carlo calculation. This propagation has been validated for iron cross sections by comparison with a reference Total Monte Carlo approach. Two gains are obtained with this technics: a strong reduction of the number of Monte Carlo calculations; and thanks to the estimation of the score difference using a single neutron path, the variance on the difference is reduced. This method appears to be very useful for integral experiments requiring long calculations with a small statistical uncertainty, such as reaction rate estimations in a reflector.

References

[1] Rochman, D. A., Bauge, E., Vasiliev, A., Ferroukhi, H. and Perret, G., EPJ Nuclear Sciences & Technologies, **4**, 7 (2018).

[2] Rief, H., Annals of nuclear energy, **11**, 9, 455–476 (1984).

[3] Leppänen, J., Pusa, M., Viitanen, T., Valtavirta, V. and Kaltiaisenaho, T., Annals of Nuclear Energy, **82**, 142–150 (2015).

[4] Laureau, A., Buiron, L. and Fontaine, B., EPJ Nuclear Sciences & Technologies, **3**, 16 (2017).

[5] Lamirand, V., Perret, G., Radman, S., Siefman, D. J., Hursin, M., Frajtag, P., Gruel, A, Leconte, P., Blaise, P. and Pautz, A., ISRD16 International Symposium on Reactor Dosimetry (2017).

[6] Lamirand, V., Laureau, A, Rochman, D., Perret, G., Gruel, A, Leconte, P., Blaise, P. and Pautz, A., International Workshop On Nuclear Data Evaluation for Reactor applications, (2018).

An Experimental Programme optimized with Uncertainty Propagation: PETALE in the CROCUS Reactor

Vincent Lamirand[1,2*], *Axel* Laureau[1], *Dimitri* Rochman[2], *Gregory* Perret[2], *Adrien* Gruel[3], *Pierre* Leconte[4], *Patrick* Blaise[3], and *Andreas* Pautz[1,5]

[1]EPFL, Laboratory for Reactor Physics and Systems behaviour (LRS), 1015 Lausanne, Switzerland
[2]PSI, Laboratory for Reactor Physics and Thermal Hydraulics (LRT), 5232 Villigen, Switzerland
[3]CEA/DEN, Experimental Physics, Safety & Instrumentation Division (SPESI), 13108 St-Paul-Lez-Durance, France
[4]CEA/DEN, Reactor Physics and Fuel Cycle Division (SPRC), 13108 St-Paul-Lez-Durance, France
[5]PSI, Nuclear Energy and Safety Research Division (NES), 5232 Villigen, Switzerland

Abstract. The PETALE experimental programme in the CROCUS reactor at EPFL intends to contribute to the validation and improvement of neutron nuclear data in the MeV energy range for stainless steel, particularly in the prospect of heavy reflector elements of pressurized water reactors. It mainly consists of several transmission experiments: first, through metallic sheets of nuclear-grade stainless steel interleaved with dosimeter foils, and, successively, through its elemental components of interest – iron, nickel, and chromium. The present article describes the study for optimizing the response of the dosimetry experiments to the nuclear data of interest.

1 Introduction

The present article describes an optimization study for the design of PETALE [1], an oncoming experimental programme in the CROCUS reactor. CROCUS is a teaching and research zero-power reactor operated by the Laboratory for Reactor Physics and Systems Behaviour (LRS) at the Swiss Federal Institute of Technology in Lausanne (EPFL). Within the framework of a collaboration with CEA Cadarache, PETALE intends to contribute to the validation efforts on neutron nuclear data in the MeV energy range for stainless steel, particularly focusing on heavy reflector elements of pressurized water reactors (PWR). It mainly consists of transmission experiments, where part of the water reflector will be replaced by successive sheets of stainless steel alloy and its elemental components – natural iron, nickel, and chromium. The metallic sheets will be set in a watertight box in air, and interleaved with thin dosimeter activation foils. In addition, reflector-reactivity-worth estimates, in-core spectra characterisations, and sheet activation mapping, when possible, will be performed. While the experiments with the alloy will be used as a reference, the successive experiments with its main elements aim at avoiding compensation effects with data assimilation methods. [2]

Because of the opposition between the low neutron flux in the dosimeters in the range of interest, requiring irradiations of maximal power and duration, and the operational constraints of the teaching activities at the CROCUS facility, an optimization of the experimental programme was carried out. Reaction rates of interest were maximised, and therefore the sensitivity to target nuclear data, while the integrated power was minimized. The work was performed by propagating uncertainties using the Total Monte Carlo (TMC) approach with ACE files from the TENDL-2017 library and the Serpent2 Monte Carlo code, in association with an acceleration method, Correlated Sampling (CS). [3-5] In this article, the design of the dosimetry experiments optimized with this methodology is presented, starting from the selection of dosimeter materials, to target uncertainties with regards to nuclear data and operation prospects.

2 Experimental setup

2.1 The CROCUS reactor

CROCUS is an experimental zero-power reactor, uranium-fuelled and water-moderated, dedicated to education in radiation and reactor physics, and to research [6-17]. A complete description of the core can be found in the International Reactor Physics Experiments Handbook (IRPhE) [18].

CROCUS is licensed for operating at a maximum power of 100 W, i.e. ~2.5×10^{9} $cm^{-2}.s^{-1}$ total neutron flux at the core centre. Criticality is controlled either by water level using a spillway (see Fig. 1), or two B_4C absorber control rods, with an accuracy of ±0.1 mm (equivalent to approximately ±0.4 pcm) and ±0.5 mm (±0.2 pcm), respectively. This precision and water level's linear trend are of interest for the estimate of reflector worth. It operates at room temperature using a controlled water loop with secondary and tertiary circuits, two heat exchangers and an electrical heater.

* Corresponding author: vincent.lamirand@epfl.ch

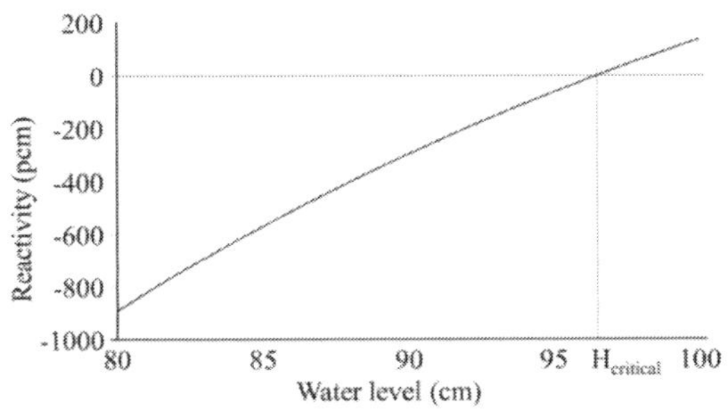

Fig. 1. Reactivity dependence on water level around criticality.

The core is located in an Al-6060 grade vessel of 130 cm in diameter, 160 cm in height, and 1.2 cm in thickness (see Fig. 2). The vessel is filled with demineralized light water which is used as both a moderator and reflector. The core active part has the approximate shape of a cylinder of 100 cm in height and about 60 cm in diameter. It consists of two interlocked fuel zones with square lattices of different pitches:

• an inner zone of 336 UO_2 rods, 1.806 wt.% enriched, and with a pitch of 1.837 cm,

• an outer zone of 172 U_{metal} rods in nominal configuration, 0.947 wt.% and 2.917 cm,

• a varying water gap between the two zones because of the two different pitches.

Both uranium fuels consist of a 1-m pile of cylindrical pellets cladded in aluminium. The rods are maintained vertically by two octagonal aluminium grid plates spaced 1 m apart. In this configuration, the grids have a 0.5-mm cadmium layer to limit axial neutron leakage to the environment, i.e. structure activation, with the active zone of the fuel starting above the lower cadmium layer.

The reactor possesses six independent shutdown mechanisms to bring it to a sub-critical state in less than one second: two cruciform Cd safety blades in the inner zone; and four expansion tanks to drain the moderator, set in the vessel corners and controlled each by a valve.

2.2 Metal reflectors

Following the feasibility study [1], each metal reflector is composed of eight 2-cm thick sheets of 30×30 cm^2. These latter ones are made of stainless steel (nuclear-grade low cobalt 304L type), natural iron (99.85 %), nickel (99.7 %), and chromium (99.95 %). Their widths were chosen to limit detection of thermal neutrons scattered in the surrounding water, while keeping a reasonable size for limiting production constraints, cost, and in-core volume. Activation foils shall be positioned between the sheets with great precision, as the flux gradient is large. The metal reflector will be placed in the northwest periphery of the core in a mobile aluminium case, which is currently under design at CEA Cadarache (see Fig. 2). It will allow reproducible positioning as well as easy insertion and extraction of the sheets and dosimeters. Moreover, the case will prevent perturbations from water slipping between the sheets.

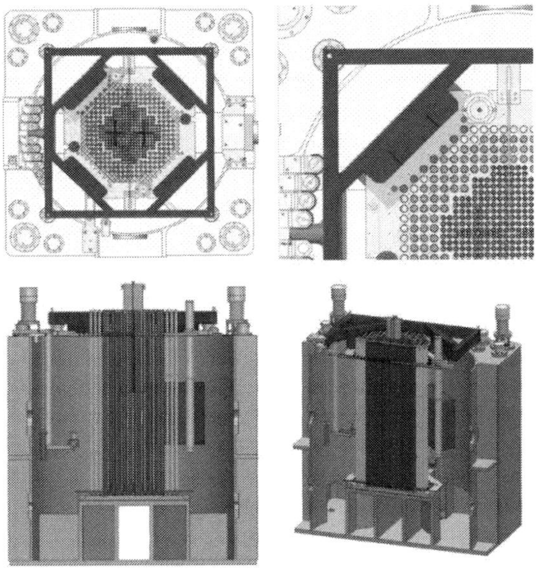

Fig. 2. Model of the metal reflector northwest positioning with respect to the core (top and side views).

2.3 Gamma spectrometry of activated foils

After irradiation, the dosimeters' activities will be measured using high-purity germanium (HPGe) gamma spectrometers to be compared with calculation results. The LRS team owns several HPGe detector stations, including a recent Canberra station (detector: GC4518/S; shielding: VG-BB-98/16D1-2), whose efficiency would suffice to measure most activation dosimeters sensitive to the thermal energy range. Regarding dosimeters with low activity, other possibilities are prospected for accurate measurements, e.g. counting in remote low-radioactivity laboratories.

3 Optimization of dosimetry experiments

As a zero-power light water reactor, CROCUS is limited in power and in fast neutron flux component. Moreover, its use for education purposes requires the accessibility to its cavity for most of the year to students or, more generally, to visitors. The shutdown configuration has no in-core moderator and, therefore, calls for limited residual dose of the core, i.e. limited integrated power. These two reasons were the main object of the feasibility study [1], while the study presented hereafter focuses on the optimization of the experiments. In addition, although it was proven that dosimeters of interest could indeed be activated, the consequent constraint on nuclear data was not quantified.

The optimization was carried out using Total Monte Carlo uncertainty propagation in association with a Correlated Sampling technique, implemented in a modified version of the Serpent2 Monte Carlo code. The TMC-CS method and the optimization results are presented hereafter.

3.1 Total Monte Carlo with Correlated Sampling

The TMC method consists of sampling distributions of quantities of interest to observe their impact on a system, and consequently propagate their uncertainties. Correlated sampling is a numerical method for Monte Carlo particle transport that allows to maintain the representativeness of two systems using a single particle track. [19] For each sampled event, its probability is calculated for a reference, and for a system modified with regards to the quantity of interest. By the use of weighting for each event, and correction of the final results, score representatives of both systems are obtained in one calculation only. An implementation of CS in the Serpent2 code is available in [20] for thermal feedback estimations. Its coupling to the TMC methods allows reduction of the computational cost.

In the case of the study presented here, the TMC-CS method was used to propagate uncertainties of the nuclear cross sections of interest to the reactions rates in the reflectors dosimeters. By sampling over cross sections distributions representative of their uncertainties, such as random ACE files provided by the TENDL-2017 library, we obtained the associated expected reaction rates distribution in the reflector foils. The development and validation of the method for the present application is presented in [21].

Two examples of reaction rates distributions are presented in Fig. 3, as a function of the dosimeter position in the reflector: position 1 corresponds to the front of the reflector, 2 to the position after the first sheet of iron, etc. For each position, distributions of absolute and relative values are shown (top and bottom). Capture (thermal range) and inelastic (fast range) reactions on indium-115 are compared (left and right sides). For each position, the distribution allows quantifying the target uncertainty of the dosimeter activation to provide feedback on the cross section. It should be significantly lower than the variation induced by the cross section. In addition, specific trends with distance are observable, which should allow analysis by relative comparison of dosimeters at different positions.

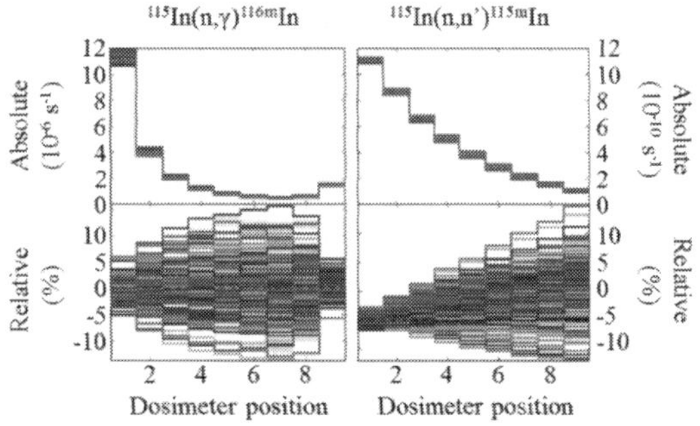

Fig. 3. Distributions of absolute and relative capture (left) and inelastic (right) reaction rates of indium dosimeters with regards to their position in the natural iron reflector. These examples were computed with Serpent2 using TMC-CS and 128 ACE files from TENDL-2017 representative of the iron cross sections.

3.2 TMC-CS for design optimization

Reaction rate distributions were calculated for all three reflector elemental materials and their associated cross sections, and all dosimetry reactions of interest. They were then compared to the expected uncertainties of the dosimeter activities for reasonable irradiation powers and durations (1 to 3 h at 50 W), and activity measurements (1 to 24 h) with the available systems. The calculations are representative of the experimental setup, and take into account self-shielding of dosimeters. Results for all reflector materials are presented in Fig. 4.

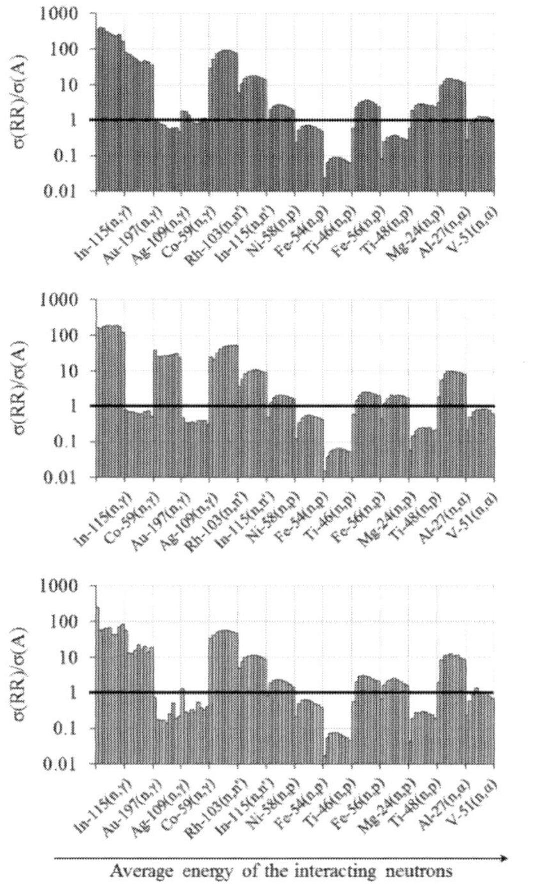

Fig. 4. Ratio of reaction rates' distributions in the elemental reflectors (Fe, Cr and Ni top to bottom), over uncertainties on activity determination for all considered reactions sorted by lethargy-averaged neutron energy, and for each reaction sorted by dosimeter position.

We observe that, for the range of envisaged irradiations, some reactions are more efficient to set constraints on the nuclear data, consistently over all cases. In a first step, the results provide orders of magnitude that help us to decide which reactions to use for each energy range, and at which position the dosimeters should be set in the reflectors for maximum sensitivity. In the final definition of the experimental programme, the results will help to dimension irradiation specifications for each dosimeter.

https://doi.org/10.1051/epjconf/201921103003

Using TMC-CS, we also quantified the correlations between the reactions' outputs. It allows for us to know which reactions carry redundant or supplementary information about the cross sections. The aim is to choose appropriate redundancy. The results for all reflector types are presented in Fig. 5, reaction by reaction, and for each dosimeter position.

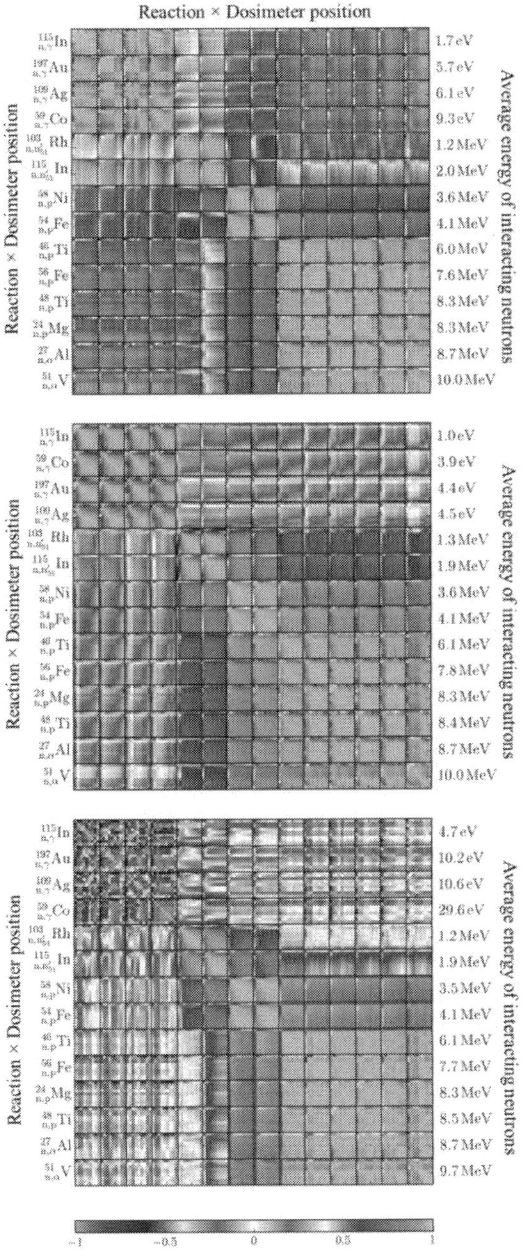

Fig. 5. Correlation matrices of reaction rates for all considered dosimetry reactions in the elemental reflectors (Fe, Cr and Ni, top to bottom), sorted by lethargy-averaged neutron energy and dosimeters position ("pixels" within each reaction).

For all three cases, but with variations in intensity, we observe four general groups divided by range of lethargy-averaged energy of the interacting neutrons:

a. eV range: In(n,γ), Au(n,γ), Ag(n,γ), Co(n,γ),
b. 1-2 MeV range: Rh(n,n'), In(n,n'),
c. 2-5 MeV range: ^{58}Ni(n,p), ^{54}Fe(n,p),
d. above 5 MeV: ^{46}Ti(n,p), ^{56}Fe(n,p), ^{48}Ti(n,p), ^{24}Mg(n,p), Al(n,α) and V(n,α).

It should be noted that we obtain a significant sensitivity in the fast range, with no less than three groups carrying information of interest. Still, in the case of the chromium reflector, correlation results present a clear separation between thermal and fast ranges only: all dosimeters in the fast range present maximal correlations, slightly less pronounced for the 1-2 MeV range. Regarding the nickel reflector, correlation oscillations in the thermal range are representative of the low dispersion in the reactions rates induced by the low nuclear data uncertainties. In the prospect of data assimilation, these correlations between dosimeters should be taken into account with great care.

3.3 Selection of dosimetry reactions

From the reactions' properties, and with the additional information deduced from the ratios and correlations presented above, we are able to define which reactions should be favoured. For consistency purposes, and as the study confirmed the possibility, we would favour the same set of reactions for all transmission experiments. The list of selected reactions and their ratio values for the iron reflector are presented in Table 1.

Table 1. Reactions of interest for all transmission experiments, with propagation results in the case of the natural iron reflector. The reactions are sorted by neutron energy range and group of interest, as defined in Section 3.2. The maximum ratios of reaction rate distribution over activity uncertainty are indicated, as well as the associated dosimeter positions in the reflector.

Group	Reaction	Half-life	Max. ratio	Max. position
a	^{115}In(n,γ)	54.3 min	400	2
a	^{197}Au(n,γ)	2.70 d	79	1
b	^{115}In(n,n')	4.49 h	18	5
c	^{58}Ni(n,p)	70.9 d	2.7	4
c	^{54}Fe(n,p)	312 d	0.7	5
d	^{56}Fe(n,p)	2.58 h	3.7	5
d	^{27}Al(n,α)	15.0 h	15	4

We observe high ratios for several reactions, both in the thermal and fast ranges, indicating space for minimisation of the irradiation power and duration. Reactions with low ratios but long half-lives, such as ^{58}Ni(n,p) and ^{54}Fe(n,p), are to be favoured for measurements in remote laboratories. In addition, we observe that some dosimetry materials used in their natural abundance, i.e. In and Fe, should provide complementary information of interest for our set of experiments by using different reactions, and without increasing the number of irradiations. As the position of maximum sensitivity varies with the reaction, irradiations with mixed dosimeter types are also to be considered. In the specific case of the ^{103}Rh(n,n') reaction, and despite its high ratio, it was discarded because of the difficulty to measure the low energy X-ray emission coupled with a short half-life (56 min), preventing measurements in remote laboratories. Another result of this study is the confirmation of the interest of nickel and iron dosimeters: it validates the principle of mapping the activity after irradiation of the corresponding reflector sheets themselves.

4 Conclusion

In the present article, we described an optimization study carried out for the design of PETALE's transmission experiments on stainless steel and its main elements. It used a new uncertainty propagation method based on Total Monte Carlo coupled with Correlated Sampling using Serpent2 and TENDL-2017. Because of the flux and operational constraints of the CROCUS facility, quantification and optimization of the impact of the experiments on target nuclear data was required. It concluded on an optimized set of dosimetry reactions – namely ^{115}In(n,γ), ^{197}Au(n,γ), ^{115}In(n,n'), ^{58}Ni(n,p), ^{54}Fe(n,p), ^{56}Fe(n,p), ^{27}Al(n,α) –, and target uncertainties for defining irradiation and measurements conditions. The next step is to propagate the uncertainties from technological parameters and the cross sections of the dosimetry reactions, in order to finalise the experimental setup and to define the experimental programme.

The authors thank Mathieu Hursin and Daniel Siefman for their assistance and feedback on uncertainty quantification and propagation.

References

1. V. Lamirand, G. Perret, S. Radman, D. Siefman, M. Hursin, P. Frajtag, A. Gruel, P. Leconte, P. Blaise, and A. Pautz, React. Dosim. 16th Int. Symp. ASTM **STP1608**, 7 (2018).
2. D. Siefman, M. Hursin, D. Rochman, S. Pelloni, and A. Pautz, Eur. Phys. J. Plus **133**, 429 (2018)
3. A. J. Koning and D. Rochman, Ann. Nucl. Energy **35**, no. 11, 2024–2030 (2008).
4. A. J. Koning and D. Rochman, Nucl. Data Sheets, **113**, no. 12, 2927–2934 (2012)
5. J. Leppänen et al., Ann. Nucl. Energy **82**, 142-150 (2015)
6. A. Rais, D. Siefman, G. Girardin, M. Hursin, and A. Pautz, *IGORR 2014* (2014)
7. M. Hursin, A. Rais, D. Siefman, G. Girardin, and A. Pautz, *International Technical Meeting on Small Reactors* (2014)
8. O. Pakari, V. Lamirand, G. Perret, D. Godat, M. Hursin, P. Frajtag, and A. Pautz, *PHYSOR 2018* (2018)
9. V. Lamirand, G. de Izarra, A. Krasa, G. Perret, O. V. Pakari, M. Hursin, P. Blaise, J. Wagemans, and A. Pautz, *PHYSOR 2018* (2018)
10. A. Rais, D. Siefman, G. Girardin, M. Hursin, and A. Pautz, Sci. Technol. Nucl. Install. (2015)
11. D. J. Siefman, G. Girardin, A. Rais, A. Pautz, and M. Hursin, Ann. Nucl. Energy **85**, 434–443 (2015)
12. A. Rais, M. Hursin, G. Perret, and A. Pautz, *PHYSOR 2016* (2016)
13. V. Lamirand, M. Hursin, G. Perret, P. Frajtag, O. Pakari, and A. Pautz, *RRFM/IGORR 2016*, 284–292 (2016)
14. A. Rais, Ecole Polytechnique Fédérale de Lausanne (EPFL) (2017)
15. V. Lamirand, A. Gruel, G. Perret, P. Leconte, P. Frajtag, M. Hursin, P. Blaise, and A. Pautz, *ISRD16* (2017)
16. O. Pakari, V. Lamirand, G. Perret, and A. Pautz, *ANIMMA 2017* (2017)
17. O. Pakari, V. Lamirand, G. Perret, P. Frajtag, and A. Pautz, IEEE Trans. Nucl. Sci. **65**, no. 9, 2456–2460 (2018).
18. U. Kasemeyer, R. Früh, J. M. Paratte, and R. Chawla, International Reactor Physics Experiments Handbook (IRPhE) **4440**, OECD, Ed. 2007, 94 (2007)
19. H. Rief, Ann. Nucl. Energy **11**, no. 9, 455–476 (1984)
20. A. Laureau, L. Buiron, and B. Fontaine, EPJ Nucl. Sci. Technol. **3**, 17 (2017)
21. A. Laureau, V. Lamirand, D. Rochman, and A. Pautz, *WONDER 2018* (2018)

Impact of delayed neutron constants on reactivity effects measured in a fast reactor

Antonín Krása[1,*], *Anatoly* Kochetkov[1], *Nadia* Messaoudi[1], *Alexey* Stankovskiy[1], *Guido* Vittiglio[1], *Jan* Wagemans[1]

[1]Belgian Nuclear Research Centre SCK•CEN, Boeretang 200, 2400 Mol, Belgium

Abstract. Delayed neutron parameters of fast VENUS-F reactor core configurations are determined with Monte Carlo calculations using various nuclear data libraries. Differences in the calculated effective delayed neutron fraction and the impact of the delayed neutron data (6- or 8-group precursors) that are applied in the experimental data analysis on the measured reactivity effects are studied. Considerable differences are found due to application of ^{235}U and ^{238}U delayed neutron data from JEFF, JENDL and ENDF evaluations.

1 Introduction

Analysis of experimental data measured in numerous nuclear reactor physics experiments require the knowledge of core averaged kinetic parameters. Deterministic [1] or Monte Carlo methods [2,3] can be used to calculate the effective delayed neutron fraction β_{eff}, decay constants λ_i and abundances β_i of the precursor groups for a specific reactor core configuration as they depend on fuel composition and neutron spectrum.

Evaluated delayed neutron data (delayed neutron multiplicity ν_d, half-lives and abundances of precursors, delayed neutron spectra χ_d) are available in general purpose nuclear data libraries (e.g. JEFF, JENDL, ENDF). The delayed neutron spectra are given in energy-group format (usually 16 groups of Hansen-Roach set [4]). The evaluated decay constants and abundances are given in time-group format (usually 8 groups [5] or 6 groups [6]). The group constants (as well as the fission yields) are evaluated at three energies: thermal, fast (fission), high (fusion). This study is focused on fast fission in ^{235}U and ^{238}U.

The influence of delayed neutron data on thermal reactor parameters has been widely investigated (e.g. [7,8]). In fast reactors the influence of delayed neutron spectra was studied [9,10], benchmarks for β_{eff} were carried out at MASURCA and FCA critical facilities [11] and used for adjustment of delayed neutron yields [12]. In this paper, we describe the impact of delayed neutron data on the analysis of reactivity measurements at the fast zero-power VENUS-F reactor located at the Belgian Nuclear Research Center SCK•CEN. VENUS-F has been extensively used for fast neutron data validation [13-17] serving as a mockup of the MYRRHA fast reactor/accelerator driven system [18,19]. Sensitivity and uncertainty analysis of β_{eff} were recently studied for the specific cases of the VENUS-F [20] and MYRRHA reactors [21].

* Corresponding author: antonin.krasa@sckcen.be

2 Delayed neutron data

The half-lives and abundances of delayed neutron precursors are represented either with 6 groups (isotope-dependent half-lives) or 8 groups (identical half-lives for all the fissionable isotopes). Each nuclear data library makes its own best estimate based on various evaluated data available for each of the fissile and fertile isotopes in three energy regions. The current status of the delayed neutron parameters of ^{235}U and ^{238}U (relevant for the present study) adopted in the main nuclear data libraries is the following.

The JEFF library (since version 3.1) [22] uses Spriggs data (both decay constants and abundances) in 8-group format [4,5]. The JENDL library (since version 3.2) [23] uses the 6-groups data: decay constants of Keepin [6] and abundances evaluated on the basis of a large set of experimental and evaluated data. The ENDF library (since version VII.0) [24] uses the 6-groups data that come from summation calculations, which use decay constants and delayed neutron emission probabilities from the evaluated experimental data file NuBase2003 [25]; if not available, the data are calculated using quasi-particle random-phase approximation and the statistical gross theory [26]. One of the most remarkable features in ENDF is the remarkably short first group half-life compared to the other two evaluations, see Table 1. Other libraries like BROND-3.1, CENDL-3.1, TENDL-2017 adopted the same decay constants as in ENDF.

There are no uncertainties of the decay constants and group abundances in the evaluated data libraries. Only the original Spriggs data [5] contain uncertainties of the group abundances (group half-lives are fixed, thus, their uncertainties are reflected in the uncertainties of the abundances), which further leads to the uncertainty of the average half-life of the delayed neutrons (9.10 s ± 1% for fast ^{235}U and 5.3 s ± 3% for fast ^{238}U fissions).

Table 1. Half-lives of precursor groups in JEFF-3.1.2, JENDL-4.0, ENDF/B-VII.1.

	Group structure	8-g	6-g			
	Isotopes	all	^{235}U		^{238}U	
Main precursors	Library	JEFF	JENDL	ENDF	JENDL	ENDF
for 8-g	Group	Half-life [s]				
^{87}Br	1	55.6	55.7	52.0	52.4	50.9
^{137}I	2	24.5	22.7	21.2	21.6	22.1
^{88}Br	3	16.3	6.22	5.74	5.00	5.62
^{138}I, ^{93}Rb, ^{89}Br	4	5.21	2.3	2.29	1.93	2.14
^{94}Rb, ^{139}I, ^{85}As, ^{98m}Y	5	2.37	0.61	0.82	0.49	0.77
^{93}Kr, ^{144}Cs, ^{140}I	6	1.04	0.23	0.24	0.17	0.23
^{91}Br, ^{95}Rb	7	0.424				
^{96}Rb, ^{97}Rb	8	0.195				

Although the group constants evaluated for thermal (25 meV) and high (14 MeV) energy have unambiguously defined energy, the mean energy of the fast spectrum specified in the evaluations is not always clear because of a wide variation in the mean energies of the experimental setups used in integral data studies. The ^{235}U evaluations in the fast spectrum correspond to a mean energy of hundreds of keV (some evaluations refer to 200 keV [4], 400 keV [22], 500 keV [24] or a range from 50 keV to 7 MeV [23]). However, the fast spectrum values for ^{238}U rather correspond to energies slightly above 1 MeV (although the evaluations again refer to 400 or 500 keV, which is not realistic due to the ^{238}U fission threshold energy). The energy dependence of the group decay constants was extensively measured [27].

In general, the delayed neutron production has a non-trivial energy dependence, especially above 1 MeV, see ^{235}U in Fig. 1. Below 10 keV, ν_d is constant and JENDL and ENDF give nubar about 2% smaller than JEFF. Between 10 and 50 keV, ν_d increases according to JENDL and ENDF, while it stays constant in JEFF evaluation. Between 50 keV and 1 MeV, ENDF gives ν_d about 2-3% bigger than JEFF and JENDL. The sharp decrease of ν_d above 4 MeV is likely to be caused by the second chance fission process. The ν_d covariances in JENDL lead to 3.5% uncertainty below 20 keV and 7 % uncertainty between 20 keV-3 MeV, which is significantly bigger than the differences between various evaluations.

The delayed neutron production in ^{238}U is about 3 times bigger than in ^{235}U with differences up to 5% between the evaluations.

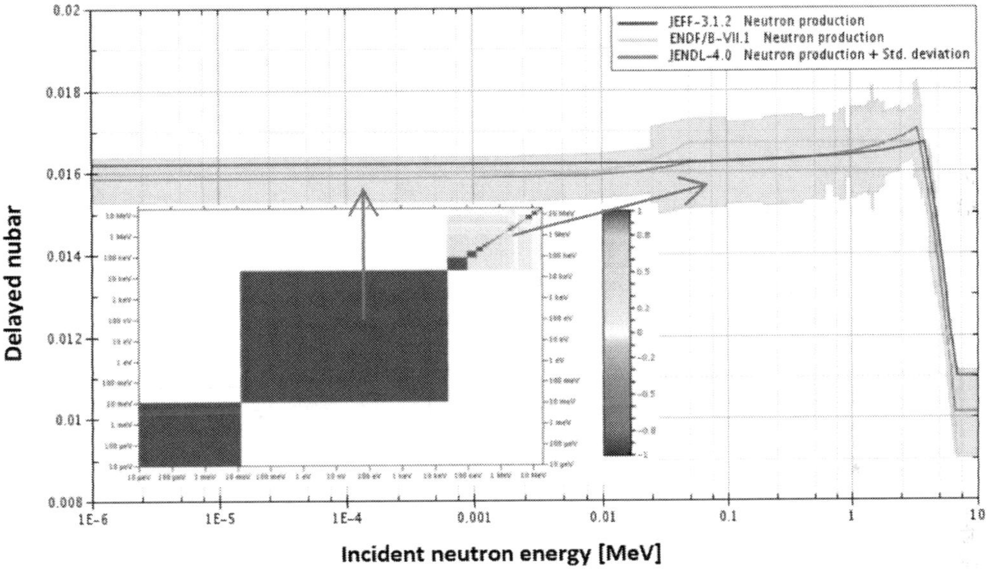

Fig. 1. The evaluated average number of delayed neutrons for ^{235}U(n,f). The inset shows covariances from JENDL-4.0 and the arrows show where they contribute to the uncertainties.

3 Experiment

VENUS-F [28] is a fast zero-power reactor loaded with 30% wt. metallic uranium fuel. The core configuration presented in this paper used solid Bi as a coolant simulator and lead as reflector, see Fig. 2 (detailed description of fuel assembly composition and the entire core can be found in [15]).

For measurements of reactivity effects, either the compensation method [29] or the MSM method [30] are used. Both require the application of the delayed neutron parameters. The former through the calibration of control rods with a period measurement (the inhour equation). The latter requires the determination of a reference sub-critical state. For that purpose, a rod drop method is applied (see further).

The VENUS-F reactor was operated in a critical state at a stable power for sufficiently long time for the neutron population and precursor concentrations to reach equilibrium (about 20 min). Fission chambers [31] inserted in experimental channels in the reflector (see Fig. 2 left) were used to record a constant count rate n_0.

Then, an absorber rod (made of B_4C powder) located in the reflector (see Fig. 2) was rapidly inserted (the rod drop took 0.4 s). Detector count rates $n(t)$ were measured as a function of time after the rod drop ($t = 0$) until counting only background counts, i.e. the complete decay of the precursors, see Fig. 3 left.

The reactivity of the sub-critical configuration was determined using the integral counting method:

$$-\frac{\rho}{\beta_{\text{eff}}} = \frac{n_0 \sum_{i=1}^{G} \frac{\alpha_i}{\lambda_i}}{\int_0^\infty n(t)dt} \tag{1}$$

that requires the knowledge of the effective delayed neutron fraction β_{eff} and the relative abundances of each precursor group $\alpha_i = \beta^i_{\text{eff}}/\beta_{\text{eff}}$ (the number of groups being $G = 6$ or 8) characterized by the decay constants λ_i (calculated using the Monte Carlo method, see next section).

The expression $\sum\alpha_i/\lambda_i$ in equation (1) represents the mean life time of a delayed neutron $\tau_d = 1/\lambda_d = T_{1/2}/\ln(2)$. Its uncertainty combined with the uncertainty of the counting statistics of n_0 and the integral of $n(t)$ after the rod drop contribute to the total uncertainty of the measured reactivity in units of \$. The uncertainty of β_{eff} needs to be added if the reactivity in units of pcm is required.

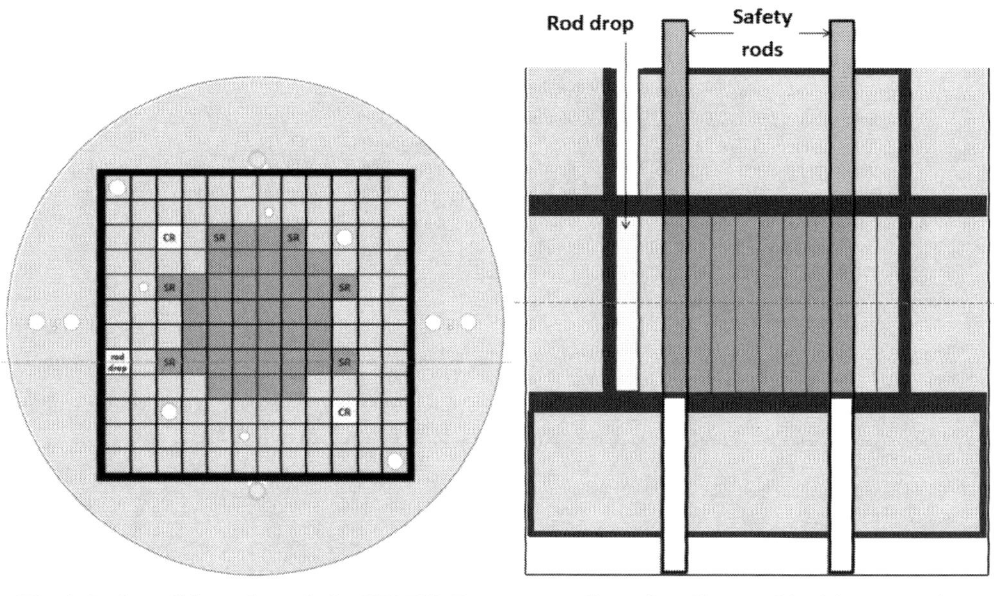

Fig. 2. Left: radial cut through the VENUS-F reactor configuration discussed in this paper. The cut passes through the core midplane (marked with a dashed line in the right-hand side figure). The assembly pitch is 8 cm, both the core diameter and height are 160 cm.
Right: axial cut through the core at the plane passing the rod drop and two safety rods (marked with a dashed line in the left-hand side figure). The depicted situation is after the rod was dropped (i.e. it is fully inserted in the core). Fuel assemblies – violet, B_4C control rods (CR) and rod drop system – yellow, B_4C safety rods (SR) - orange, lead reflector – green, experimental channels with fission chambers – white circles, stainless steel casing – black.

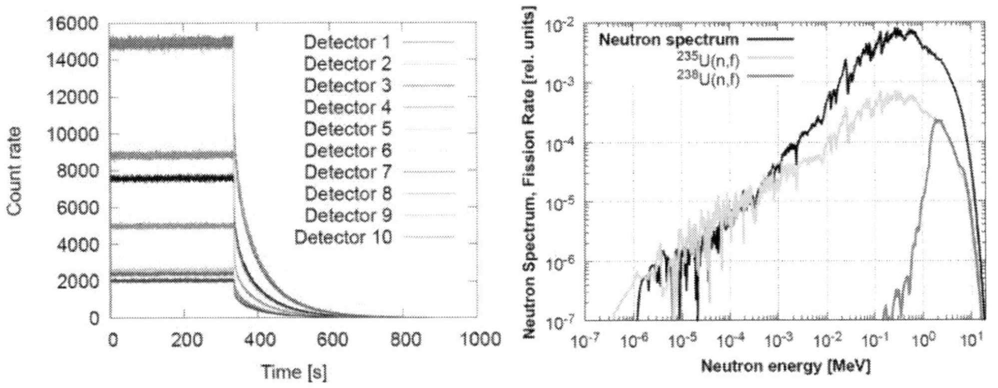

Fig. 3. Left: Detector count rates during a stable critical operation, rod drop and decay.
Right: MCNP simulation of neutron spectrum and energy dependence of the fission rates in fuel.

4 Calculation tools

The MCNP5-1.60 code [32] was used to calculate the kinetic parameters averaged over the VENUS-F core (i.e. spectrum and isotope averaged) with the iterated fission probability (IFP) approach [33]. The KOPTS card [34] with the number of cycles in blocks for adjoint weighting equal to 10 was used. Each criticality calculation consisted of 1850 active cycles preceded by 50 inactive, each of them having 10^6 neutrons. Such high number of histories was chosen to suppress the statistical uncertainties of the calculated kinetic parameters. The uncertainty of the calculated β_{eff} is 2 pcm (i.e. 0.3%) in all cases.

It was tested at various VENUS-F core configurations that the IFP approach gives the same result of β_{eff} as the prompt method [2]. An extensive verification of the Serpent-2 code against MCNP performed with the first VENUS-F critical core [35] provided a confidence in consistency of results, among others, β_{eff} results of both codes agree within 1σ.

Cross-sections from the following evaluated nuclear data libraries were used in the MCNP calculations:

- JEFF-3.1.2 (released in 2012 [36]) that is taken as a reference library in the next section; the same delayed neutron data were adopted in the newest release JEFF-3.3 (Nov 2017 [37]);
- JENDL-4.0 (released in 2010 [23]);
- ENDF/B-VII.1 (released in 2011 [24]) – the same delayed neutron data were adopted in the newest release ENDF/B-VIII.0 (Feb 2018 [38]), the only difference being the added ν_d covariances.

5 Results

First, the core averaged kinetic parameters calculated with different libraries are compared. Then their influence on the measured reactivity effects is discussed and the experimental results are compared with the MCNP calculations of the reactivity effects.

5.1 Effective delayed neutron fraction

The calculation with JEFF-3.1.2 gives β_{eff} = (741±2) pcm, while JENDL-4.0 and ENDF/B-VII.1 give 1% (i.e. 3-4 σ) smaller values. In one of the previous VENUS-F cores the kinetic parameters were measured [39]. All the calculations agree with the experimental value β_{eff} = (730±11) pcm within the uncertainties.

That is the situation if cross sections and delayed neutron parameters of all the isotopes are used from a single library. In order to distinguish between the impact of both uranium isotopes and to investigate whether the cross sections have any impact on the core averaged kinetic parameters, new calculations were performed having JEFF-3.1.2 as a reference library and altering either the cross sections and the delayed neutron data or only the delayed neutron data of JEFF-3.1.2 with those of ENDF/B-VII.1 and JENDL-4.0 for ^{235}U and ^{238}U isotopes separately, see Table 2.

Replacing any of the JEFF-3.1.2 ^{235}U or ^{238}U cross-sections or the delayed neutron data with those of JENDL-4.0 (while keeping everything else from JEFF-3.1.2) leads to the β_{eff} value decreases by 1 %. No statistically significant influence of the ^{235}U and ^{238}U cross sections on β_{eff} was observed.

If the delayed neutron data from ENDF/B-VII.1 are used for ^{235}U or ^{238}U, then the β_{eff} value increases by 1 % or decreases by 2%, respectively. Such a strong influence of ^{238}U can be explained by the fast neutron spectrum (about 10% of fissions in the fuel happens in ^{238}U, see Fig. 3 right) and the bigger delayed neutron multiplicity than ^{235}U. The opposite effects of ^{235}U and ^{238}U partially cancel out and lead to a β_{eff} value that is 1 % smaller when ENDF/B-VII.1 is used for all the isotopes compared to JEFF-3.1.2.

Table 2. Effective delayed neutron fraction calculated with MCNP using JEFF-3.1.2, JENDL-4.0, ENDF/B-VII.1 libraries for cross sections and delayed neutron parameters.

Cross-sections ^{235}U	^{238}U	all other isotopes	DN data ^{235}U	^{238}U	β_{eff} [pcm]	Diff from reference
JEFF	JEFF	JEFF	JEFF	JEFF	741 ± 2	0.0%
ENDF	ENDF	ENDF	ENDF	ENDF	733 ± 2	-1.1%
JENDL	JENDL	JENDL	JENDL	JENDL	734 ± 2	-0.9%
ENDF	JEFF	JEFF	**ENDF**	JEFF	747 ± 2	0.8%
JEFF	JEFF	JEFF	**ENDF**	JEFF	748 ± 2	0.9%
JEFF	**ENDF**	JEFF	JEFF	**ENDF**	726 ± 2	-2.0%
JEFF	JEFF	JEFF	JEFF	**ENDF**	727 ± 2	-1.9%
JENDL	JEFF	JEFF	**JENDL**	JEFF	734 ± 2	-0.9%
JEFF	JEFF	JEFF	**JENDL**	JEFF	731 ± 2	-1.3%
JEFF	**JENDL**	JEFF	JEFF	**JENDL**	735 ± 2	-0.8%
JEFF	JEFF	JEFF	JEFF	**JENDL**	733 ± 2	-1.1%

5.2 Precursor groups

Core averaged abundances of delayed neutron precursors calculated using the three main libraries are compared in Fig. 4. Only the 1st and the 4th group from ENDF and JENDL have similar values. The 2nd and the 3rd group in JENDL have abundances bigger by 20% than in ENDF, while the 5th and the 6th group in ENDF have abundances bigger than JENDL by 20% and 40%, respectively. ENDF has thus more weight on the shorter-living precursors and, as a consequence, the average delayed neutron half-life is shorter: 7.05 s for ENDF, 8.22 s for JENDL, 8.18 s for JEFF.

Mixing the delayed neutron data for fissionable isotopes in 6 and 8 groups in one calculation, as it was done in the previous section, can lead to absurd core averaged group half-lives. JEFF-3.1.2 used as a reference has the same half-lives of the 8 groups for both ^{235}U and ^{238}U. If ENDF or JENDL data in 6 groups are applied for one of the isotopes, then the core averaged half-lives of the 1st-6th groups change, while those of the 7th and the 8th groups remain the same. That can cause the 6th group having shorter half-life than the 7th group, see example in Table 3.

In fact, such mixing is not an artificial thing. In JEFF, delayed neutron data in the 8 groups format are used for the well-known fissionable isotopes, however, for some other isotopes the delayed neutron data are evaluated in the 6 groups format [37]. Even in one of the test versions of JEFF-3.3, the 6 groups format was considered for ^{235}U and the 8 groups format for ^{238}U [37].

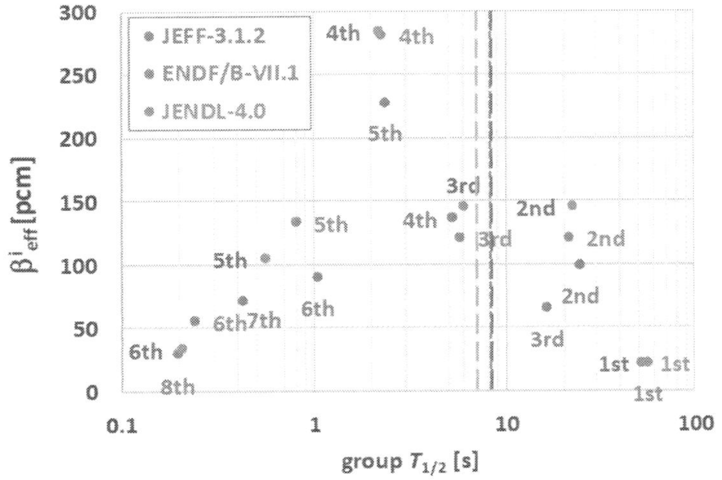

Fig. 4. Core averaged abundances of delayed neutron precursors in dependence of the half-life (in log-scale). Calculated with MCNP using JEFF-3.1.2 (8 groups), JENDL-4.0 and ENDF/B-VII.1 (6 groups) libraries. The uncertainties are ≤ 1 pcm in all cases. The average half-life of each of the three data sets is shown with a dashed line.

Table 3. Core averaged group half-lives calculated with MCNP using 8 groups format for both fissionable isotopes (first row) or combining 8 groups for ^{238}U and 6 groups for ^{235}U (second row). For the latter, the 6th group has shorter half-life than the 7th group.

DN data library		DN group							
		1st	2nd	3rd	4th	5th	6th	7th	8th
235U	238U	Half-life [s]							
JEFF (8-g)	JEFF (8-g)	55.6	24.5	16.3	5.21	2.37	1.04	0.424	0.195
ENDF (6-g)	JEFF (8-g)	52.2	21.6	5.96	2.41	1.05	0.374	0.424	0.195

5.3 Reactivity effects

The evaluated delayed neutron data have impact on the measured rod drop reactivity, see equation (1). The reactivity (in $) is linearly proportional to the mean delayed neutron life time τ_d. As shown in the previous section, τ_d of the investigated VENUS-F core has almost the same values (difference of 0.6 %) when calculated with 8 groups of JEFF and 6 groups of JENDL delayed neutron precursor data. Therefore, the application of JEFF and JENDL lead to the same measured reactivity in $ and ENDF gives about 14% smaller reactivity in $. For reactivity in pcm, the difference between JEFF and the other two libraries increases by 1 %, see section 5.1.

MCNP calculations of the rod drop are compared with the experimental results in Table 4. A small variation (not more than 2 %) in the calculated rod drop reactivity was found when different libraries were used in the calculations.

When delayed neutron data from JEFF or JENDL were applied in the analysis of experimental data, then the calculations underestimate such experimental results (by 6-8 %). In case of ENDF, the calculations overestimate the experiment by 6 %.

The core averaged delayed neutron data from mixed libraries (discussed in sections 5.1 and 5.2) were also applied to the experimental reactivity determination, see Table 4. Replacing the JEFF delayed neutron data of ^{235}U by JENDL does not change the reactivity value, while replacing by ENDF leads to a significant change and overestimation of the experiment by 9 %. On the contrary, replacing the JEFF delayed neutron data of ^{238}U by either JENDL or ENDF leads to a perfect agreement between experiment and calculation.

Table 4. Comparison of the experimental and calculated rod worth when JEFF-3.1.2, JENDL-4.0 and ENDF/B-VII.1 libraries are utilized for experimental data analysis (eq. (1)) as well as for the MCNP calculations.

XS library	DN data library		Rod worth [pcm]		
	235U	238U	EXP	MCNP	C/E
JEFF	JEFF		194 ± 4	178 ± 2	0.92 ± 0.02
ENDF	ENDF		164 ± 3	174 ± 2	1.06 ± 0.02
JEFF	ENDF	JEFF	165 ± 3	179 ± 2	1.09 ± 0.02
JEFF	JEFF	ENDF	175 ± 4	174 ± 2	0.99 ± 0.02
JENDL	JENDL		192 ± 4	181 ± 2	0.94 ± 0.02
JEFF	JENDL	JEFF	188 ± 4	179 ± 2	0.95 ± 0.02
JEFF	JEFF	JENDL	176 ± 4	177 ± 2	1.00 ± 0.02

6 Conclusion

The rod drop experiment performed at the fast VENUS-F core was analysed using core averaged delayed neutron parameters calculated with MCNP using the JEFF-3.1.2 (8-group precursors), JENDL-4.0 and ENDF/B-VII.1 (6-group precursors) libraries. Considerably discrepant experimental results (up to 15 % difference) were observed.

Application of the JEFF-3.1.2 or JENDL-4.0 delayed neutron data leads to C/E < 1, while ENDF/B-VII.1 leads to C/E > 1. Similar trends (up to 10-15 %) were observed in other VENUS-F cores (the same fuel, slightly modified neutron spectrum).

Although the best agreement of experiments and calculations was reached with ^{235}U delayed neutron data from JEFF-3.1.2 combined with ^{238}U delayed neutron data from either JENDL-4.0 or ENDF/B-VII.1, mixing 6-group and 8-group time structure can lead to unphysical distribution of decay constants among the precursor groups.

As a next step, the impact of delayed neutron data on the reactivity effects in the MYRRHA core with MOX fuel will be investigated.

The first author acknowledges D. Foligno for fruitful discussion on the topic of delayed neutron measurements and evaluations.

References

1. M. Carta et al., Sci. Technol. Nucl. Ins., Article ID 584256 (2011)

2. R. K. Meulekamp, S. C. van der Marck, Nucl. Sci. Eng. **152**, 142-148 (2006)
3. B. Kiedrowski, F. Brown, P. Wilson, LA-UR-09-06786 (2009)
4. G. Rudstam et al., NEA/WPEC-6, Int. Evaluation Cooperation, Vol. **6** (2002)
5. G. D. Spriggs, J.M. Campbell, V. M. Piksaikin, Prog. Nucl. Energy **41**, 223-251 (2002)
6. G. R. Keepin, T. F. Wimett, R. K. Zeigler, Phys. Rev. **107**, 1044 (1957)
7. B. Geslot, C. Jammes, B. Gall, Ann. Nucl. Energy **34**, 652–660 (2007)
8. A. Zoia et al., Ann. Nucl. Energy **110**, 11–24 (2017)
9. E. Kiefhaber, Nucl. Sci. Eng. **111**, 197-204 (1992)
10. S. Das, Nucl. Sci. Eng. **122**, 344-358 (1996)
11. S. Okajima et al., Prog. Nucl. Energy **41**, 285-301 (2002)
12. T. Sakurai, S. Okajima, J. Nucl. Sci. Technol. **39**, 19-30 (2002)
13. A. Kochetkov et al., EPJ Web of Conferences **146**, 06007 (2017)
14. A. Krása et al., EPJ Web of Conferences **146**, 06019 (2017)
15. A. Krása et al., ASTM **STP1608**, 286–295 (2018)
16. E. Fridman, A. Kochetkov, A. Krása, Ann. Nucl. Energy **108**, 239–252 (2017)
17. M. Sarotto et al., Ann. Nucl. Energy **121**, 626–637 (2018)
18. G. Van den Eynde et al., Proc. TCADS-3, 141–150 (2016)
19. A. Stankovskiy et al., Nucl. Data Sheets **118**, 513–515 (2014)
20. H. Iwamoto, A. Stankovskiy, L. Fiorito, G. Van den Eynde, J. Nucl. Sci. Technol. **55**, 539-547 (2018)
21. I. A. Kodeli, Ann. Nucl. Energy **113**, 425-435 (2018)
22. A. Santamarina, D. Bernard, Y. Rugama, JEFF Report **22** (2009)
23. K. Shibata et al., J. Nucl. Sci. Technol. **48**(1), 1-30 (2011)
24. M.B. Chadwick, Nuclear Data Sheets **112**, 2887-2996 (2011)
25. G. Audi et al., Nucl. Phys. **A729** 3-129 (2003)
26. P. Moller, B. Pfeiffer, K.-L. Kratz, Phys. Rev. C. **67** 055802 (2003)
27. V. M. Piksaikin et al., , Prog. Nucl. Energy **41**, 203-222 (2002)
28. A. Kochetkov et al., EPJ Web of Conferences **106**, 06004 (2016)
29. A. Kochetkov et al., Proc. PHYSOR-2018, Cancun, Mexico
30. J. L. Lecouey et al., Ann. Nucl. Energy **83**, 65–75 (2015)
31. J. Wagemans et al., EPJ Web of Conferences **170**, 04027 (2018)
32. X-5 Monte Carlo Team, LA-UR-03-1987 (2003)
33. B. C. Kiedrowski, F. B. Brown, P. P. H. Wilson, Nucl. Sci. Eng. **168**, 226–241 (2011)
34. B. C. Kiedrowski et al., LA-UR-10-06217 (2010)
35. E. Fridman, A. Kochetkov, A. Krása, Ann. Nucl. Energy **108**, 239–252 (2017)
36. A. J. Koning et al., J. Korean Phys. Soc. **59**, 1057-1062 (2011)
37. A. J. M. Plompen et al., to be published in EPJ
38. D. A. Brown et al., Nucl. Data Sheets **148**, 1-142 (2018)
39. X. Doligez et al., IEEE Nucl. Sci. Symp. Conf. Rec, Proc. ANIMMA-2015 (2015)

Neutron inelastic scattering of ^{232}Th: measurements and beyond

Eliot Party[1,*], *Catalin* Borcea[2], *Philippe* Dessagne[1], *Xavier* Doligez[3], *Grégoire* Henning[1], *Maëlle* Kerveno[1], *Alexandru* Negret[2], *Markus* Nyman[4], *Adina* Olacel[2], and *Arjan* Plompen[4]

[1]Université de Strasbourg, CNRS, IPHC UMR 7178, F-67000 Strasbourg, France
[2]Horia Hulubei National Institute for Physics and Nuclear Engineering (IFIN-HH), Reactorului 30, Magurele – Bucharest, 077125, Romania
[3]Institut de Physique Nucléaire, CNRS-IN2P3, Univ. Paris-Sud, Université Paris-Saclay, 91406 Orsay Cedex, France
[4]European Commission, Joint Research Centre, Directorate G – Nuclear Safety and Security, Unit G.2, Retieseweg 111, 2440 Geel, Belgium

Abstract. Gamma production cross sections have been obtained for 81 transitions in ^{232}Th from (n,n'γ), 11 in ^{231}Th from (n,2nγ) and 7 in ^{230}Th from (n,3nγ) reactions using prompt gamma spectroscopy. Incident neutron energies were determined using the neutron time-of-flight technique. Sources of uncertainty have been examined and their correlations have been computed. Total uncertainty on cross sections ranges from 4 to 20%. Obtained cross sections are in agreement with prior experiments, but are not well reproduced by the TALYS-1.8 reaction code using default parameters. During analysis, discrepancies between our findings and the Evaluated Nuclear Structure Data File (ENSDF) were noted. Future work related to the present experiment includes: improving theoretical models, quantifying the influence of the ^{232}Th inelastic neutron cross section on reactor core parameters, and conducting additional measurements.

1 Introduction

Today, the lack of precision in nuclear data affects our capability to predict accurately some phenomena in nuclear reactors. For example, inelastic scattering off ^{238}U is the main source of uncertainty due to nuclear data in power distribution of large reactor [1]. To answer to the need of lower uncertainties in nuclear data for applications, new precise measurements of cross sections are necessary.

In this paper, measurements of neutron inelastic scattering off ^{232}Th using prompt gamma spectroscopy are presented. ^{232}Th is an alternative nuclear fuel using the ^{232}Th/^{233}U cycle. Inelastic scattering off ^{232}Th may play the same role in this fuel as the one off ^{238}U in usual fuel. Cross section measurements referenced in the EXFOR database [2], are mostly anterior to 1990, and used mainly direct neutron detection. Two studies used gamma spectroscopy to determine partial cross sections in 1980's, one by Dave et al. [3], the other by Filatenkov et al. [4]. Only a few of these measurements, and none using gamma detection, extended beyond 3 MeV in incident neutron energy, as shown on Fig. 1. This work provides new and precise measurements on a range of incident neutron energy extending to 20 MeV.

*e-mail: eliot.party@etu.unistra.fr

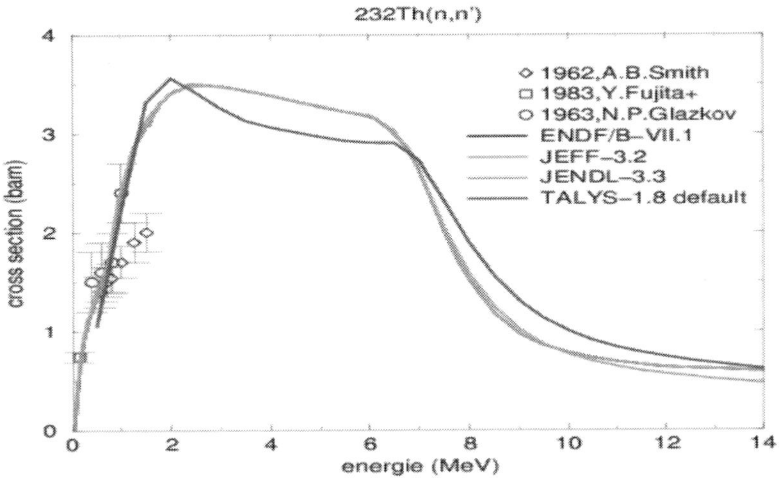

Figure 1. (color online) Previous experimental and evaluated data of the total inelastic neutron scattering cross section of ^{232}Th.

2 Method used and experimental setup

To study inelastic neutron scattering off ^{232}Th we detected γ-rays emitted by the deexciting ^{232}Th nuclei, coupled with the time-of-flight technique, in which the incident neutron energy is deduced from the known distance to the neutron source and the measured flight time. This method can be used for all reactions where a gamma ray is emitted, such as the (n,2n) and (n,3n) reactions. Reactions producing one specific gamma transition from a residual nucleus, after emission of x neutrons, will be called (n,xnγ) reactions in the following. The goal is to obtain angle-integrated (n,xnγ) reaction cross sections for as many transitions as possible, on a neutron energy range pertinent to reactor applications.

Neutron source used was GELINA (Geel Electron LINear Accelerator) [5], operated by the European Commission Joint Research Centre (JRC) in Geel, Belgium. It generates neutron bursts at 800 Hz frequency by colliding electron pulses from the accelerator on a uranium target, where neutrons are produced by bremsstrahlung-induced reactions. Both moderated and unmoderated neutron beams are available on several flight paths of different lengths. The unmoderated neutron spectrum used in the present experiment is similar to the fission neutron spectrum in a nuclear reactor.

Our experimental setup, named GRAPhEME (GeRmanium Array for Actinides PrEcise MEasurements) [6] is placed at the 30 meter station of flight path 16. The time of flight range of interest spreads from 0.5 to 6 microseconds, corresponding to neutron energies from 20 MeV to 100 keV, respectively.

γ-rays are detected using 4 planar HPGe counters placed at 110° and 150° from beam axis. Incident neutrons are monitored using a Fission Chamber (FC) containing ^{235}U foils installed upstream of the target. Output signals from the detectors are processed by TNT-2 100 MHz digital acquisition cards. Measurements have been conducted over three campaigns in 2009, 2010 and 2013, totalling 800 hours of beam time. Only events within 10 μs before to 50 μs after a neutron pulse emission, are registered, storing the instant of detection and the signal amplitude. This large time gate allows observation of background radiation. HPGe counters are set to observe γ-rays of low energy, with a maximum efficiency for energies between 150 and 200 keV.

3 Analysis

3.1 Data extraction

In order to determine the number of events in the fission chamber and the HPGe counters as a function of neutron time of flight, the data are sorted into two-dimensional time vs. energy histograms (as Fig. 2).

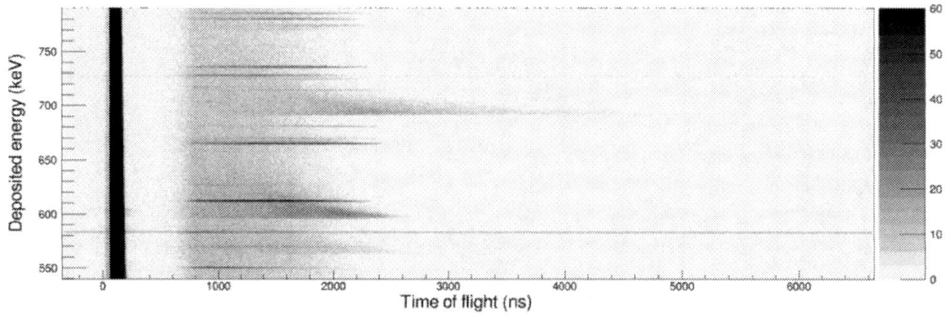

Figure 2. An example of a time-of-flight vs. deposited energy matrix, from data collected with one of the GRAPhEME HPGe detectors.

Before a neutron pulse emission, only background events are observed, mainly from ^{235}U foil alpha radioactivity in FC, and ^{232}Th target radioactivity γ-rays. This background is constant and present at every time. Then events caused by the bremsstrahlung radiation produced in the primary target are detected. The bremsstrahlung produces only some photofission in the FC, but their scattering by the ^{232}Th target makes up 95% of events in HPGe counters, and induces significant pile up. After about 500 ns time-of-flight, corresponding to an incident neutron energy of 20 MeV, the first neutron-induced γ-rays are detected. As time passes, few neutrons reach target anymore, and only background is detected until the next neutron pulse.

In HPGe counters, we observe a smooth background and well resolved peaks in energy, present on a limited time of flight range. These peaks are identified with a transition in an isotope using mainly their centroid energy. If needed, their time range of presence and their area may act as hints. For peaks originating from (n,xnγ) reactions, area under peak N_γ on each HPGe counter is obtained for each neutron energy (time of flight) bin. Correction factor τ_{pup} accounting for pile-up is computed using acquisition cards tagging of pile-up events. And the counter efficiency ϵ_γ at given gamma energy is computed by MCNP [7], and verified using calibrated ^{152}Eu source.

Number of incident neutrons impinging on the target for the same neutron energy bin is obtained using the number of fission events N_{fiss} detected in the corresponding time-of-flight bin. Other terms include ^{235}U foil atomic density per unit area $\zeta_{^{235}U}$, ^{235}U standard fission cross section $\sigma_{f,^{235}U}$ from reference [8], as well as FC detection efficiency ϵ_{FC} and correction factor τ_{att} for beam attenuation between FC and ^{232}Th target [9]. Multiple scattering do not need to be corrected as thin target was used.

Then, for each energy bin, the angle-differential (n,xnγ) cross section is given by equation (1), with ζ_{Th} being target atomic density per unit area.

$$\frac{d^2\sigma_{(n,xn\gamma)}}{dEd\theta}(E_n) = \frac{N_\gamma(E_n)\,\tau_{pup}\,\sigma_{f,^{235}U}(E_n)\,\epsilon_{FC}\,\zeta_{^{235}U}}{\zeta_{Th}\,\epsilon_\gamma\,N_{fiss}(E_n)\,\tau_{att}} \quad (1)$$

Choice of neutron energy bins is a compromise between neutron energy resolution and an increased uncertainty on the cross section caused by reduced statistics in smaller bins.

3.2 Covariances

To combine measurements of all detectors and to propagate the uncertainties, we need to compute covariances. Covariances of cross sections are obtained from covariances of each nine factors in Eq. (1) using uncertainty propagation in linear approximation. As some factors values are common to all energy bins (like target density ζ_{Th}), positive correlation is induced. A correlation matrix obtained for the 681 keV (n,xnγ) cross section is shown in Fig. 3. Only positive correlations are present, so the color palette ranges from 0 to 1. There are positive correlations between the cross section values over the neutron energy range covered here. Such a positive correlation between different neutron energies implies a lower uncertainty on the shape of the cross section than if no correlation was assumed. Thus, adequate treatment of covariances is necessary for the correct use of our results, for example in an evaluation.

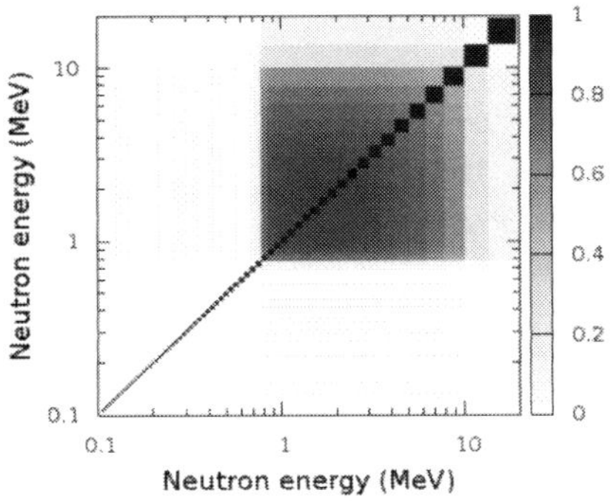

Figure 3. (color online) Correlation matrix of the measured 681 keV (n,n'γ) cross section.

4 Results

4.1 Some remarks on the measured cross sections

81 cross sections from ^{232}Th(n,n'γ) reactions have been extracted. Additionally 11 ^{232}Th(n,2nγ)^{231}Th and 7 ^{232}Th(n,3nγ)^{230}Th cross sections have been obtained. For 28 out of the 30 lowest lying excited states of ^{232}Th, at least one deexciting transition was observed and its γ-ray production cross section measured. The remaining two were weakly populated due to their high spins (7^+ and 12^+). Four (n,n'γ) cross sections are shown in Fig. 4. The two on the left are the 49 keV transition from 2^+ state and the 223 keV transition from 8^+ state of ground state band, and the two on the right are from other bands head states. Main component of the total uncertainty of the cross section is either statistics or detection efficiencies, depending on the transition. Efficiencies and other systematic uncertainties result in a correlated uncertainty of 3.5% on the measured cross sections. Best precision obtained is 4% for the measured (n,n'γ) cross sections.

A large number of (n,n'γ) cross sections had been published by Dave et al. (1985) [3], while other studies only published deduced level population cross section. As Dave et al. measurements only extend up to 2.2 MeV, energy ranges in Fig. 4 have been reduced to 10 MeV to ease comparison. Our results are in good agreement: from the 57 cross sections measured in both experiments (no transitions below 550 keV in Dave et al.), only 5 show significant differences, but they are weakly populated or are contaminated by nearby peaks.

Obtained cross sections have also been compared with predictions of the reaction code TALYS 1.8 [10], using the default calculation parameters. Differences are observed in shape or in magnitude for most cross sections. They may be explained to some extent by incomplete nuclear structure knowledge, such as unknown spins, parities, or branching ratios, which can lead to over- or under-estimation of the cross section in TALYS calculations.

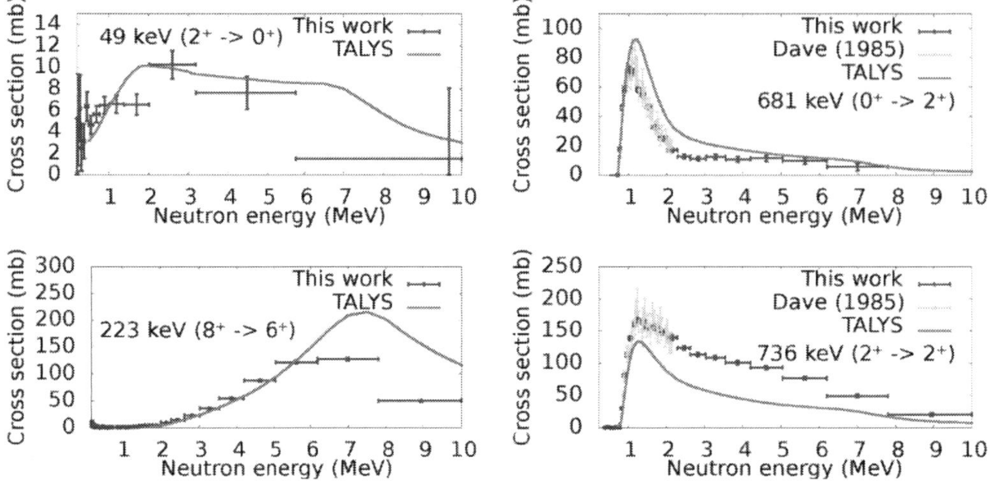

Figure 4. (color online) The (n,n'γ) cross sections are chosen as representative of shapes observed. This work is compared with Dave et al. [3] and predictions of TALYS-1.8 with default parameters.

4.2 Shortcomings in the nuclear structure knowledge

Some discrepancies in the ENSDF [11] nuclear structure database were found during data analysis. First, we found an erroneous Relative Intensity (RI in the following) in ENSDF for the level 7^- at 1043 keV in ^{232}Th. 486 keV and 159 keV transitions from this level are reported with RIs of 65 and 100 respectively. Yet, our measurements suggests RIs of 100 and 18.5(30). Indeed, RIs of 4.5 and 6.9 (relative to another transition) in Nuclear Data Sheets [12] were instead reported to be 45 and 6.9 in original reference [13], equivalent to RIs of 100 and 15.3. The original values are compatible with our measurements.

Second, there are ten levels in ENSDF for which at least one γ-ray observed in our measurement has an unknown RI. Yet, these are not necessarily weak. For the 1121 keV(2^+) level of ^{232}Th, while RI are only reported for transitions at 959, 407 and 347 keV as 100%, 37% and 30%, our measurement suggests RI of 300% (relative to 959 keV transition) for the 1072 keV transition. Other studies [14–16] confirm the high RI for the 1072 keV transition.

Finally, new levels and transitions for ^{232}Th were proposed by Demidov et al. (2008) [16]. Some of these transitions were observed in our measurements, in particular a strong transition at 1075.5 keV included in our results. This recent proposition, as well as uncertain spin-parity assignements, shows that the level scheme of ^{232}Th is not definitive at high excitation energies covered in our measurement.

Our study uses nuclear structure, both in analysis, as γ-rays are identified from transitions registered in ENSDF or XUNDL [11] (eXperimental Unevaluated Nuclear Data List), and in theoretical interpretation, as prediction from theory is affected by deexcitation paths. Consequently, these shortcomings interfere with our study and improved knowledge of the ^{232}Th level scheme is required.

5 Outlook of related activities

5.1 Collaboration on theory

To obtain cross sections pertinent to reactor applications, like the total (n,n') cross section, only from the (n,xnγ) cross sections is not possible if all deexcitation paths are not observed.

Our approach is to use theoretical models constrained with experimental values in order to fill the gaps. This work is done in collaboration with theoreticians from CEA (Commissariat à l'Energie Atomique et aux énergies alternatives). Collaboration on the ^{232}Th cross section is only beginning, but results have already been obtained on ^{238}U [18]. This paper points that adjustment of theory parameters (spin distribution, gamma strength...) could reproduce several features observed in measurements. For example, adjustment of the spin distribution from an ad hoc exciton model to a QRPA calculated distribution enabled a better agreement with measurement, especially for transitions in the ground state band. The measured ^{232}Th(n,xnγ) cross sections should put further constraints on the models.

5.2 Sensitivity studies

To quantify the influence on reactors of inelastic scattering of ^{232}Th, compared to other cross-sections, it was decided to propagate uncertainties and to calculate sensitivity on three different parameters of nuclear reactors (neutron multiplication factor k_{eff}, delayed neutron fraction β_{eff} and radial power distribution $P(r)$). Recent developments in the Monte-Carlo transport codes MCNP [7] and SERPENT [19] allow sensitivity calculations of k_{eff} to nuclear data thanks to the generalized perturbation theory. Then, from those sensitivity coefficients, it is possible to calculate uncertainties if the covariance matrix of nuclear data used is known. On the other hand, the Total Monte Carlo method (TMC) [20] consists in sampling randomly nuclear parameters in order to build, for one isotope of interest, hundreds of evaluated cross section file from TENDL [21]. The sampling is made to be representative of uncertainties on nuclear data. Then a MCNP or SERPENT calculation is performed for each file from this library of nuclear data and the distribution of calculated parameters gives directly the uncertainty due to the nuclear data of one particular isotope.

These methods were applied to a Pressurized Water Reactor fueled with a ^{232}Th/^{233}U fuel. The TMC method applied for k_{eff} calculations gave uncertainties up to 6%, justifying the strong needs of improvement for thorium cross sections. We took the opportunity of TENDL to build covariance matrices for ^{232}Th and ^{233}U cross-section for several energy binning. We combined them with the sensitivity coefficients calculated with SERPENT and MCNP to compare the perturbation theory with the TMC method. We showed that a high number of energy bins for the covariance matrix is necessary to obtain results in perfect agreement. This condition allows us to reduce drastically the huge computational power needed for the TMC and we proved that reliable covariance of nuclear data may be obtained with TENDL. Further studies are ongoing for β_{eff} and $P(r)$ on other reactor types, including sodium cooled reactors for fast neutron spectrum.

5.3 Experimental developments

The GRAPhEME spectrometer is continuously being upgraded. It has been used for ten years to measure (n,xnγ) cross sections on several isotopes. HPGe counters have been added in 2014 (one being segmented to overcome difficulties induced by very high activity isotopes ^{233}U and ^{239}Pu) and in 2018 the acquisition system has been changed to the FASTER system [22].

Apart from this gradual evolution, we plan to adress one of the problems of gamma spectroscopy: unobserved level deexcitation by internal conversion. As an example, 99.7% of deexcitation of the ^{232}Th first excited level at 49 keV occurs by internal conversion, decreasing a prominent 3 b deexcitation cross section to a barely detectable 10 mb gamma emission cross section, with consequent uncertainties (see Fig. 4 upper left corner). Tests for a proof of concept of DELCO (for Detecteur d'ELectron de COnversion), a new detection setup aiming at detecting conversion electrons using silicon detectors, have been started. If its completion is successful, it will allow the detection of deexcitation paths still unobserved in our experiment.

As lack of nuclear structure information for some isotopes affects the interpretation of our measurements, it would also be of interest to perform nuclear structure measurements on the isotopes concerned, primarily ^{238}U. This could be carried out by coupling GRAPhEME with GAINS [23, 24], another HPGe spectrometer in our experimental collaboration.

6 Conclusion

The main result of this work is the measurement of 81 ^{232}Th(n,n'γ) cross sections, on a large incident neutron energy range, up to 20 MeV. These cross sections show and cover the deexcitation of most low-lying levels. Covariances associated to each cross section have been derived to ensure their correct uses. The present measurement is a significant improvement from preceding measurements by other teams, in which the incident neutron energy did not exceed 3 MeV, the uncertainties were higher and first transitions in the ground state band were not reported. Upcoming work will include the integration of measured cross section in EXFOR, as well as collaboration to derive theoretical interpretation, sensitivity studies in reactor applications, and development of new equipment to account for conversion electrons.

References

[1] M. Salvatores et al., International Evaluation Co-operation Vol. 26, NEA No .6410 (2008)
[2] V. Zerkin et al., Nucl. Instrum. Methods A **888**, p. 31 (2018)
[3] J.H. Dave et al., Nuclear Science and Engineering **91**, p. 187 (1985)
[4] A.A. Filatenkov et al., USSR report to the I.N.D.C. **298**, p. 63 (1989)
[5] D. Ene et al., Nucl. Instrum. Methods A **618**, p. 54 (2010)
[6] M. Kerveno et al., European Physical Journal A **51**:167, p. 1-18 (2015)
[7] J.T. Goorley et al., Initial MCNP6 Release Overview - MCNP6 version 1.0 , LANL, LA-UR-13-22934 (2013)
[8] A.D. Carlson et al., Nucl. Data Sheets **148**, p. 143-188 (2018)
[9] J.C. Thiry, Ph.D. thesis, Univ. de Strasbourg (2010), http://scd-theses.u-strasbg.fr/2016/
[10] A.J. Koning et al., "TALYS-1.0" in Int. Conf. on Nuc. Data for Sci. and Tech., EDP Sciences, p. 211 (2008)
[11] Evaluated Nuclear Structure Data File, available from http://www.nndc.bnl.gov/ensdf
[12] E. Browne, Nucl. Data Sheets **107**, p. 2579 (2006)
[13] C. Gunther et al., Phys. Rev. C **61**, 064602 (2000)
[14] W.R. McMurray et al., Zeitschrift für Physik **253**, p. 289 (1972)
[15] F.K. McGowan et al., Nuc. Phys. A**562**, p. 241 (1993)
[16] A.M. Demidov et al., Phys. At. Nucl. **71**, p. 1839 (2008)
[17] Experimental Unevaluated Nuclear Data List, available from www.nndc.bnl.gov/xundl
[18] M. Dupuis et al., European Physical Journal A **51**:168, p. 1-16 (2015)
[19] J. Leppänen et al., Ann. Nuc. Energy **82**, p. 142-150, (2015)
[20] D. Rochman et al., Ann. Nuc. Energy **38** p. 942-952 (2011)
[21] A.J. Koning, D. Rochman, Nucl. Data Sheets **113** p. 2841- 2934 (2012)
[22] FASTER project—IN2P3, available from: http://faster.in2p3.fr/
[23] D. Deleanu et al., Nucl.Instrum. Methods A **624**, p. 130 (2010)
[24] A. Negret et al., Nucl. Data Sheets **119**, p. 179 (2014)

Fission program at n_TOF

L. Tassan-Got[1,2,3], *N.* Colonna[4], *M.* Diakaki[2], *Z.* Eleme[5], *A.* Manna[6,7], *A.* Sekhar[8], *A.* Stamatopoulos[2], *O.* Aberle[1], *V.* Alcayne[9], *S.* Amaducci[10,11], *J.* Andrzejewski[12], *L.* Audouin[3], *V.* Babiano-Suarez[13], *M.* Bacak[1,14,15], *M.* Barbagallo[1,4], *S.* Bennett[8], *E.* Berthoumieux[15], *D.* Bosnar[16], *A. S.* Brown[17], *M.* Busso[18,19], *M.* Caamaño[20], *L.* Caballero[13], *M.* Calviani[1], *F.* Calviño[21], *D.* Cano-Ott[9], *A.* Casanovas[21], *F.* Cerutti[1], *E.* Chiaveri[1,8], *G. P.* Cortés[21], *M. A.* Cortés-Giraldo[22], *L.* Cosentino[10], *S.* Cristallo[18,23], *L. A.* Damone[4,24], *P. J.* Davies[8], *M.* Dietz[25], *C.* Domingo-Pardo[13], *R.* Dressler[26], *Q.* Ducasse[27], *E.* Dupont[15], *I.* Durán[20], *B.* Fernández-Domíngez[20], *A.* Ferrari[1], *I.* Ferro-Gonçalves[28], *P.* Finocchiaro[10], *V.* Furman[29], *R.* Garg[25], *A.* Gawlik[12], *S.* Gilardoni[1], *K.* Göbel[30], *E.* González-Romero[9], *C.* Guerrero[22], *F.* Gunsing[15], *S.* Heinitz[26], *J.* Heyse[31], *D. G.* Jenkins[17], *U.* Jiri[26], *A.* Junghans[32], *Y.* Kadi[1], *F.* Käppeler[33], *A.* Kimura[34], *I.* Knapová[35], *M.* Kokkoris[2], *Y.* Kopatch[29], *M.* Krtička[35], *D.* Kurtulgil[30], *I.* Ladarescu[13], *C.* Lederer-Woods[25], *J.* Lerendegui-Marco[22], *S.-J.* Lonsdale[25], *D.* Macina[1], *T.* Martínez[9], *A.* Masi[1], *C.* Massimi[6,7], *P. F.* Mastinu[36], *M.* Mastromarco[1], *E.* Maugeri[26], *A.* Mazzone[4,37], *E.* Mendoza[9], *A.* Mengoni[38,6], *V.* Michalopoulou[1,2], *P. M.* Milazzo[39], *M. A.* Millán-Callado[22], *F.* Mingrone[1], *J.* Moreno-Soto[15], *A.* Musumarra[10,11], *A.* Negret[40], *F.* Ogállar[41], *A.* Oprea[40], *N.* Patronis[5], *A.* Pavlik[42], *J.* Perkowski[12], *C.* Petrone[40], *L.* Piersanti[18,23], *E.* Pirovano[27], *I.* Porras[41], *J.* Praena[41], *J. M.* Quesada[22], *D.* Ramos Doval[3], *R.* Reifarth[30], *D.* Rochman[26], *C.* Rubbia[1], *M.* Sabaté-Gilarte[22,1], *A.* Saxena[43], *P.* Schillebeeckx[31], *D.* Schumann[26], *A. G.* Smith[8], *N.* Sosnin[8], *P.* Sprung[26], *G.* Tagliente[4], *J. L.* Tain[13], *A. E.* Tarifeño-Saldivia[21], *B.* Thomas[30], *P.* Torres-Sánchez[41], *A.* Tsinganis[1], *S.* Urlass[1,32], *S.* Valenta[35], *G.* Vannini[6,7], *V.* Variale[4], *P.* Vaz[28], *A.* Ventura[6], *D.* Vescovi[18,44], *V.* Vlachoudis[1], *R.* Vlastou[2], *A.* Wallner[45], *P. J.* Woods[25], *T. J.* Wright[8], and *P.* Žugec[16]

[1]European Organization for Nuclear Research (CERN), Switzerland
[2]National Technical University of Athens, Greece
[3]IPN, CNRS-IN2P3, Univ. Paris-Sud, Université Paris-Saclay, F-91406 Orsay Cedex,France
[4]Istituto Nazionale di Fisica Nucleare, Bari, Italy
[5]University of Ioannina, Greece
[6]Istituto Nazionale di Fisica Nucleare, Sezione di Bologna, Italy
[7]Dipartimento di Fisica e Astronomia, Università di Bologna, Italy
[8]University of Manchester, United Kingdom
[9]Centro de Investigaciones Energéticas Medioambientales y Tecnológicas (CIEMAT),Spain
[10]INFN Laboratori Nazionali del Sud, Catania, Italy
[11]Dipartimento di Fisica e Astronomia, Università di Catania, Italy
[12]University of Lodz, Poland
[13]Instituto de Física Corpuscular, CSIC - Universidad de Valencia, Spain
[14]Technische Universität Wien, Austria
[15]CEA Saclay, Irfu, Université Paris-Saclay, Gif-sur-Yvette, France
[16]Department of Physics, Faculty of Science, University of Zagreb, Croatia
[17]University of York, United Kingdom
[18]Istituto Nazionale di Fisica Nazionale, Perugia, Italy
[19]Dipartimento di Fisica e Geologia, Università di Perugia, Italy
[20]University of Santiago de Compostela, Spain
[21]Universitat Politècnica de Catalunya, Spain

[22]Universidad de Sevilla, Spain
[23]Istituto Nazionale di Astrofisica - Osservatorio Astronomico d'Abruzzo, Italy
[24]Dipartimento di Fisica, Università degli Studi di Bari, Italy
[25]School of Physics and Astronomy, University of Edinburgh, United Kingdom
[26]Paul Scherrer Institut (PSI), Villigen, Switzerland
[27]Physikalisch-Technische Bundesanstalt (PTB), Bundesallee 100, 38116 Braunschweig, Germany
[28]Instituto Superior Técnico, Lisbon, Portugal
[29]Joint Institute for Nuclear Research (JINR), Dubna, Russia
[30]Goethe University Frankfurt, Germany
[31]European Commission, Joint Research Centre, Geel, Retieseweg 111, B-2440 Geel,Belgium
[32]Helmholtz-Zentrum Dresden-Rossendorf, Germany
[33]Karlsruhe Institute of Technology, Campus North, IKP, 76021 Karlsruhe, Germany
[34]Japan Atomic Energy Agency (JAEA), Tokai-mura, Japan
[35]Charles University, Prague, Czech Republic
[36]Istituto Nazionale di Fisica Nucleare, Sezione di Legnaro, Italy
[37]Consiglio Nazionale delle Ricerche, Bari, Italy
[38]Agenzia nazionale per le nuove tecnologie, l'energia e lo sviluppo economico sostenibile (ENEA), Bologna, Italy
[39]Istituto Nazionale di Fisica Nazionale, Trieste, Italy
[40]Horia Hulubei National Institute of Physics and Nuclear Engineering (IFIN-HH),Bucharest
[41]University of Granada, Spain
[42]University of Vienna, Faculty of Physics, Vienna, Austria
[43]Bhabha Atomic Research Centre (BARC), India
[44]Gran Sasso Science Institute (GSSI), L'Aquila, Italy
[45]Australian National University, Canberra, Australia

Abstract. Since its start in 2001 the n_TOF collaboration developed a measurement program on fission, in view of advanced fuels in new generation reactors. A special effort was made on measurement of cross sections of actinides, exploiting the peculiarity of the n_TOF neutron beam which spans a huge energy domain, from the thermal region up to GeV. Moreover fission fragment angular distributions have also been measured. An overview of the cross section results achieved with different detectors is presented, including a discussion of the ^{237}Np case where discrepancies showed up between different detector systems. The results on the anisotropy of the fission fragments and its implication on the mechanism of neutron absorption, and in applications, are also shown.

1 Introduction

The fission measurements at n_TOF have been carried out with different detection systems: ionisation chambers, μMegas detectors, parallel plate avalanche counters (PPAC). Although the cross section was the main topic, other quantities as the prompt γ emission and the angular distribution of fission fragments have also been measured. The scope of this presentation is to illustrate some of the measurements done recently.

2 The n_TOF facility

C. Rubbia proposed to use the CERN/PS synchrotron, accelerating protons up to 20 GeV/c, to produce an intense neutron source based on the spallation of lead [1]. A long flight path of 185 m and a proton bunch width of σ = 7 ns made possible the tagging of the neutron energy with high accuracy. The n_TOF facility was commissioned in 2001 [2, 3] with the long horizontal flight path leading to experimental area EAR1.

In 2014 a new experiment station EAR2, with a vertical flight path 20 m long, was also made available for experiments with low mass samples requesting higher neutron flux [4].

Figure 1 shows the neutron spectrum in each area, integrated over the neutron beam area. In the case of EAR1 the spot size is defined with a cylindical collimator of diameter 2 cm for capture and 8 cm for fission.

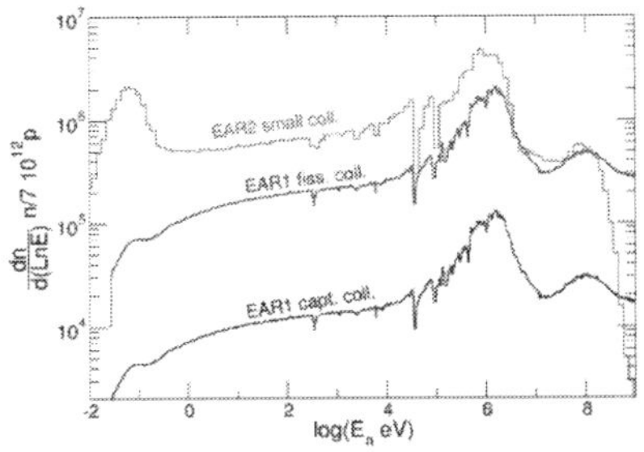

Figure 1: Flux integrated over the neutron beam spot defined by a capture and a fission collimator in EAR1 and by a small collimator in EAR2.

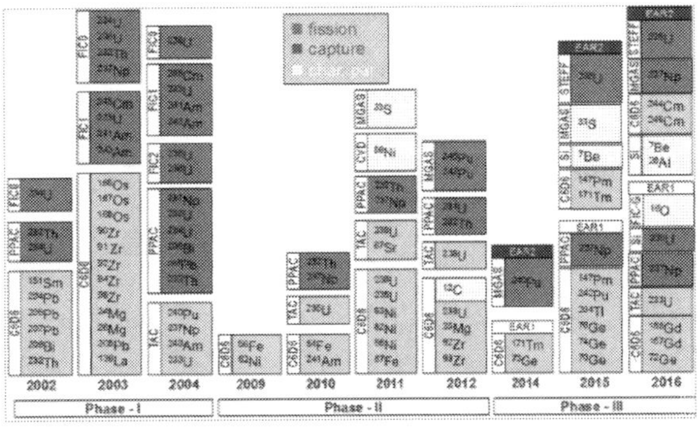

Figure 2: Measurements performed in the 2002-2016 period, including radiative capture and fission in both areas [5].

The energy spectrum spans 11 order of magnitude, reaching 1 GeV in EAR1 and thermal energy in both areas. In the case of EAR1 a borated water moderator has been used for the capture measurements to reduce the photon background produced by capture in the water. This wide energy range is very suitable for fission measurements.

3 Measurements

An overview of the measurements performed in the period 2002-2016 is given in figure 2 [5]. A significant fraction of the program was dedicated to fission of actinides. In the latter case several types of detectors have been used: fast ionisation chamber (FIC), parallel plate avalanche counters (PPAC), micromegas detectors (μMegas) and the STEFF ensemble. They included samples covering entirely the neutron beam spot, defined in EAR1 by the large fission collimator (8 cm in diameter), and the small collimator ($\approx$3 cm) in EAR2. The FIC was used in EAR1 until 2004 for cross section measurements [6]. Thenceforth μMegas has been used.

In the following section we show examples of measurements performed in the last past years.

4 Excerpt of fission measurements

4.1 Fission of Pu isotopes with μMegas

The fission cross sections of ^{240}Pu ($T_{1/2}$=6.6 ky) and ^{242}Pu ($T_{1/2}$=375 ky) have been measured. ^{235}U and ^{238}U samples were used as references for the accurate determination of the neutron flux using ENDF/B-VII.1.

A μMegas cell is made of a drift gap where ionisation by the fission fragments takes place and the collected electrons are multiplied in the μMegas detector, as shown in figure 3 [8]. The ^{242}Pu measurement has been carried out in EAR1 with 4 samples and a total mass of 3.1 mg [7]. Figure 4 shows the comparison with previous measurements and evaluations.

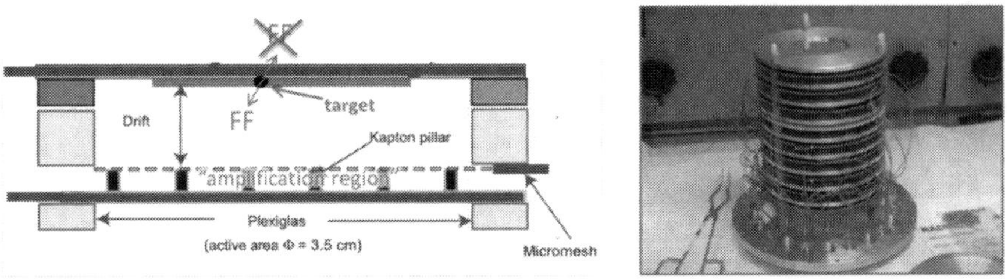

Figure 3: μMegas detector for fission (left hand side) and stack of couples of sample/detector (right hand side).

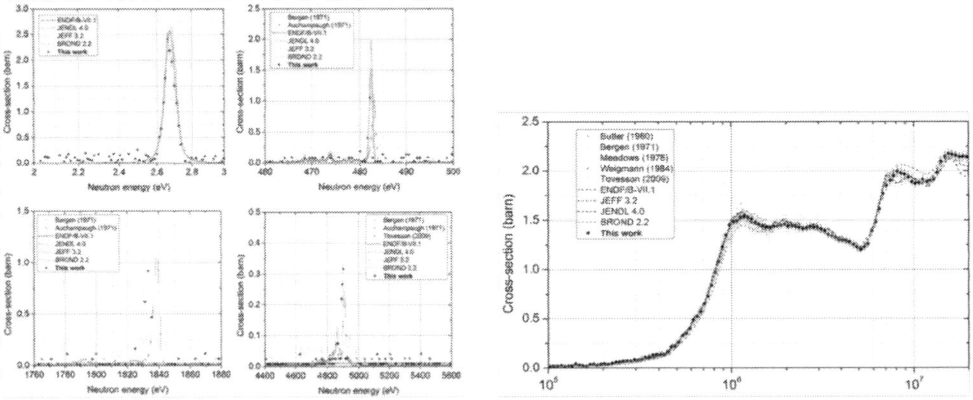

Figure 4: ^{242}Pu(n,f) cross section in the resonance region (left hand side), and in threshold region (right hand side) [7]

In the resonance region (left hand side) the already reported resonances are found again, but shifted in energy and we recall that our energy determination is very accurate thanks to the long flight path. It is worth noticing that the evaluations stop the resolved resonance domain below 1.9 keV, although they could be extended at least up to 10 keV as the density of resonances is still low. In the threshold region up to the third chance opening (right hand side of the figure) the agreement with previous data and evaluations is good, although a slight overestimation shows up with respect to Tovesson's data [9].

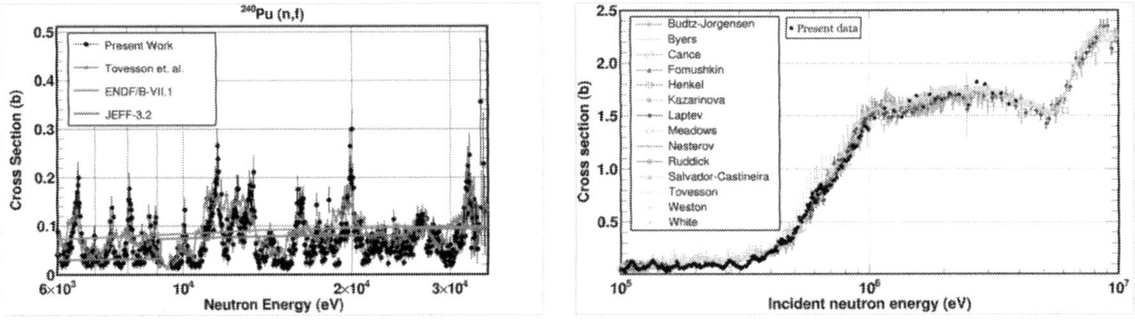

Figure 5: ^{240}Pu(n,f) cross section in the resonance region (left hand side), and in threshold region (right hand side) [8]

https://doi.org/10.1051/epjconf/201921103006

The ^{240}Pu measurement was one of the first experiments performed in EAR2. Figure 5 shows the results, in the 10 keV region (left hand side) and in the threshold region (right hand side). Compared to Tovesson's data in the resonance region [9], the high energy resolution of the n_TOF facility is clearly visible. In the plateau above the threshold the data are consistent with Tovesson, the fluctuation is induced by the dead time correction which is significant in this domain due to the high neutron flux, and the sample mass.

4.2 γ emission in fission with STEFF

The STEFF ensemble is made of arms equiped with Bragg chambers and fast detectors for measuring the time of flight of the fission fragments (figure 6), thereby allowing the application of the *2v-2E* method. STEFF was mounted in EAR2 with a large collimator to define a beam spot covering the 8 cm size of the sample made of a deposit of ^{235}U on a thin Al foil 0.7 μm thick (figure 6) to allow the detection of the 2 fission fragments in 2 opposite arms. In coincidence with fission events in the Bragg chambers the NaI scintillators were recording the emitted γ's [10].

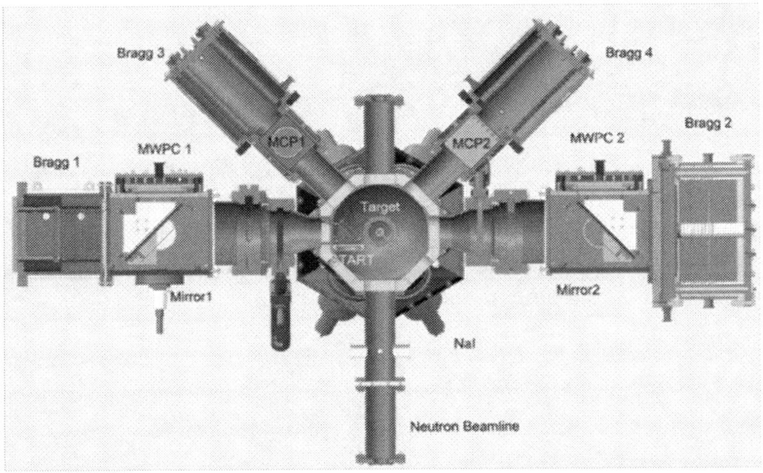

Figure 6: Scheme of the STEFF ensemble (left hand side) showing the 2 horizontal arms terminated by Bragg chambers and the fast timing detectors for the time of flight measurement [10]. In green around the target, the NaI scintillators for detection of the photons emitted in fission. On the right hand side the ^{235}U target used in the experiment.

The recorded photon multiplicity and the sum energy distributions are shown figure 7. The left column displays the actually measured distributions. They are highly dependent on the detection efficiency which has to be corrected for. This is achieved by calculating the matrix response of the system by a Monte Carlo method. The matrix is inverted to get the emitted distributions which are displayed in the right column of figure 7. From those distributions the average multiplicity and energy sum can be obtained and displayed in table 1 in comparison with previous measurements. Although previous measurements are done with thermal neutrons, STEFF data span the interval thermal-1eV. The γ energy threshold is 160 keV for STEFF, higher than in previous measurements where it lies between 10 and 150 keV.

It can be seen that a higher energy sum is found with a lower multiplicity. In this experiment a larger 0-fold is found in comparison with previous STEFF experiments, possibly indicating a higher background in the 2017 measurement. Therefore a better estimation of the in-beam background has to be done before a firm conclusion.

 https://doi.org/10.1051/epjconf/201921103006

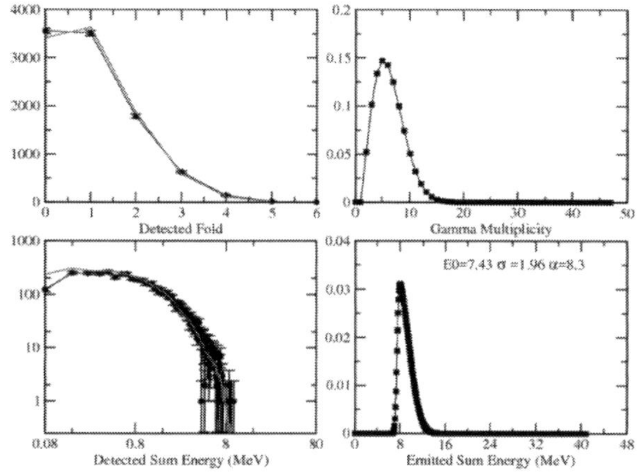

Figure 7: Distributions of multiplicity and total energy for γ emission in ^{235}U(n,f). Left column: raw data (black) and result of a modelling (red). Right column: distributions after efficiency correction [10].

Table 1: Comparison of average multiplicity and total energy with previous experimental data. In STEFF (2017) the energy range of neutrons is thermal-1 eV.

Experiment	$< M >$	E_{tot} (MeV)
STEFF (2017)	6.3 ± 0.2	9.0 ± 0.1
DANCE (2015)	7.35 ± 0.35	8.35 ± 0.4
Oberstedt (2014)	8.19 ± 0.11	6.92 ± 0.09
Verbinski (1973)	6.70 ± 0.30	6.51 ± 0.30
Pleasonton (1972)	6.51 ± 0.30	6.43 ± 0.30
Peelle (1971)	7.45 ± 0.35	7.18 ± 0.26

4.3 The ^{237}Np(n,f) problem

The neutron-induced cross section of ^{237}Np has already been measured at n_TOF with a PPAC system, over a broad energy range (0.7 eV to 1 GeV), giving values in average 7 % higher than the existing data and the evaluations (ENDF/B-VII.1 and JEFF3.1) [11]. Nevertheless the insertion of this cross section in the simulation of a critial benchmark indicated that this higher cross section could not be ruled out [12].

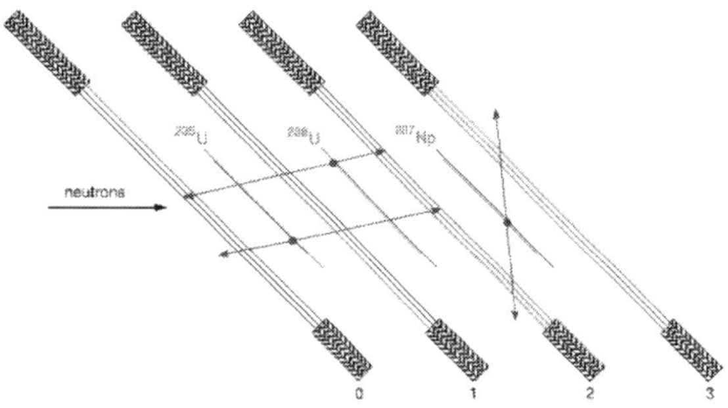

Figure 8: Ensemble of PPAC detectors and samples tilted at 45° in respect to the neutron beam.

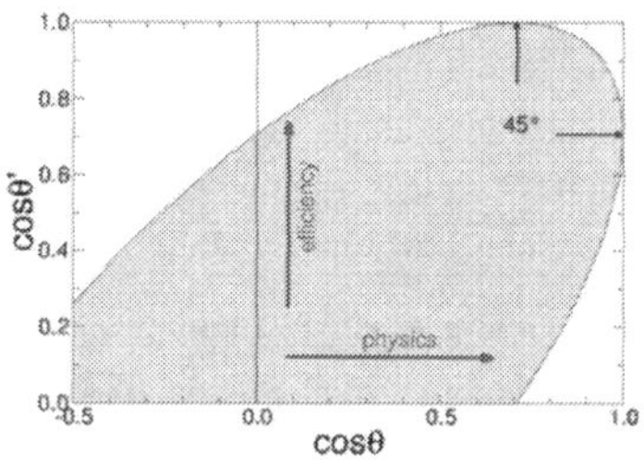

Figure 9: Mapping of the available angles $\cos\theta$ relative to beam and $\cos\theta'$ to the normal to detectors. The hatched area marks the geometrically available zone.

This motivated a new experiment at n_TOF with a setup where a good control on the detection efficiency is possible. In the PPAC system each of the 9 samples is inserted between two PPACs and both fission fragments are detected in coincidence. The localisation of the fragment crossing point in each surrounding detector provides the fission direction, provided a back to back emission is assumed in the laboratory system, as sketched in figure 8. Although this assumption is not valid due to the boost imparted by the incoming neutron, a Monte Carlo simulation shows that the deviation between laboratory and center of mass fission angles does not exceed a few degrees [13]. This is due to the saturation of the deposited linear momentum above a few tens of MeV of incoming energy [14]. From the fission reconstruction the fission angle is extracted and the fission fragment angular distribution (FFAD) is obtained, in addition to the cross section. In a configuration where the detectors and samples are tilted by 45° relative to the beam, as sketched in figure 8, all fission angles can be accessible, and the detection efficiency is almost independent of the angular distribution. Most importantly this configuration makes possible a better control on the detection efficiency.

Any fission trajectory has an angle θ relative to the beam and an angle θ' relative to the normal to detectors and samples. The detection efficiency $\epsilon(\cos\theta')$ depends only on θ' and not on θ, whereas the angular distribution $W(\cos\theta)$ depends only on θ and not on θ', so that the number of detected fissions is: $N_f = N\,W(\cos\theta)\,\epsilon(\cos\theta')$. The key point is that a given θ can be reached with a range of θ', having different azimuthal angle ϕ', as illustrated in figure 9. The θ' dependence of the counting just reflects the efficiency. The latter can be obtained by pieces corresponding to different θ. Knowing that the efficiency is 1 when the fragment trajectory is perpendicular to the detectors and samples, one gets the absolute detection efficiency as a function of θ'.

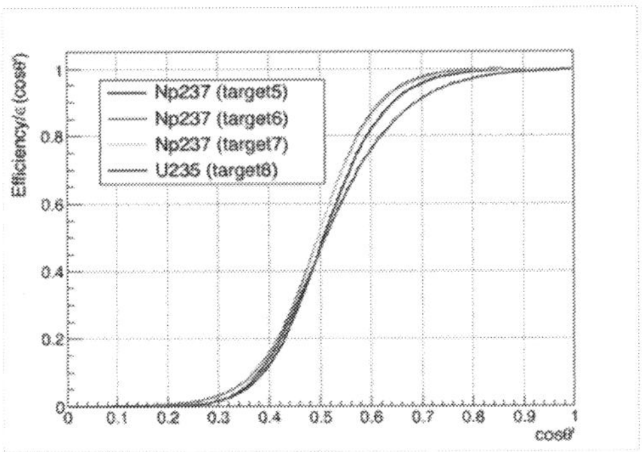

Figure 10: Fit of the angle dependence of the efficiency for each sample. Three ^{237}Np samples are represented and one ^{235}U used as the cross section reference for determining the neutron flux.

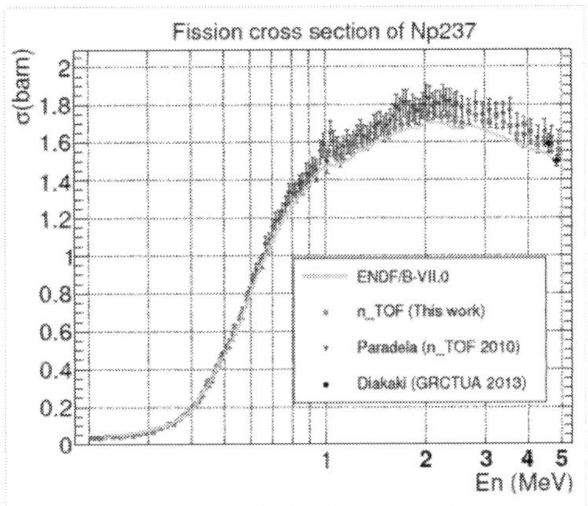

Figure 11: ^{237}Np(n,f) cross section compared with the previous PPAC measurement [11], and ENDF/B-VII.0

The figure 10 shows fits of the efficiency dependence for the three ^{237}Np samples and the ^{235}U sample used as reference. The dependences follow the expected shape, with a drop at large angle due to the thicker dead layers (backing, electrodes) stopping the fission fragments. The three ^{237}Np samples are consistent with each other, and different from the ^{235}U sample, for which the integrated efficiency is lower. If this difference in efficiency is taken into account the extracted ^{237}Np cross section is lowered as shown in figure 11 where it is compared to the older PPAC measurement and to ENDF/B-VII.0. The element dependence of the electroplating process is probably the reason of the efficiency difference, the uranium deposits having a rougher surface and being more loaded with oxygen atoms and water molecules, as diagnosed by Rutherford backscattering spectrometry (RBS). This elemental effect explains the higher cross section measured in [11], where the efficiency could not be extracted experimentally.

4.4 Fission angular distribution in ^{232}Th(n,f)

The fission trajectories have been reconstructed for the ^{232}Th(n,f) reaction, by using the PPAC detection system, assuming again a back to back emission.

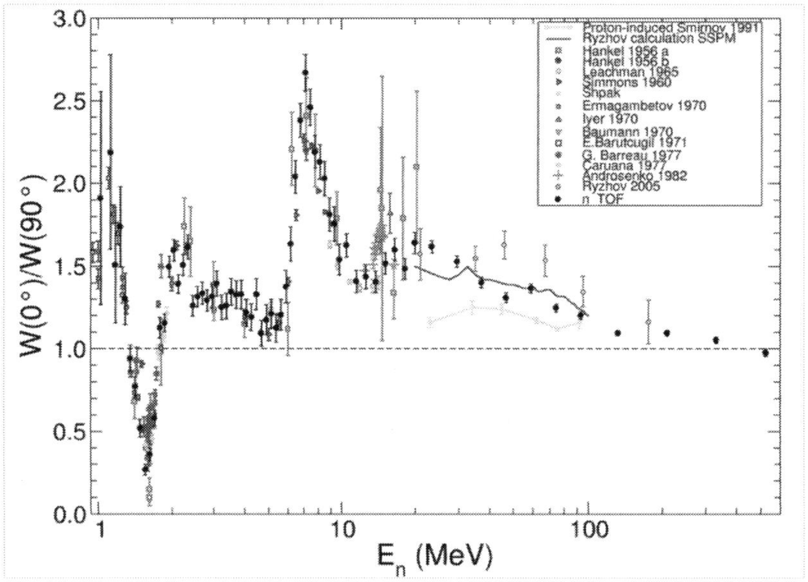

Figure 12: Fragment anisotropy for ^{232}Th(n,f) measured with the PPAC system (black solid points) compared to previous measurements. The blue line is a modelling from [15], and the turquoise one is the systematics of ^{232}Th(p,f) [16].

From the fission fragment angular distribution (FFAD) the anisotropy is defined as the ratio of the emission rate at 0° to that at 90°. This anisotropy is plotted in figure 12 as a function of the energy of the incoming neutron. For the first time the anisotropy is measured between the fission threshold and 500 MeV. Below 10 MeV the agreement is very good with older data and at the third chance opening around 14 MeV the n_TOF data are more accurate due to the time of flight technique. Above 30 MeV the only existing data were obtained with a monoenergetic beam [17] and the anisotropy was found much higher than the n_TOF data.

The FFAD is governed by the spin J of the fissioning nucleus and its projection K along the fissioning axis [18]. When they are comparable the anisotropy is lower than 1 (sideward peaking), as seen in the 1.6 MeV region. If $K \ll J$ it is higher than 1 (backward-forward peaking). When the excitation at the saddle-point is more than a few MeV, many levels contribute to fission, so that K distributes statistically with an average: $K_0 = \sqrt{I_{eff}T/\hbar^2}$ where $I_{eff}$is the moment of inertia and T the temperature. The lower K_0 the higher the anisotropy. The peak at every chance-fission opening is explained by the drop of K_0 due to the temperature decrease after neutron emission. Similarly the lower anisotropy in case of proton-induced fission is explained by the lower barrier of the fissioning nucleus (higher fissility), so that the temperature at the saddle point is higher. The fading of the difference between proton- and neutron-induced at high energy is expected, as the incoming nucleons does not get absorbed anymore so that the fissioning nucleus is the same for both entrance channels. Conversely the difference below 30 MeV demonstrates that the incoming nucleus is essentially trapped.

A recent measurement up to 200 MeV [19] confirmed our results above 30 MeV, contradicting again [17], and in good agreement as well with modelling [15].

5 Conclusion

The fission program at n_TOF addressed the measurement of cross sections, γ emission and fission fragment angular distributions. It exhibited the effect of the micro-structure and elemental composition of the sample on the detection efficiency in the case of the PPAC system. The fission fragment angular distribution gives important information on the fission mechanism. As neutrons in fission are essentally emitted by the fragments, this anisotropy induces also an anisotropy in neutron emission whose practical effects might not be negligible in some cases, such as small size critical benchmarks, and should be taken into account in simulations.

The refurbishment of the spallation target, within the next two years, will provide an even higher flux in EAR2 with a better energy resolution. The ongoing developments in detectors electronics will also allow measurements to higher energies.

References

[1] C. Rubbia et al., Tech. rep., CERN (1998), CERN/LHC/98-002-EET
[2] C. Borcea et al., Nucl. Instr. Meth. Res. A **513**, 524 (2003)
[3] n_TOF collaboration, *CERN n_TOF facility: Performance report* (2010), on-line access from http://cdsweb.cern.ch/collection/n_TOF%20Public%20Notes
[4] C. Weiß et al., Nucl. Instr. Meth. Res. A **799**, 90 (2015)
[5] F. Gunsing et al., *The measurement programme at the neutron time-of-flight facility nTOF at CERN*, in *ND2016 International Conference on Nuclear Data for Science and Technology 2016, Bruges, Belgium, 11-16 Sept 2016*, edited by EPJ Web of Conferences (2017), Vol. 146, p. 11002, DOI: 10.1051/epjconf/201714611002
[6] F. Belloni et al., Eur. Phys. J. A **47**, 2 (2011), DOI: 10.1040/epja/i2011-11002-y
[7] A. Tsinganis et al., *private communication*
[8] A. Stamatopoulos et al., *Measurement of the ^{240}Pu(n,f) cross-section at the CERN n_TOF facility: First results from experimental area II (EAR-2)*, in *ND2016 International Conference on Nuclear Data for Science and Technology 2016, Bruges, Belgium, 11-16 Sept 2016*, edited by EPJ Web of Conferences (2017), Vol. 146, p. 04030, DOI: 10.1051/epjconf/2017146040030
[9] F. Tovesson et al., Phys. Rev. C **79**, 014613 (2009)
[10] J.A. Ryan, Ph.D. thesis, School of Physics and Astronomy, University of Manchester (2017)
[11] C. Paradela et al., Phys. Rev. C **82**, 034601 (2010)
[12] L.S. Leong et al., Ann. Nucl. Energy **54**, 36 (2013)
[13] Diego Tarrío Vilas, Ph.D. thesis, PhD thesis, Universidade de Santiago de Compostela (2012)
[14] M. Fatyga et al., Phys. Rev. C **32**, 1496 (1985)
[15] I. Ryzhov et al., Nucl. Phys. A **760**, 19 (2005), EXFOR #22898 ^{232}Th: .002, ^{238}U: .003
[16] A. Smirnov, I. Gorshkov, A. Prokofiev, V. Eismont, *Total and differential cross sections of 10 to 95 MeV proton-induced fission of actinide nuclei*, in *21th International Symposium on Nuclear Physics, Gaussig, Germany* (1991), p. 214, EXFOR #O0606.013
[17] G.A. Tutin et al., Nucl. Instr. Meth. Res. A **457**, 646 (2001)
[18] R. Vandenbosch, J.R. Huizenga, *Nuclear Fission* (Academic Press, New York, 1973)
[19] A.S. Vorobeyev et al., *Angular distributions and anisotropy of fission fragments from neutron-induced fission in intermediate energy range 1-200 MeV*, in *ND2016 International Conference on Nuclear Data for Science and Technology 2016, Bruges, Belgium, 11-16 Sept 2016*, edited by EPJ Web of Conferences (2017), Vol. 146, p. 04011, DOI: 10.1051/epjconf/201714604011

Preliminary results on the ^{233}U capture cross section and alpha ratio measured at n_TOF (CERN) with the fission tagging technique

M. Bacak[1,2,3,*], M. Aiche[4], G. Bélier[5], E. Berthoumieux[3], M. Diakaki[3], E. Dupont[3], F. Gunsing[3,1], J. Heyse[6], S. Kopecky[6], B. Laurent[5], H. Leeb[2], L. Mathieu[4], P. Schillebeeckx[6], O. Serot[7], J. Taieb[5], V. Vlachoudis[1], O. Aberle[1], J. Andrzejewski[8], L. Audouin[9], J. Balibrea[10], M. Barbagallo[11], F. Bečvář[12], J. Billowes[13], D. Bosnar[14], A. Brown[15], M. Caamaño[16], F. Calviño[17], M. Calviani[1], D. Cano-Ott[10], R. Cardella[1], A. Casanovas[17], F. Cerutti[1], Y. H. Chen[9], E. Chiaveri[1,13,18], N. Colonna[11], G. Cortés[17], M. A. Cortés-Giraldo[18], L. Cosentino[19], L. A. Damone[11,20], C. Domingo-Pardo[21], R. Dressler[22], I. Durán[16], B. Fernández-Domínguez[16], A. Ferrari[1], P. Ferreira[23], P. Finocchiaro[19], V. Furman[24], K. Göbel[25], A. R. García[10], A. Gawlik[8], S. Gilardoni[1], T. Glodariu†[26], I. F. Gonçalves[23], E. González-Romero[10], E. Griesmayer[2], C. Guerrero[18], H. Harada[27], S. Heinitz[22], D. G. Jenkins[15], E. Jericha[2], F. Käppeler[28], Y. Kadi[1], A. Kalamara[29], P. Kavrigin[2], A. Kimura[27], N. Kivel[22], I. Knapova[12], M. Kokkoris[29], M. Krtička[12], D. Kurtulgil[25], E. Leal-Cidoncha[16], C. Lederer[30], J. Lerendegui-Marco[18], S. Lo Meo[31,32], S. J. Lonsdale[30], D. Macina[1], A. Manna[32,33], J. Marganiec[8,34], T. Martínez[10], A. Masi[1], C. Massimi[32,33], P. Mastinu[35], M. Mastromarco[11], E. A. Maugeri[22], A. Mazzone[11,36], E. Mendoza[10], A. Mengoni[31], P. M. Milazzo[37], F. Mingrone[1], A. Musumarra[19,38], A. Negret[26], R. Nolte[34], A. Oprea[26], N. Patronis[39], A. Pavlik[40], J. Perkowski[8], I. Porras[41], J. Praena[41], J. M. Quesada[18], D. Radeck[34], T. Rauscher[42,43], R. Reifarth[25], C. Rubbia[1], J. A. Ryan[13], M. Sabaté-Gilarte[1,18], A. Saxena[44], D. Schumann[22], P. Sedyshev[24], A. G. Smith[13], N. V. Sosnin[13], A. Stamatopoulos[29], G. Tagliente[11], J. L. Tain[21], A. Tarifeño-Saldivia[17], L. Tassan-Got[9], S. Valenta[12], G. Vannini32,[33], V. Variale[11], P. Vaz[23], A. Ventura[32], R. Vlastou[29], A. Wallner[45], S. Warren[13], C. Weiss[2], P. J. Woods[30], T. Wright[13], P. Žugec[14,1]

[1]European Organization for Nuclear Research (CERN), Switzerland
[2]Technische Universität Wien, Austria
[3]CEA Irfu, Université Paris-Saclay, F-91191 Gif-sur-Yvette, France
[4]CENBG, CNRS/IN2P3-Université de Bordeaux, Gradignan, France
[5]CEA, DAM, DIF, F-91297 Arpajon, France
[6]European Commission, Joint Research Centre, Geel, Retieseweg 111, B-2440 Geel, Belgium
[7]CEA, DEN, Cadarache, France
[8]University of Lodz, Poland
[9]Institut de Physique Nucléaire, CNRS-IN2P3, Univ. Paris-Sud, Université Paris-Saclay, F-91406 Orsay Cedex, France
[10]Centro de Investigaciones Energéticas Medioambientales y Tecnológicas (CIEMAT), Spain
[11]Istituto Nazionale di Fisica Nucleare, Sezione di Bari, Italy
[12]Charles University, Prague, Czech Republic
[13]University of Manchester, United Kingdom
[14]Department of Physics, Faculty of Science, University of Zagreb, Zagreb, Croatia
[15]University of York, United Kingdom

[*]corresponding author: michael.bacak@cern.ch

[16]University of Santiago de Compostela, Spain
[17]Universitat Politècnica de Catalunya, Spain
[18]Universidad de Sevilla, Spain
[19]INFN Laboratori Nazionali del Sud, Catania, Italy
[20]Dipartimento di Fisica, Università degli Studi di Bari, Italy
[21]Instituto de Física Corpuscular, CSIC - Universidad de Valencia, Spain
[22]Paul Scherrer Institut (PSI), Villingen, Switzerland
[23]Instituto Superior Técnico, Lisbon, Portugal
[24]Joint Institute for Nuclear Research (JINR), Dubna, Russia
[25]Goethe University Frankfurt, Germany
[26]Horia Hulubei National Institute of Physics and Nuclear Engineering, Romania
[27]Japan Atomic Energy Agency (JAEA), Tokai-mura, Japan
[28]Karlsruhe Institute of Technology, Campus North, IKP, 76021 Karlsruhe, Germany
[29]National Technical University of Athens, Greece
[30]School of Physics and Astronomy, University of Edinburgh, United Kingdom
[31]Agenzia nazionale per le nuove tecnologie (ENEA), Bologna, Italy
[32]Istituto Nazionale di Fisica Nucleare, Sezione di Bologna, Italy
[33]Dipartimento di Fisica e Astronomia, Università di Bologna, Italy
[34]Physikalisch-Technische Bundesanstalt (PTB), Bundesallee 100, 38116 Braunschweig, Germany
[35]Istituto Nazionale di Fisica Nucleare, Sezione di Legnaro, Italy
[36]Consiglio Nazionale delle Ricerche, Bari, Italy
[37]Istituto Nazionale di Fisica Nucleare, Sezione di Trieste, Italy
[38]Dipartimento di Fisica e Astronomia, Università di Catania, Italy
[39]University of Ioannina, Greece
[40]University of Vienna, Faculty of Physics, Vienna, Austria
[41]University of Granada, Spain
[42]Department of Physics, University of Basel, Switzerland
[43]Centre for Astrophysics Research, University of Hertfordshire, United Kingdom
[44]Bhabha Atomic Research Centre (BARC), India
[45]Australian National University, Canberra, Australia

Abstract. ^{233}U is of key importance among the fissile nuclei in the Th-U fuel cycle. A particularity of ^{233}U is its small neutron capture cross-section, which is on average about one order of magnitude lower than the fission cross-section. The accuracy in the measurement of the ^{233}U capture cross-section depends crucially on an efficient capture-fission discrimination, thus a combined set-up of fission and γ-detectors is needed. A measurement of the ^{233}U capture cross-section and capture-to-fission ratio was performed at the CERN n_TOF facility. The Total Absorption Calorimeter (TAC) of n_TOF was employed as γ-detector coupled with a novel compact ionization chamber as fission detector. A brief description of the experimental set-up will be given, and essential parts of the analysis procedure as well as the preliminary response of the set-up to capture are presented and discussed.

1 Introduction

The Th-U fuel cycle [1,2] has been proposed as an alternative to the U-Pu fuel cycle for nuclear power. As one of the key nuclei, ^{233}U influences many parts of the design of a nuclear power plant like neutronics performance, economics, nuclear safety, etc. The available data for the ^{233}U capture cross section are scarce, as shown in Fig. 1, because the measurement is challenging due to the competing fission reaction which is on average one order of magnitude more likely than the capture cross section, see Fig. 2. The first measurement [3] at the n_TOF facility [4] at CERN was successfully performed but proved to be challenging due to the need to accurately distinguish between capture and fission γ-rays without any additional discrimination tool.

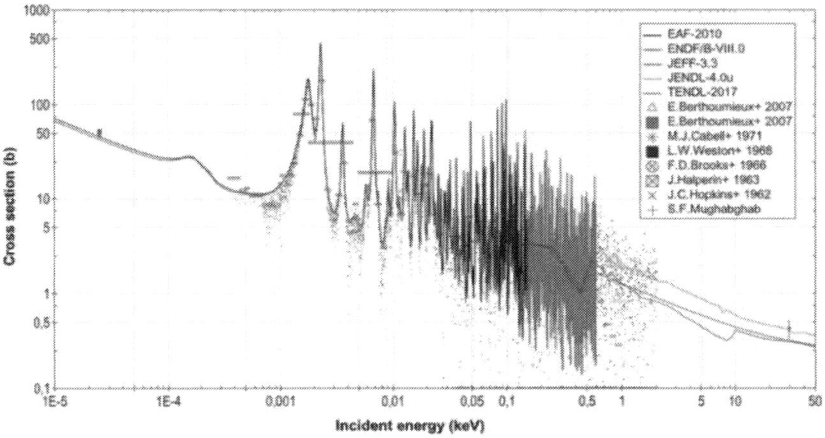

Fig. 1. Existing data sets and evaluations for the ^{233}U capture cross section.

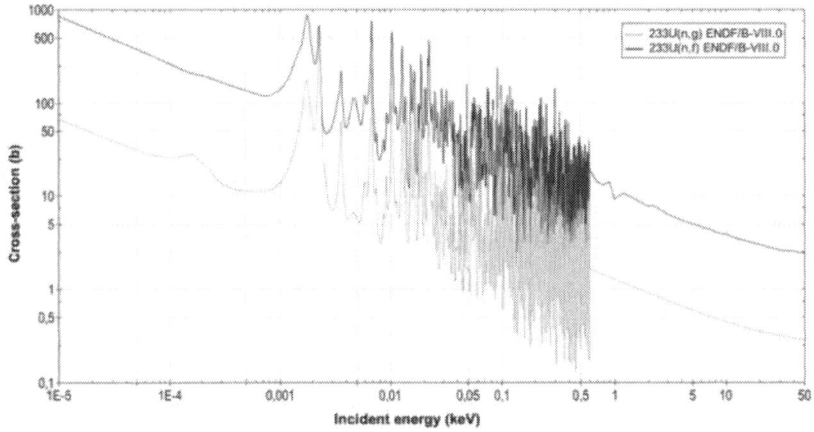

Fig. 2. ^{233}U fission and capture cross section in comparison.

A new measurement [5] was proposed at n_TOF and performed at the end of 2016, aiming to provide a higher level of discrimination between the competing capture and fission channels and to obtain more precise and accurate data up to 10 keV neutron energy. As the accuracy in the measurement of the ^{233}U capture cross-section essentially relies on an efficient capture-fission discrimination, a combined set-up of γ-detectors and fission detectors was used, described in sections 2.3 and 2.4, respectively. A similar methodology of extracting the capture cross-section of fissile actinides has been under development at the DANCE facility at the Los Alamos National Laboratory [6,7].

2 Experimental Setup

2.1 The n_TOF facility

At the n_TOF facility neutrons are produced by spallation reactions of 20 GeV/c protons provided by CERN's Proton Synchrotron impinging on a lead target. The target is surrounded by water acting as a coolant and moderator for the initially fast neutron spectrum resulting in a white neutron beam with neutron energies ranging from sub-thermal up to GeV, see Fig. 3.

The high intensity pulsed proton beam with a maximum frequency of 0.83 Hz results in a high instantaneous neutron flux in the experimental area. The measurement was performed at the experimental area 1 (EAR1), located at 185 m from the neutron-producing target. The neutron beam intensity in EAR1 is constantly monitored with four out-of-beam silicon detectors (*SiMon*) [8] measuring the outgoing particles of the standard reaction ^{6}Li(n,t) reaction from a 600 μm/cm^2 thick ^{6}Li foil intersecting the beam.

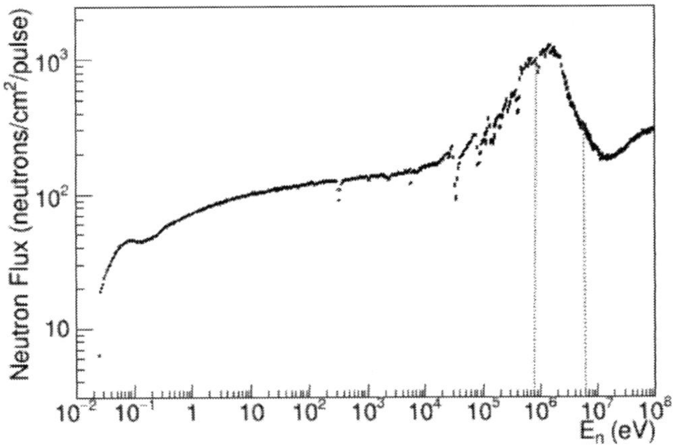

Fig. 3. Neutron Flux at n_TOF EAR1. The dashed lines indicate the region of 800 keV < E_n < 7 MeV corresponding to 5 μs < TOF < 15 μs (see discussion in section 3.1).

2.2 ^{233}U targets

The ^{233}U deposits for the measurement have been prepared by JRC-Geel by molecular plating. The base material was 99.936 % enriched in ^{233}U with the largest contaminant being 0.0496 % ^{234}U (Lot 2146, TP2015-10). Samples 4 cm in diameter covering the full neutron beam were deposited on 10 μm thick aluminium backings. A total mass of 46.5 mg ^{233}U was distributed over 14 samples with an average areal density of about 250 μg/cm^2 and an average activity of about 1 MBq per sample.

2.3 Gamma-Detector

The Total Absorption Calorimeter (TAC) [9] is designed as a high efficiency calorimeter to detect the complete prompt γ-ray cascade emitted in nuclear reactions. It consists of 40 BaF_2 crystals forming a hollow sphere with an inner radius of 10.6 cm covering almost 4π solid angle minus two opposite channels which are left open for entrance and exit of the neutron beam, see Fig. 4. The fission detector was placed at the centre of the TAC surrounded by the so-called *absorber*, a spherical neutron shielding aiming to reduce the background introduced by scattered neutrons into the crystals. The absorber is made out of polyethylene and 7.5 % natural lithium salt.

The digitised waveforms are analysed offline with a dedicated pulse shape analysis routine [10], and the individual signals are grouped together in events by setting an adequate coincidence window of 12 ns between signals. Each of those events is characterised by its time-of-flight *TOF*, determining the neutron energy E_n, the total deposited energy in the TAC E_{Sum}, and the number of hit crystals m_{cr}. One of the main advantages of the TAC is the use of the quantities m_{cr} and E_{Sum} to discriminate between different types of reactions. For example, background events from the sample activity or ambient background usually deposit a few MeV of energy in the TAC and are characterised by small crystal multiplicities. On the other hand, electromagnetic cascades from neutron capture events will deposit energies up to the neutron separation energy of the compound nucleus with larger average multiplicities. Hence, selecting the right subset of events with respect to m_{cr} and E_{Sum} will result in an improved capture-to-background ratio.

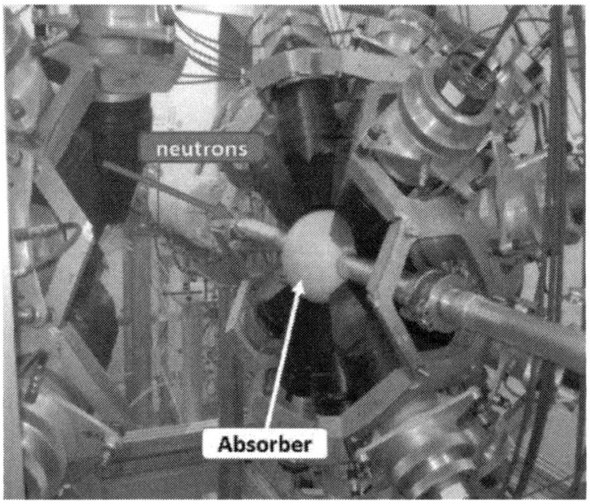

Fig. 4. The n_TOF Total Absorption Calorimeter.

2.4 Fission Detector

A previous experiment [11] at n_TOF using MicroMegas (MGAS) detectors as fission detector showed that the copper mesh of the MicroMegas is a significant source of background for neutron energies above 100 eV. Thus, a novel fission detector was designed, see Fig. 5, respecting the main constraints namely a) the restricted space inside the absorber of the TAC (maximum diameter of 10 cm); b) the fast response needed to reduce pile-up due to the high α-activity of the ^{233}U targets; c) the hosting of the maximum number of ^{233}U targets possible for sufficient statistics in a reasonable beam time. The housing of the fission chamber *FICH* is made of 1.5 mm thick aluminium with an outer diameter of 66 mm and a length of 78 mm. Two stacks of simple axial ionization cells are mounted directly on their respective motherboards and are inserted from each end of the chamber. The 3 mm inter-electrode gap in each cell is biased with 420 V and filled with the fast ionizing gas CF_4. Pre-amplifier and shaper modules are directly mounted on the motherboards of each stack to reduce signal attenuation and to improve the signal to noise ratio.

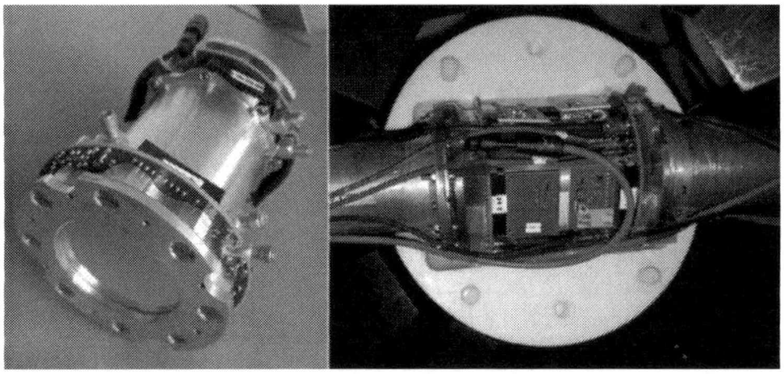

Fig. 5. The fission chamber: prototype in the lab (left panel); fully operational chamber mounted inside the white absorber in the TAC (right panel).

3 Performance of the FICH

3.1 Discrimination between fission and alpha activity

Fig. 6 shows the performance of the chamber in terms of separation between fission fragments (FF) and α-particles for various conditions on TOF, hence E_n. Looking at the full neutron energy range of interest the separation is limited due to the poor FF/α-ratio. This ratio is drastically improved in resonances of the fission cross section, for example in the energy range 1.6 eV < E_n < 1.9 eV, corresponding to the first large resonance dominated by fission. The ratio can be further improved when exploiting the shape of the neutron flux at n_TOF, namely the region between 800 keV < E_n < 7 MeV or 5 μs < TOF < 15 μs respectively. In this region the neutron flux has a maximum, see Fig. 3, further improving the fission rate, hence the FF/α-ratio.

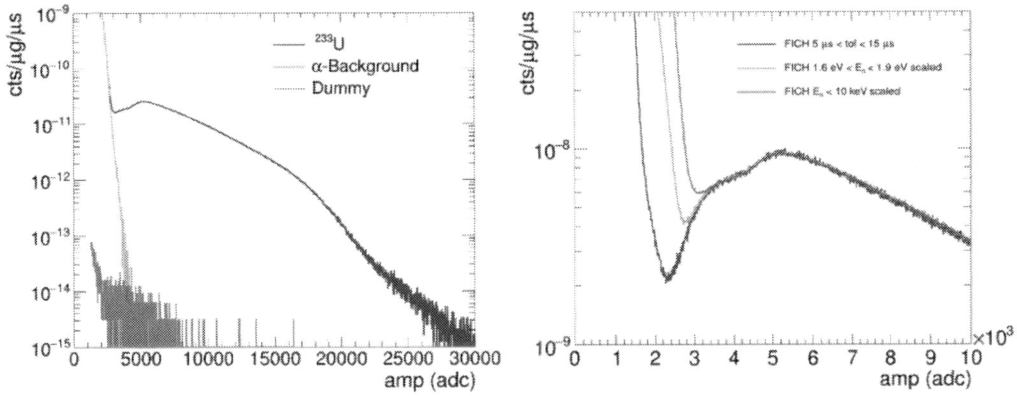

Fig. 6. Amplitude spectra of the fission chamber: various contributions for 0.02 eV < E_n < 10 keV (left panel); α-FF separation for various conditions on E_n or TOF (right panel).

3.2 Fission Tagging

The purpose of the fission chamber is to allow identification of the TAC response to fission events by looking at coincidences between FICH and TAC. In Fig. 7 the comparison between the amplitude spectra of the FICH without and with coincidence (tagged) is shown. The tagged spectrum follows the fission fragment shape perfectly even below amplitudes of 3000 channels of the ADC where the main contribution is from α-particles. In the top left panel of Fig. 8 the response of the TAC to fission events is shown.

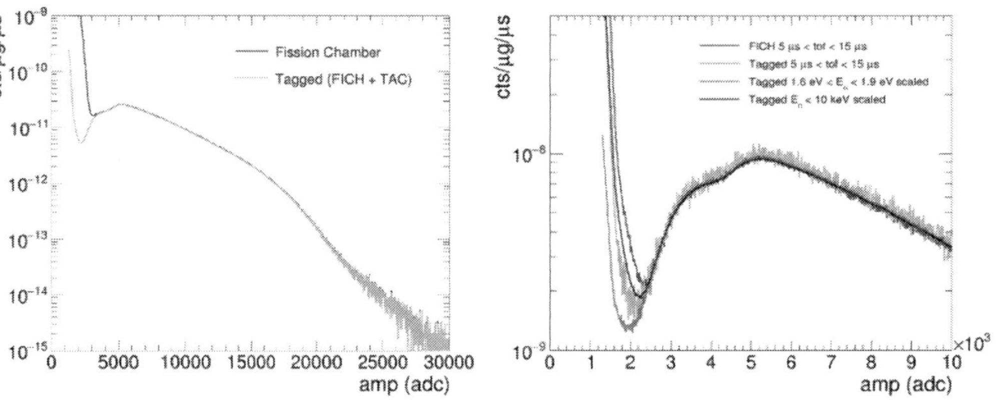

Fig. 7. Comparison of the amplitude spectra of the FICH and tagged counts: for 20 meV < E_n < 10 keV (left panel), α-FF separation for various conditions on E_n or TOF (right panel).

4 TAC response

4.1 Sources of Background

In Fig. 8 the contributions to the total measured spectrum are shown separately. It can be seen that the background consists mostly of low energy events with $E_{Sum} < 3$ MeV and $m_{cr} < 3$. The γ-rays from fission are the main source of background in the region of interest for the capture measurement, i.e. from 3 MeV < E_{Sum} < 7.5 MeV ($S_n(^{234}U)$ = 6.84 MeV). Thus, a precise subtraction of this component is crucial to minimise the uncertainty in the capture cross section.

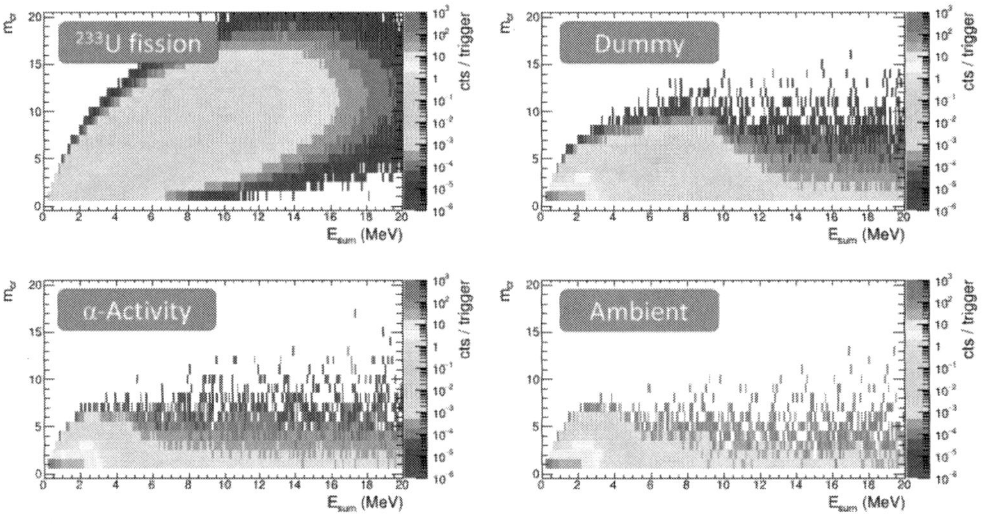

Fig. 8. TAC response for various background contributions.

4.2 FICH efficiency

The efficiency of the fission detector can be determined under the assumption that mostly fission events deposit energies higher than 10 MeV in the TAC. With the extra assumption that a fission event is detected independently by the TAC and the FICH the tagging efficiency ε_{Tagg} equals the fission chamber efficiency ε_{FICH}

$$\varepsilon_{Tagg} = \frac{c_{Tagg}}{c_{TAC}} = \varepsilon_{FICH} \tag{1}$$

with c_{Tagg} and c_{TAC} the tagged and total TAC counts respectively with $10 < E_{Sum} < 20$ MeV. The removal of the background contributions can be further improved by gating on high crystal multiplicities. The calculated fission chamber efficiency for fission amplitudes bigger than 3000 channels of the ADC is shown in Fig. 8 for various neutron energy regions. It can be seen that the data points are in a good agreement with a constant value of 0.867 ± 0.002, see fitted red line in Fig. 9.

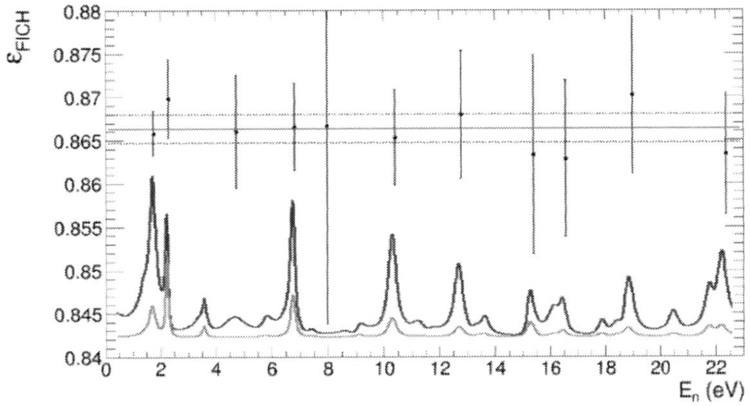

Fig. 9. Efficiency of the fission chamber as a function of neutron energy for events with fission amplitudes bigger than 3000 channels of the ADC and 10 MeV $< E_{Sum} <$ 20 MeV and $m_{cr} > 6$. The red line is a fitted constant function and the blue dashed line indicates the error band of the fit. The solid blue and green line at the bottom illustrate the shape of the ENDF/B-VIII.0 fission and capture cross section respectively.

4.3 Response to Capture

In Fig. 10 the total measured deposited energy spectrum compared to the various contributions can be seen in the case of neutrons absorbed in the large capture resonance at 2.1 eV $< E_n <$ 2.5 eV. The lower cut-off is defined by the 200 keV threshold set per individual detector and a minimum measured multiplicity $m_{cr} > 2$. It is evident that the largest contribution are the fission events in the region of interest. The peak in the capture response at the neutron separation energy $S_n(^{234}U) = 6.84$ MeV is clearly visible and followed by a smooth bump towards lower energies. Below 2.5 MeV the background subtraction has to be investigated further, because a smooth shape going to zero is expected in the capture response. Above 7.5 MeV there is still some remaining background possibly due to capture of scattered neutrons in the TAC.

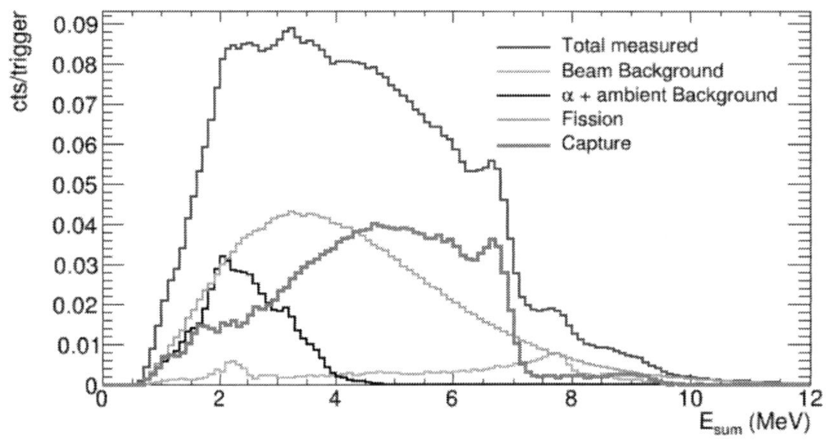

Fig. 10. Contributions to the deposited energy spectra in the TAC in the capture resonance at 2.1 eV < E_n < 2.5 eV for 2 < m_{cr} < 7.

5 Summary & Conclusion

The experiment aiming to measure the ^{233}U capture cross section has been performed successfully at the n_TOF facility (CERN) using the Total Absorption Calorimeter for the detection of the γ-rays. A fission detector was used in veto to properly remove the contribution from the fission γ-rays. Preliminary analysis of capture results have been presented and are promising.

Two background contributions, namely the scattered neutrons by the ^{233}U layers and the response of the experimental set-up to fission neutrons are still under investigation. In the next step extensive Monte Carlo simulations will be used to calculate the detection efficiency for the applied analysis cuts regarding deposited energy and crystal multiplicity. For this purpose, the capture cascades will be generated with DICEBOX [12] and fed into Geant4 where the full detector geometry has been reproduced. Furthermore, dead time corrections and pile-up effects [13, 14] will be taken into account.

References

1. V.G. Pronyaev, IAEA Report, INDC (NDS) **408** (1999)
2. The Generation IV International Forum, http://www.gen-4.org/ (2013)
3. C. Carrapico et al., Nucl. Instr. Meth. A **704**, 60 (2013)
4. C. Guerrero, and the n TOF collaboration, Eur. Phys. J. A **48**, 29 (2012).
5. C. Carrapico, E. Berthoumieux and the n_TOF Collaboration, CERN-INTC-2013-041 / INTC-P-397 (2013)
6. M. Jandel et al., Phys. Rev. Lett. **109**, 202506 (2012)
7. S. Mosby et al., Phys. Rev. C **89**, 034610 (2014)
8. S. Marrone, et al., Nucl. Instr. and Meth. A **517**, 389 (2004)
9. C. Guerrero and the n_TOF Collaboration, Nucl. Instr. Meth. A **608**, 424 (2009)
10. E. Berthoumieux, Preliminary report on BaF_2 Total Absorption Calorimeter test measurement, Rap. Tech., CEA-Saclay/DAPNIA/SPhN, (2004)
11. J. Balibrea and the n_TOF Collaboration, Nuclear Data Sheets, **119**: 10 (2014)
12. F. Bečvář, Nucl Instr Meth A. **417**, 434 (1998)
13. E. Mendoza, et al., Nucl. Instr. and Meth. A **768**, 55 (2014)
14. C. Guerrero, et al., Nucl. Instr. and Meth. A **777**, 63 (2015)

Measurement of the ^{244}Cm and ^{246}Cm neutron-induced capture cross sections at the n_TOF facility

V. Alcayne[1], A. Kimura[2], E. Mendoza[1], D. Cano-Ott[1], T. Martínez[1], O. Aberle[3], J. Andrzejewski[4], L. Audouin[5], V. Bécares[1], M. Bacak[3,6,7], M. Barbagallo[3,8], F. Bečvář[9], G. Bellia[10,11], E. Berthoumieux[7], J. Billowes[12], D. Bosnar[13], A. Brown[14], M. Busso[8,15,16], M. Caamaño[17], L. Caballero-Ontanaya[18], F. Calviño[19], M. Calviani[3], A. Casanovas[19], F. Cerutti[3], Y. H. Chen[5], E. Chiaveri[3,12,20], N. Colonna[8], G. Cortés[19], M. A. Cortés-Giraldo[20], L. Cosentino[10], S. Cristallo[8,15,21], L. A. Damone[8,22], M. Diakaki[23,3], M. Dietz[24], C. Domingo-Pardo[18], R. Dressler[25], E. Dupont[7], I. Durán[17], Z. Eleme[26], B. Fernández-Domínguez[17], A. Ferrari[3], P. Finocchiaro[10], V. Furman[27], K. Göbel[28], A. Gawlik[4], S. Gilardoni[3], T. Glodariu[29], I. F. Gonçalves[30], E. González-Romero[1], C. Guerrero[20], F. Gunsing[7], H. Harada[2], S. Heinitz[25], J. Heyse[31], D. G. Jenkins[14], F. Käppeler[32], Y. Kadi[3], T. Katabuchi[33], N. Kivel[25], I. Knapova[9], M. Kokkoris[23], Y. Kopatch[27], M. Krtička[9], D. Kurtulgil[28], I. Ladarescu[18], C. Lederer-Woods[24], J. Lerendegui-Marco[20], S. Lo Meo[34,35], S. J. Lonsdale[24], D. Macina[3], A. Manna[35,36], A. Masi[3], C. Massimi[35,36], P. Mastinu[37], M. Mastromarco[3], F. Matteucci[38,39], E. A. Maugeri[25], A. Mazzone[8,40], A. Mengoni[34], V. Michalopoulou[23], P. M. Milazzo[38], F. Mingrone[3], A. Musumarra[10,11], A. Negret[29], R. Nolte[41], F. Ogállar[42], A. Oprea[29], N. Patronis[26], A. Pavlik[43], J. Perkowski[4], L. Persanti[8,15,21], I. Porras[42], J. Praena[42], J. M. Quesada[20], D. Radeck[41], D. Ramos-Doval[5], T. Rauscher[44,45], R. Reifarth[28], D. Rochman[25], M. Sabaté-Gilarte[3,20], A. Saxena[46], P. Schillebeeckx[31], D. Schumann[25], S. Simone[10], A. G. Smith[12], N. V. Sosnin[12], A. Stamatopoulos[23], G. Tagliente[8], J. L. Tain[18], T. Talip[25], A. Tarifeño-Saldivia[19], L. Tassan-Got[3,23,5], A. Tsinganis[3], J. Ulrich[25], S. Urlass[3,47], S. Valenta[9], G. Vannini[35,36], V. Variale[8], P. Vaz[30], A. Ventura[35], D. Vescovi[8,15], V. Vlachoudis[3], R. Vlastou[23], A. Wallner[48], P. J. Woods[24], T. Wright[12], and P. Žugec[13]

[1]Centro de Investigaciones Energéticas Medioambientales y Tecnológicas (CIEMAT), Spain
[2]Japan Atomic Energy Agency (JAEA), Tokai-mura, Japan
[3]European Organization for Nuclear Research (CERN), Switzerland
[4]University of Lodz, Poland
[5]Institut de Physique Nucléaire, CNRS-IN2P3, Univ. Paris-Sud, Université Paris-Saclay, F-91406 Orsay Cedex, France
[6]Technische Universität Wien, Austria
[7]CEA Irfu, Université Paris-Saclay, F-91191 Gif-sur-Yvette, France
[8]Istituto Nazionale di Fisica Nucleare, Sezione di Bari, Italy
[9]Charles University, Prague, Czech Republic
[10]INFN Laboratori Nazionali del Sud, Catania, Italy
[11]Dipartimento di Fisica e Astronomia, Università di Catania, Italy
[12]University of Manchester, United Kingdom
[13]Department of Physics, Faculty of Science, University of Zagreb, Zagreb, Croatia
[14]University of York, United Kingdom
[15]Istituto Nazionale di Fisica Nucleare, Sezione di Perugia, Italy
[16]Dipartimento di Fisica e Geologia, Università di Perugia, Italy
[17]University of Santiago de Compostela, Spain
[18]Instituto de Física Corpuscular, CSIC - Universidad de Valencia, Spain

[19]Universitat Politècnica de Catalunya, Spain
[20]Universidad de Sevilla, Spain
[21]Istituto Nazionale di Astrofisica - Osservatorio Astronomico di Teramo, Italy
[22]Dipartimento di Fisica, Università degli Studi di Bari, Italy
[23]National Technical University of Athens, Greece
[24]School of Physics and Astronomy, University of Edinburgh, United Kingdom
[25]Paul Scherrer Institut (PSI), Villingen, Switzerland
[26]University of Ioannina, Greece
[27]Joint Institute for Nuclear Research (JINR), Dubna, Russia
[28]Goethe University Frankfurt, Germany
[29]Horia Hulubei National Institute of Physics and Nuclear Engineering, Romania
[30]Instituto Superior Técnico, Lisbon, Portugal
[31]European Commission, Joint Research Centre, Geel, Retieseweg 111, B-2440 Geel, Belgium
[32]Karlsruhe Institute of Technology, Campus North, IKP, 76021 Karlsruhe, Germany
[33]Tokyo Institute of Technology, Japan
[34]Agenzia nazionale per le nuove tecnologie (ENEA), Bologna, Italy
[35]Istituto Nazionale di Fisica Nucleare, Sezione di Bologna, Italy
[36]Dipartimento di Fisica e Astronomia, Università di Bologna, Italy
[37]Istituto Nazionale di Fisica Nucleare, Sezione di Legnaro, Italy
[38]Istituto Nazionale di Fisica Nucleare, Sezione di Trieste, Italy
[39]Dipartimento di Astronomia, Università di Trieste, Italy
[40]Consiglio Nazionale delle Ricerche, Bari, Italy
[41]Physikalisch-Technische Bundesanstalt (PTB), Bundesallee 100, 38116 Braunschweig, Germany
[42]University of Granada, Spain
[43]University of Vienna, Faculty of Physics, Vienna, Austria
[44]Department of Physics, University of Basel, Switzerland
[45]Centre for Astrophysics Research, University of Hertfordshire, United Kingdom
[46]Bhabha Atomic Research Centre (BARC), India
[47]Helmholtz-Zentrum Dresden-Rossendorf, Germany
[48]Australian National University, Canberra, Australia

Abstract. The neutron capture reactions of the ^{244}Cm and ^{246}Cm isotopes open the path for the formation of heavier Cm isotopes and heavier elements such as Bk and Cf in a nuclear reactor. In addition, both isotopes belong to the minor actinides with a large contribution to the decay heat and to the neutron emission in irradiated fuels. There are only two previous ^{244}Cm and ^{246}Cm capture cross section measurements: one in 1969 using a nuclear explosion [1] and the most recent data measured at J-PARC in 2010 [2]. The data for both isotopes are very scarce due to the difficulties in performing the measurements: high intrinsic activity of the samples and limited facilities capable of providing isotopically enriched samples.

We have measured both neutron capture cross sections at the n_TOF Experimental Area 2 (EAR-2) with three C_6D_6 detectors and also at Area 1 (EAR-1) with the TAC. Preliminary results assessing the quality and limitations (background subtraction, measurement technique and counting statistics) of this new experimental datasets are presented and discussed.

1 Introduction

Accurate neutron capture cross section data for minor actinides (MAs) are required to estimate the production and transmutation rates of MAs in light-water reactors (LWR) with a high burnup, critical fast reactors like Gen-IV systems and other innovative devices such as

accelerator driven systems (ADS) [3]. The ^{244}Cm ($T_{1/2}$=18.1 years) and ^{246}Cm ($T_{1/2}$=4730 years) isotopes are among the most important MAs due to the difficulties in their transmutation and their contribution to the radiotoxicity of the irradiated nuclear fuels. In particular, even after three years of cooling, ^{244}Cm shares nearly 50% of the total actinide decay heat in irradiated reactor fuels with a high burnup. In addition, both of them are in the path of the creation of any heavier Cm isotopes and heavier elements like Bk and Cf.
Only two previous capture measurements were done before the n_TOF measurements. The first one, done in 1969 [1], used the neutrons produced in an under-ground nuclear explosion. The ^{244}Cm (n,γ) cross section was measured in a range from 20 eV to 1 keV and for the ^{246}Cm(n,γ) from 20 eV to 400 eV. The second capture measurement was performed in 2010 with a large coverage Ge-array in the Accurate Neutron Nucleus Reaction Measurement Instrument (ANNRI) at J-PARC [2]. In this second measurement the data obtained for both isotopes range from 2 eV to 300 eV. The resonance analysis was done up to 30 eV. The scarcity of the available data and the many experimental challenges involved in the two previous measurements motivated an additional measurement under different conditions, i.e., in a different facility, with different detectors and monitors and with a different methodology.

2 Experiment

The same samples used at J-PARC have been measured at n_TOF. In this facility, the neutron beam is generated through spallation of 20 GeV/c protons, which are extracted in pulses from the CERN Proton Synchrotron and impinging on a lead target. The pulses have a nominal intensity of 7×10^{12} protons and a time spread of 7 ns (rms). The neutrons travel along two beam lines towards the two experimental areas along : EAR-1 flight length of 185 m [4] (horizontal) and EAR-2 flight length of 19 m [5] (vertical). The neutron flux is larger in the EAR-2 and the energy resolution is better in the EAR-1.

The samples have been measured in both experimental areas. The measurement in EAR-2 has been done with three C_6D_6 detectors and the Total Energy Detection (TED) [6] technique, and the measurement in EAR-1 with the n_TOF Total Absorption Calorimeter (TAC) [7], designed to detect the complete γ-ray cascade. Measuring in both experimental areas with different detectors and techniques will allow to crosscheck results and, presumably, to reduce the final uncertainties.

The targets used for the experiment consist in two samples of ^{244}Cm and one of ^{246}Cm. There were 0.4 mg of ^{244}Cm in each of the ^{244}Cm samples and 1.1 mg of ^{246}Cm in the ^{246}Cm sample. The isotopic abundances of the different actinides are presented in Table 1.

Table 1. Isotopic composition (% atoms) of the ^{244}Cm and ^{246}Cm samples.

	^{244}Cm sample	^{246}Cm sample
^{244}Cm	60.3±1.1	20.3±0.5
^{245}Cm	2.4±0.3	1.1±0.3
^{246}Cm	6.3±0.6	57.7±1.5
^{247}Cm	-	2.8±0.4
^{248}Cm	-	8.8±0.2
^{240}Pu	31.0±0.6	9.3±0.2

2.1 Measurement at the EAR-2

The three C_6D_6 detectors were placed at 5 cm from the sample and perpendicular to the beam. In addition, three additional detectors were used for monitoring the beam. Two of them measured the intensity of the proton beam. The third one was the SiMon [8], an array of four silicon detectors facing a thin enriched lithium fluoride foil, for monitoring the neutron beam.

To obtain accurate weighting functions required for the PHWT [9] technique, a very detailed description of the experimental set-up has been implemented in the Geant4 toolkit [10]. The simulated response functions have been validated with experimental data obtained with several calibration sources (^{133}Ba, ^{137}Cs, ^{60}Co, ^{88}Y, AmBe and CmC). The geometry implemented in Geant4 and one of the simulated response functions are shown in Figure 1. The data obtained at n_TOF are processed with the Pulse Shape Analysis (PSA) routine [11].

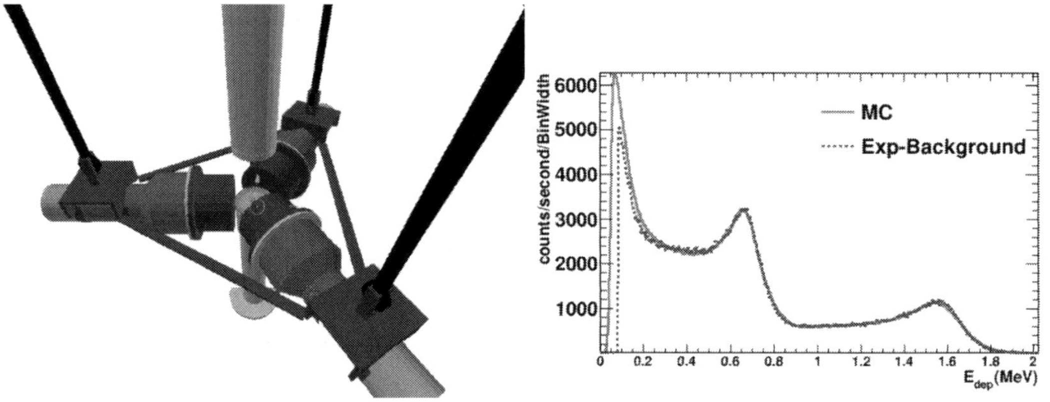

Figure 1. Geometry of the EAR-2 setup as implemented in Geant4 (left). Comparison between simulated (MC) and experimental response function to an ^{88}Y source (right).

The results of the fits are stored in ROOT files [12]: signal amplitudes, signal areas, times, etc.. Also the conversion from time-of-flight to neutron energy is done.

Accurate amplitude-to-energy calibrations and gain stability checks were performed on a weekly basis for the three C_6D_6 detectors using ^{133}Ba, ^{137}Cs, ^{60}Co, ^{88}Y, AmBe and CmC calibration sources. Small gain shifts (8%) are observed and corrected as a function of time.

The total number of counts as a function of neutron energy measured with the Cm samples in place are shown in Figure 2. Also shown are the estimated total background, the beam related background, and the no-beam related background. These backgrounds have been obtained from dedicated measurements.

A preliminary but rather complete analysis of the data measured at the EAR-2 has been performed and two preliminary capture yields (no background subtracted) have been obtained, one for the ^{244}Cm samples and the other for the ^{246}Cm sample. Both yields (no background subtracted) are presented in Figure 3 together with the yields obtained from the JEFF-3.3 [13] cross sections with the experimental background and the characteristics of the EAR-2 neutron beam (neutron flux + *resolution function*)[14]. The relative differences between the experimental and the evaluated yields (in %) are shown at the bottom of each panel, integrated for each resonance. The results have been normalized to the first large resonance of ^{240}Pu at 1 eV.

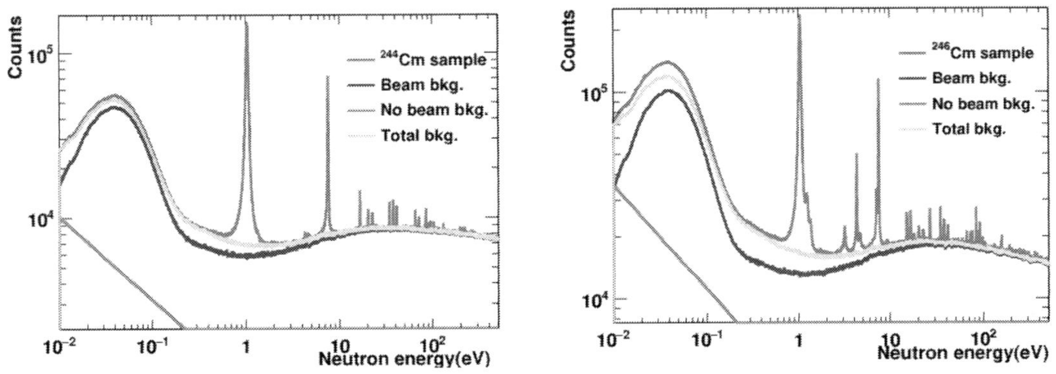

Figure 2. Total number of counts (750 bins per decade) and estimated backgrounds registered in the ^{244}Cm (left) and ^{246}Cm (right) EAR-2 measurements.

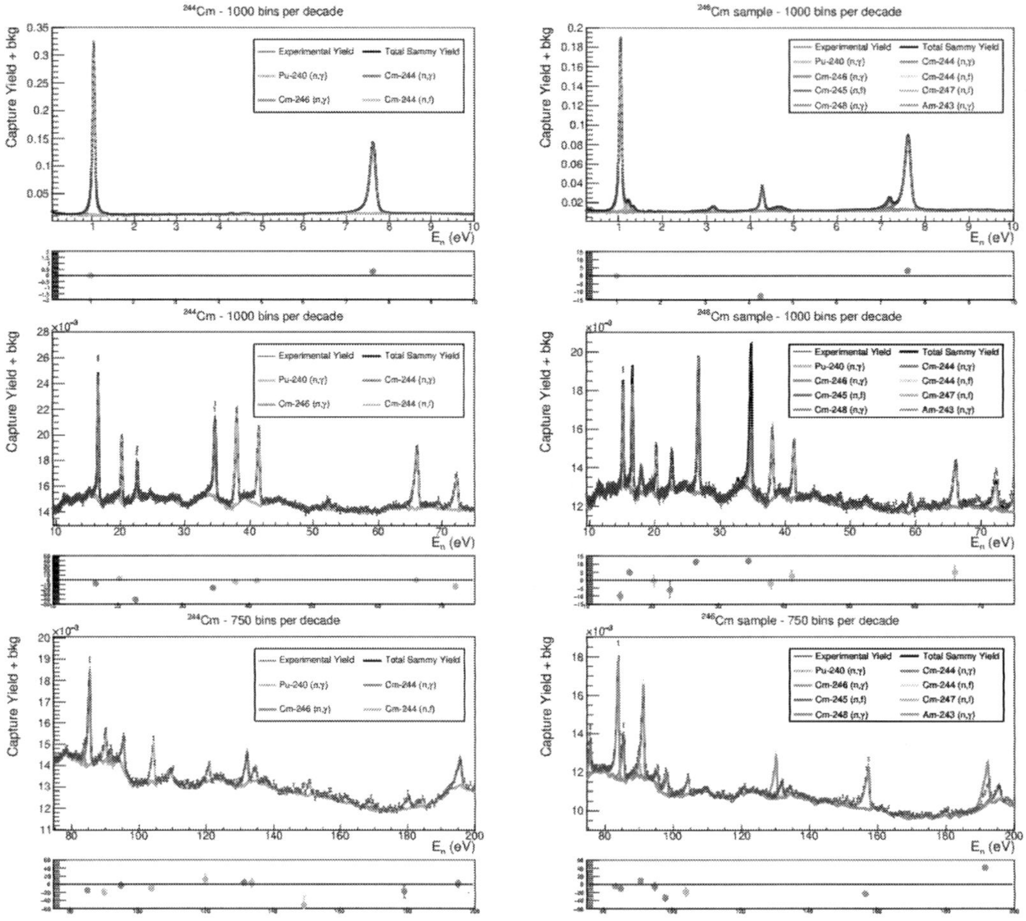

Figure 3. Preliminary experimental yields (no background subtracted) of the measured ^{244}Cm (left) and ^{246}Cm (right) samples. Together with the experimental data points, we show an estimation of the contribution of the capture and fission reactions in each isotope present in the samples. These contributions have been obtained using the SAMMY computer code to calculate the reaction yields taking into account the experimental conditions such as Doppler and resolution broadening, and taking the reaction cross sections from JEFF-3.3. The resulting yields were then normalized to fit the experimental results and added to the background, obtained from dedicated measurements.

2.2 Measurement at the EAR-1

The measurement in EAR-1 was performed with the TAC, which is an array of 40 BaF_2 crystals designed to detect the full capture γ-ray cascades. The data analysis is ongoing and it will follow similar procedures than the ones performed in previous TAC experiments [15] [16]. Signals are grouped into TAC events with a 20 ns coincidence window, and cuts in total deposited energy and detection multiplicity allow to improve the signal to background ratio.

The total number of counts as a function of neutron energy measured with the ^{244}Cm samples in place are shown in Figure 4, together with the estimated backgrounds. The better energy resolution and smaller instantaneous neutron fluence in EAR-1 results in a narrower resonant structure and a larger contribution of the no-beam related background.

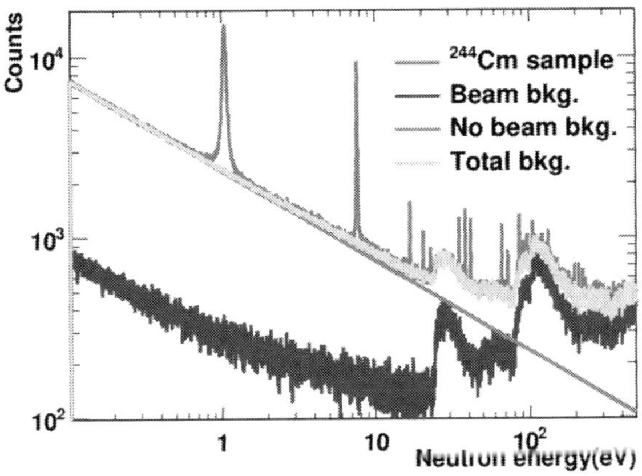

Figure 4. Total number of counts (1000 bins per decade) and estimated backgrounds registered in the ^{244}Cm EAR-1 measurement.

3 Conclusions

The capture cross sections of ^{244}Cm and ^{246}Cm are required to estimate the production and transmutation rates of MAs in LWR and also for new reactor types. There are only two previous capture measurements of these two isotopes, both of them with many experimental difficulties.Therefore, a new measurement has been performed at n_TOF using the two experimental areas and preliminary capture yields have been obtained.

References

[1] M. S. Moore et. al., Phys. Rev. C, 3, 1656 (1971)
[2] A. Kimura et. al., Jour. Nucl. Sc. Tech. 49, 708 (2012)
[3] G. Aliberti et. al., , Ann. Nucl. Ener. 33, 700 (2006)
[4] C. Guerrero et al., Eur. Phys. J. A 49, 27 (2013)
[5] C. Weiss et al., Nucl. Instrum. Meth. A 799, 90 (2015)
[6] R.L. Macklin and J.H. Gibbons,Phys. Rev. 159, 1007 (1967)
[7] C.Guerrero et al., Nucl. Instrum. Meth. A 608, 424 (2009)
[8] S. Marrone et al., Nucl. Instrum. Meth. A 517, 389 (2004)

https://doi.org/10.1051/epjconf/201921103008

[9] Abondano et al., Nucl. Instrum. Meth. A 521, 454 (2004)
[10] S. Agostinelli et al., Nucl. Instrum. Meth. A 506, 250 (2003)
[11] P. Žugec et al., Nucl. Instrum. Meth. A 812, 134 (2016)
[12] https://root.cern.ch/
[13] A. Koning et al., J. Korean Phys. Soc. 59, 1057 (2011); The JEFF library is available from https://www.oecd-nea.org/dbdata/jeff
[14] M. Sabaté-Gilarte et al., Eur. Phys. J 53, 10 (2017)
[15] C. Guerrero et al., Phys. Rev. C 85, 044616 (2012)
[16] E. Mendoza et al., Phys. Rev. C 90, 034608 (2014)

Absolute cross section measurements of ^{238}U(n,f) and ^{237}Np(n,f) in the neutron energy range 1-2.4 MeV

Paula Salvador-Castiñeira[1,*], *Franz-Josef* Hambsch[2], *Alf* Göök[2], *Marzio* Vidali[2], *Nigel P.* Hawkes[1], *Stephan* Oberstedt[2], *Neil J.* Roberts[1], *Graeme C.* Taylor[1], and *David J.* Thomas[1]

[1]National Physical Laboratory, Hampton Road, Teddington, Middlesex, TW11 0LW, United Kingdom
[2]EC-JRC-Directorate G-Nuclear Safety & Security, Unit G.2, Retieseweg 111, 2440 Geel, Belgium

Abstract. New standard (n,f) cross sections other than ^{235}U are important to study the relevant cross sections for Generation-IV power plants. A specific need for such standards is for performing new experiments with quasi-monoenergetic neutron beams, such as those produced by Van de Graaf accelerators. Neutrons down-scattered to low energies in the experimental environment, so called room-return, become relevant for this type of measurements. Hence, a standard (n,f) cross section with a fission threshold is of great interest, in order to suppress the contribution from room-return background. For this reason we have performed two experiments at the VDG of the National Physical Laboratory to measure absolutely the (n,f) cross sections of ^{235}U, ^{238}U and ^{237}Np in the fast neutron energy region. Our preliminary results are in agreement with the most up-to-date evaluations.

1 Introduction

Modelling of Generation-IV nuclear power plants requires highly accurate values of cross sections in the fast energy region. Cross section measurements are usually performed relative to the primary standard ^{235}U(n,f), but in environments where the thermal and epi-thermal neutron background is non-negligible, such as the target hall of a Van de Graaff accelerator, other isotopes with a fission threshold should be preferred. Up to now, none of the isotopes with a fission threshold has been considered a primary standard. Two isotopes, for which the experimental database is sufficient, could potentially fill this role: ^{238}U(n,f) and ^{237}Np(n,f). ^{238}U(n,f) has a fission threshold at around $E_n = 1.6$ MeV and is a secondary standard from $E_n = 2$ MeV. Although the JEFF 3.2 evaluation showed discrepancies up to 7% in the range 1.5 MeV $< E_n <$ 5 MeV with respect to ENDF/B-VIII.0, the new JEFF 3.3 evaluation is in agreement with ENDF/B-VIII.0. ^{237}Np(n,f) would be more suitable as a standard because of its lower fission threshold ($E_n = 0.5$ MeV) and its higher cross section above $E_n = 1.0$ MeV, but some discrepancies between measurements have been found in the last years [1].

To address these issues, two experimental campaigns have been performed at the Van de Graaff accelerator of the National Physical Laboratory (NPL,UK) under the European CHANDA project. The Nuclear Metrology group at NPL is known worldwide for its capability to provide very accurate and precise neutron flux measurements by using a well characterized long counter. Within their facilities there is also a large low-scatter target hall ideal for

*e-mail: paula.salvador.castineira@npl.co.uk

Table 1. Properties of the samples used during the experiment.

Isotope	Sample ID	Mass (μg)	Purity
^{235}U	^{235}U$_{2016}$	562 ±2.0%	99.83%
^{235}U	^{235}U$_{2017}$	701 ±0.6%	99.93%
^{238}U	^{238}U	681 ±2.6%	>99.99%
^{237}Np	^{237}Np	489.5±0.5%	>99.99%
^{242}Pu	^{242}Pu	671 ±0.9%	>99.97%

cross section measurements. A twin Frisch-grid ionization chamber was used as fission fragment detector. Measurements were done absolutely, by using the long counter, and relatively, by placing two samples in a back-to-back configuration in the fission chamber. The first campaign was performed in January 2016 and the isotopes measured were ^{235}U(n,f), ^{238}U(n,f) and ^{237}Np(n,f). For the second campaign, performed in January 2017, measurements were performed for ^{235}U(n,f), ^{237}Np(n,f) and ^{242}Pu(n,f).

Results from both campaigns are presented, although results on the ^{242}Pu(n,f) cross section will be reported in a future publication as they are outside the scope of the present work. This paper presents a follow-up on the experiment reported on [2].

2 Experimental setup

The experimental setup consisted of a long counter (LC) and a twin Frisch-grid ionization chamber (TFGIC). The measurements were done in the following chronological order: first, a LC measurement was performed with the desired neutron energy to determine the neutron fluence per monitor count. The monitor is measuring the current on the neutron producing target can produced by the incident ion beam. Then, a shadow cone was used to subtract the contribution from scattered neutrons to the neutron fluence measured by the LC. Finally, the TFGIC was placed in front of the neutron producing target.

Two proton-induced reactions were used during the experiment: ^{7}Li(p,n)^{7}Be for producing E_n = 0.565 MeV and ^{3}H(p,n)^{3}He for producing E_n = 0.9-2.4 MeV.

2.1 Long counter

The LC consists of a BF_3 tube surrounded by a moderating layer and an outer shield. It has a high sensitivity, a negligible gamma response and a nearly flat fluence response from 1 keV up to 7 MeV [3]. Figure 1 (left) shows the LC placed in front of the beam line and with the shadow cone in place.

2.2 Fission Fragment detector

Two different TFGICs were employed as fission fragment detector with very similar dimensions, one in 2016 and another one in 2017. For the measurements reported here the configuration was the same as in [2] (see Fig. 1 (right)).

2.3 Fissile samples

During both campaigns five different samples were used. Their properties are listed in table 1.

Figure 1. Left: NPL low-scatter area with the long counter and shadow cone aligned with the beam line. Right: TFGIC aligned with the beam line and neutron producing target.[2]

3 Measurements

Two measurement campaigns have been performed, one in 2016 and one in 2017. In 2016 the focus was on the ability to reproduce the ^{235}U(n,f) cross section, thus proving our experimental setup and our data analysis. In addition, the ^{238}U(n,f) and the ^{237}Np(n,f) cross section were measured. As it will be shown in Sect. 5, the results of the 2016 measurement on the ^{235}U(n,f) cross section were systematically higher than predicted by the evaluations. For this reason, the same cross section was measured in 2017 using two different samples, together with the ^{237}Np(n,f) and ^{242}Pu(n,f).

4 Data analysis

4.1 Neutron fluence at the sample positions

The neutron fluence determined by the LC needs to be corrected by several effects that are related with the fact that we could not perform a shadow cone measurement with the TFGIC. In addition, the code used to obtain the neutron fluence at the sample positions from the LC measurement treats the system as a point-to-point problem, it considers the neutron producing target as a point in the space and the fission samples as a point. However, this was no longer applicable when we measured with the TFGIC because the edge of the fission chamber was only around 3 cm away from the neutron producing target, therefore the total distance between the neutron producing target and the fission samples was only around 9 cm. For this reason MCNP6 simulations [4] were performed to allow for the neutron fluence emitted by the disk-like neutron producing target and transmitted through the disk-like samples. This correction is around 2-4% depending on the incident neutron energy.

4.1.1 Target can scattering

The target can scattering was taken into account as this is subtracted in the shadow cone measurement performed with the long counter, but could not be taken into account in the TFGIC measurement due to the close distance between the TFGIC and the neutron producing target. The calculation was made using MCNP6 and the correction is of 2-3% depending on the incident neutron energy.

4.1.2 Neutron attenuation on the TFGIC

Finally, the attenuation of neutrons not only on the front face of the TFGIC, but on the first anode and grid, needs to be considered as these layers will be contributing to the neutron slowing down and, therefore, change the spectrum of the neutrons impinging on the samples in the cathode. The calculation was made using MCNP6 and the correction is of 1.5-2% depending on the incident neutron energy.

4.2 Fission fragment determination

The fission fragments were detected using the TFGIC. The signal used in our experiments corresponded to the anodes and grid only. A 2D grid vs anode pulse height was used to clean the anode pulse height (PH) distribution from any α-particle background. In addition, the anode projected PH was extrapolated to lower PH values to account for FFs of lower energy than the level of the electronic threshold, this was done using a straight line. This correction is about 2-5%.

4.2.1 Reaction rate due to lower energy neutrons

An MCNP6 simulation was needed to simulate the target hall to account for all scattered neutrons that will be impinging on the samples with lower energy than expected. These neutrons will be inducing fissions in the samples as a function of their corresponding (n,f) cross section. Therefore, this correction will have more impact on the ^{235}U samples as its cross section increases as the neutron energy decreases. The correction is of 4-10% depending on the sample and the incident neutron energy.

4.3 Cross section determination

The cross section has been determined as:

$$\sigma(E_n) = \frac{C_{\text{corr}} \cdot k_{\text{FF,low}}}{\epsilon} \frac{A}{m \cdot N_A} \frac{1}{\Phi_n(E_n) \cdot k_{\text{PP-DD}} \cdot k_{\text{TS}} \cdot k_{\text{AttFC}}}, \tag{1}$$

where C_{corr} are the corrected counts for the electronic threshold, ϵ accounts for the FF loss within the sample, A is the atomic number of the sample, m its mass, N_A is Avogadro's number and $\Phi_n(E_n)$ is the absolute fluence. The correction factor $k_{\text{FF,low}}$ accounts for the fission fragments produced by neutrons of lower energy than the main peak, $k_{\text{PP-DD}}$ corrects for the point-to-point to disk-to-disk difference when extracting the absolute fluence, k_{TS} corrects for the neutrons scattered in the target can and k_{AttFC} corrects for the attenuation of neutrons in the front face of the TFGIC as well as the anode and grid facing the neutron beam.

5 Preliminary results and discussion

The preliminary results are presented in Fig. 2. All results presented include both statistical and systematic uncertainties. In all case, the most significant ones are the counting statistics and the uncertainty related to the neutron fluence determination at the position of the samples.

5.1 ^{235}U(n,f)

The ^{235}U(n,f) cross section was measured in both campaigns, 2016 and 2017. In 2016, only one sample was employed, namely $^{235}U_{2016}$, whilst in 2017 we measured with $^{235}U_{2016}$ and with a new sample, namely $^{235}U_{2017}$. The main reason to measure with different samples was the increased cross section values found in 2016. In order to confirm that the issue was due to the sample, in 2017 we repeated the same measurement with the same sample and measured as well with another sample. The results we have obtained for the ^{235}U(n,f) cross section are in very good agreement with the present evaluations when using the $^{235}U_{2017}$ sample (red squares). However, there was clearly an issue with the $^{235}U_{2016}$ sample (green and blue squares). This unidentified issue increased the measured cross section by around 5% in the investigated neutron energy range. Although additional experiments are planned to further study this effect, the most probable hypothesis is that the high-Z backing of the $^{235}U_{2016}$ sample could have an impact on the amount of backscattered FFs detected.

5.2 ^{238}U(n,f)

The ^{238}U(n,f) absolute cross section measured in this experiment agrees within uncertainties with the present evaluations.

5.3 ^{237}Np(n,f)

In the case of ^{237}Np(n,f) measurements were performed in 2016 and 2017 with some neutron energies repeated, mainly due because in 2017 the ^{242}Pu(n,f) cross section was measured at the same time. The ^{237}Np(n,f) absolute cross section measured in this experiment agrees within uncertainties with the present evaluations and the newest data from Diakaki [5]. Therefore, does not confirm the 5% increase seen by Paradela [1]. (The ratio $\sigma_{\mathrm{ENDF}}/\sigma_{\mathrm{meas}}$ is not shown as its value is out of scale.)

6 Conclusions

The neutron-induced fission cross section has been measured absolutely for ^{235}U, ^{238}U and ^{237}Np. The results show a very good agreement with the present evaluations for ^{238}U(n,f) and ^{237}Np(n,f). In the case of ^{235}U(n,f) our data agree with evaluation when using one of the samples ($^{235}U_{2017}$), but we obtain a higher cross section when using the $^{235}U_{2016}$ sample. However, we suspect that further investigations will be able to explain this mismatch. Our uncertainties are dominated by the statistical uncertainty due to the beam current limitation of the facility as well as the fluence uncertainty. A follow up experiment is planned to resolve the discrepancy obtained with the $^{235}U_{2016}$ sample. Our preliminary results agree with the latest evaluations in the neutron energy ranges studied in this work and we believe that they could be used, when finalized, along with other high quality data in a possible evaluation of ^{238}U(n,f) and ^{237}Np(n,f) as standards in these energy ranges.

This work was partly funded by the CHANDA project (EC-FP7-Fission-2013).

References

[1] Paradela, C., *et al.*, Phys. Rev. C **82(3)**, 034601 (2010)

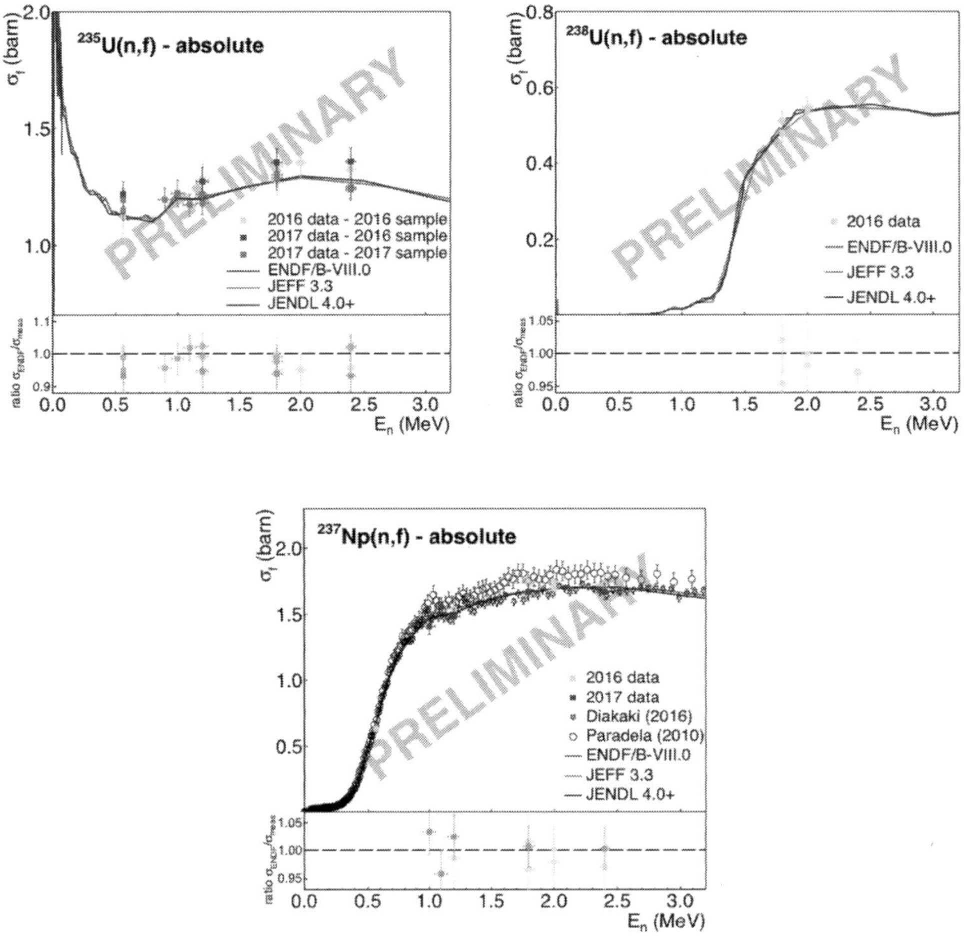

Figure 2. Preliminary results obtained via the absolute measurement for the ^{235}U, ^{238}U and ^{237}Np. The data is compared to the ENDF/B-VIII.0, the JEFF 3.3 and the JENDL 4.0+ evaluation. In addition, the ^{237}Np is compared to the experimental data from Paradela *et al.* [1] and Diakaki *et al.* [5]. For each case, the ratio between the measured data and the ENDF/B-VIII.0 evaluation is also shown. See text for further details.

[2] Salvador-Castiñeira, P., Hambsch, F.-J., Göök, A., Vidali, M., Hawkes, N.P., Roberts, N.J., Taylor, G.C. and Thomas, D.J., EPJ Web of Conferences **146**, 04050 (2017)

[3] Tagziria, H., and Thomas, D.J., Nucl. Instr. Meth. **A452**, 470-483 (2000)

[4] Goorley, T., *et al.*, Nuclear Technology, **180**, 298-315 (2012).

[5] Diakaki, M., *et al.*, Phys. Rev. C **93(3)**, 034614 (2016)

Development of a Gaseous Proton-Recoil Detector for neutron flux measurements between 0.2 and 2 MeV neutron energy

P. Marini[1,*,**], *L.* Mathieu[1], *M.* Aïche[1], *T.* Chiron[1], *P.* Hellmuth[1], *J.-L.* Pedroza[1], *G.* Kessedjian[2], *S.* Czajkowski[1], *B.* Jurado[1], and *I.* Tsekhanovich[1]

[1]CENBG, CNRS/IN2P3-Université de Bordeaux, Chemin du Solarium B.P. 120, 33175 Gradignan, France
[2]LPSC, Université Grenoble-Alpes, CNRS/IN2P3, F-38026 Grenoble Cedex, France

Abstract. Absolute measurements of neutron fluence are an essential prerequisite of neutron-induced cross section measurements, dosimetric investigations and neutron beam lines characterisation. Independent and precise neutron flux measurements can be performed with respect to the H(n,p) elastic cross section. However, the use of silicon proton recoil detectors is not straightforward below incident neutron energy of 1 MeV, due to a high background in the detected proton spectrum. A new gaseous proton-recoil detector has been designed to answer the challenge. The detector is described in details and results of the commissioning tests are presented.

1 Introduction

A precise absolute measurements of neutron flux is essential for neutron-induced cross section measurements as well as for neutron-beam lines characterisation and dosimetric investigations. In particular, high quality heavy-nuclei neutron-induced cross section data are needed for the development of Generation IV nuclear systems [1, 2]. However, despite the experimental efforts, large discrepancies and uncertainties exist nowadays on these data in the actinide region [1, 3].
The neutron flux is typically determined with respect to fission standard reactions (^{235}U(n,f), ^{238}U(n,f), ^{237}Np(n,f)), whose accuracy ranges from 0.5% to 10% [4, 5]. Independent and more accurate measurements can be carried out with respect to the ^{1}H(n,p) elastic scattering cross section, which is evaluated with a precision better than 0.5% from 1 meV to at least 20 MeV [4, 5]. If silicon-based proton-recoil telescopes are the preferred method for absolute determination of monoenergetic neutron flux for neutron energies above 1 – 2 MeV [6, 7], the extension of the σ(n,p)-based method at energies relevant for the development of Generation IV nuclear systems, namely between 100 keV to 1 MeV [8], is not trivial. Indeed it requires the detection of low energy protons ($\lesssim$ 1 MeV) in a neutron and γ-dense environment, thus a clean discrimination of the signal from the detector background, which is technically difficult [9].

*e-mail: paola.marini@cea.fr
**Present address: CEA, DAM, DIF, F-91297 Arpajon, France

Ref.[9] has shown that the most significant contributions to the low energy background arises from scattered neutrons impinging on H-contaminated surrounding materials, and from Compton electrons scattered by low energy γ-rays produced by the neutron source. Proton-recoil detectors based on silicon detectors are very sensitive to these low energy background sources and thus not suited for measurements on a large dynamical range below 1 MeV.

Plastic scintillators using time-of-flight technique [10] and proton-recoil proportional counters [11–13] are typically used in this low energy region. They present, however, significant limitations: scintillators light output is known to be non-linear, the detector efficiency is difficult to calculate accurately (better than $1-2\%$) and $^{12}C(n,x)$ reactions must be discriminated from (n,p) reactions [14–17]. On the other hand, proportional counters must relying on simulations to account for the dependence on the ionizing particle energy of the gas ionizing potential, Quenching corrections, important wall effects and interferences from photons and carbon recoils [11, 18, 19].

A double-stage gas-based telescope, Gasseous Proton Recoil Telescope (GPRT), for neutron flux measurments at energies below 1 MeV was therefore developed by the ACEN group of CENBG. In Secs.2 and 3 the design of the detector and its performances are presented. Sec.4 concludes with the perspectives for further developments of the GPRT.

2 Detector design

The lay-out of the GPRT telescope is shown in Fig.1. The detector is constituted by a H-rich deposit (polypropylene, mylar foil or tristearine deposit), and a segmented ΔE-E ionization chamber (4x4x12 cm^3), read by a 64 pads Micromegas-based detection plane (Fig.1).

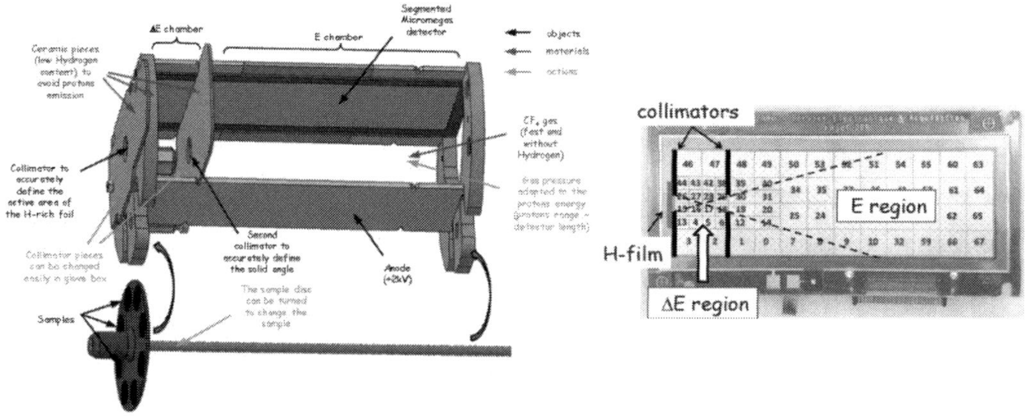

Figure 1. Schematic layout of the detector (left) and of the detection plane (right). The sample disk is shown separately for clarity purpose.

Seven different H deposit with different thicknesses can be mounted on a turning disk. The possibility to select the sample thickness adapted to the neutron incident energy is crucial to obtain precise measurements at low neutron energy since it allows to keep the proton energy loss in the H-foil below 10% and thus to reduce their energy spread. An empty slot is available for background measurements.

Two ceramic collimators, one placed right after the sample, the other between the two ionization chambers, provide a well-defined geometry, essential for precise neutron flux measurements, and allow a physical separation of the ΔE and E sides of the ionization chamber.

The detection plane is a Micromegas, segmented in 64 pads, as shown in Fig.1. The aim is to reduce the electronic noise associated to the pad capacitance, and to perform a raw track analysis in order to reject false coincidences with abnormal trajectories.

The chamber is filled with gas (CF_4 or N_2-CO_2) at an adjustable pressure, from few tens of mbar to few tens of 100 mbar, to reduce as much as possible the detector sensitivity to Compton electrons. Indeed the proton range is adapted to the detector length, which, on the contrary, corresponds to a very small fraction of electrons range. The coincidence between the ΔE and E sides of the ionization chamber allow a first discrimination between direct and scattered neutrons. A field cage placed among the electrods (not shown in Fig.1) assures the uniformity of the electric field applied to the ionization chamber.

Special care was devoted during the project phase to minimize as much as possible the amount of materials to reduce neutron scattering. Finally, the gas and the ceramic materials have been chosen for their very low or absent hydrogen content, so that no additional protons can be emitted by the gas or the detector structure.

More details on the detector design can be found in Refs.[20, 21]. A detailed description of the detector associated electronics, based on the GET system [22], will be presented in a forthcoming publication [23].

3 Performances

Measurements to evaluate the detector performances were performed at the AIFIRA facility [24] with quasi-monoenergetic neutron beam, with energies from 0.3 to 1 MeV. Both CF_4 and N_2-CO_2 (70% N_2 and 30%CO_2) gasses were used during the tests. The gas pressure was varied from 100 to 30 mbar, depending on the neutron energy, and the applied voltage was adapted consequently. Tests were also performed with and without the field cage to study its impact.
Measurements were performed to:

- **Evaluate the direct vs scattered neutron discrimination.** Direct and scattered neutrons could be discriminated i) imposing a ΔE-E coincidence, ii) via trajectory reconstruction. In Fig.2 are shown the signal amplitudes measured on each hit pad by a proton generated by 1 MeV incident neutron, and its reconstructed 2D trajectory. It should be noted that the energy calibration of the detector is not straightforward, since the charge collection strongly depends on the gas pressure and on the applied voltage. The direct vs scattered neutron discrimination is therefore based on the analysis on the lenght of the trajectory: scattered neutrons have lower energy than direct ones, inducing a shorter track of the recoil proton. A further improvement in the discrimination could be obtained if a 3D trajectory reconstruction is possible. This could be achieved using a gas with a slower mobility than CF_4, as the N_2-CO_2 gas, provided that background events have a small rate. An example of 3D trajectory reconstruction is shown in Fig.2, where the drift time of electrons on each pad was measured.
- **Study the detector sensitivity to:** i) **e^- and/or γ-rays:** no signal associated to electrons and γ-rays were detected when irradiating the detector with neutrons but removing the H-foil. The measurement was performed imposing a minimum multiplicity of three pads hit in the same event; ii) **cosmic rays:** easily discriminated based on the reconstructed trajectory, despite the fact that several of them were observed during 6 days beam. An example of cosmic ray track is shown in Fig.3 iii) **a direct neutron irradiation of the Micromegas**: no signals associated to these events were detected. These observations indicate that the event rate is mainly due to proton events and therefore that a slower gas than CF_4 can be used.

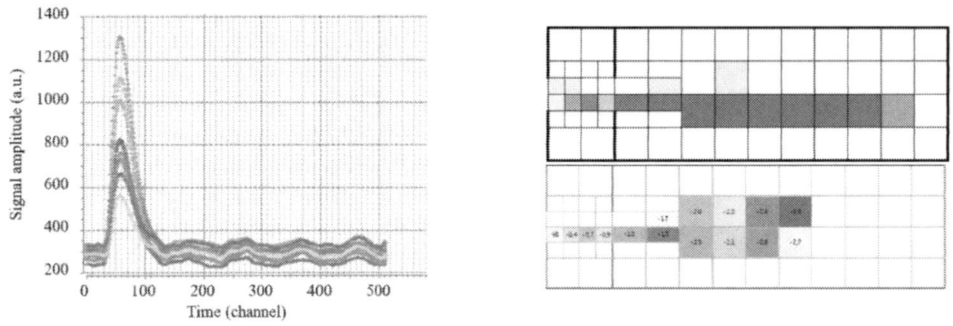

Figure 2. Signal amplitudes for a given event on the different pads (left), and 2D (top) and 3D (bottom) tracks reconstruction for two proton events generated by 1 MeV neutrons (right). The values on the 3D track reconstructions are expressed in cm and represents the proton evolution in the z-axis.

- **Evaluate the lowest neutron energy**. Proton traces at neutron energies as low as 200 keV have been obtained during the test using N_2−CO_2 gas, as shown in Fig.3, fixing the lowest neutron energy at which the detector can be used. Efforts are ongoing to lower this limit.

Figure 3. Trajectory reconstruction of a cosmic ray event (left) and of a proton scattered by a 300 keV neutron.

- **Validate the pressure-voltage working regime** Electrostatics discharges were observed for gas pressure of few tens of mbar and grid voltages of the order of −500 V during the measurements. Indeed, according to Paschen's law, this regime corresponds to a lowering of the breakdown voltage for the used gasses [25]. To work with (p,V) values close to the breakdown voltage requires a very good control over the gas pressure in the detector, i.e. reduced pressure fluctuations, and a very low oxygen contamination, i.e. a very good detector chamber sealing.
- **Evaluate the uniformity of the electrostatic field**. Two main issues were encountered during the tests without the field cage. First of all, smaller amplitude signals were observed on the external pads of the detection plane. This effect, strongly suppressed when using the field cage, was investigated with the code OPERA [26], where the whole detector was simulated, as shown in Fig.4.

  In Fig.5 the equipotential lines obtained without and with the field cage are shown. In the first case, the field is distorted at the entrance and at the exit of the detector, due to the presence of grounded structures (motor axis, reaction chamber...), translating in a degraded charge collection for the external pads of the detection plane. Simulations with a field cage show a better uniformity of the electric field, as confirmed by the experimental results.

  Moreover, a charge accumulation on the insulator pieces under beam irradiation was observed. This induced a progressive reduction of the signal amplitude on the pads closest to the insulator pieces (i.e. on the first raw of pads). The effect of the charge accumulation is

 https://doi.org/10.1051/epjconf/201921103010

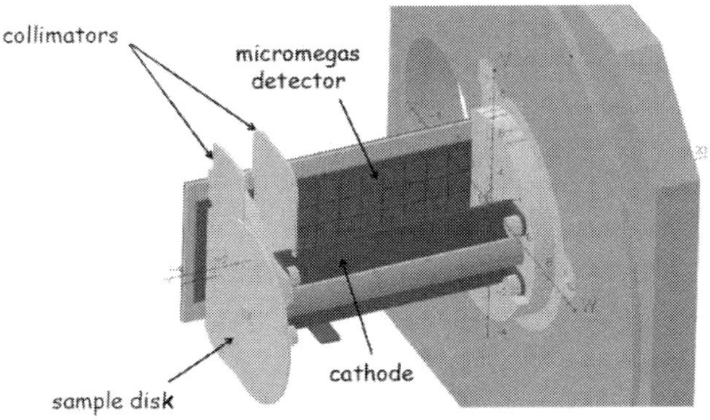

Figure 4. Detector geometry implemented in OPERA simulations.

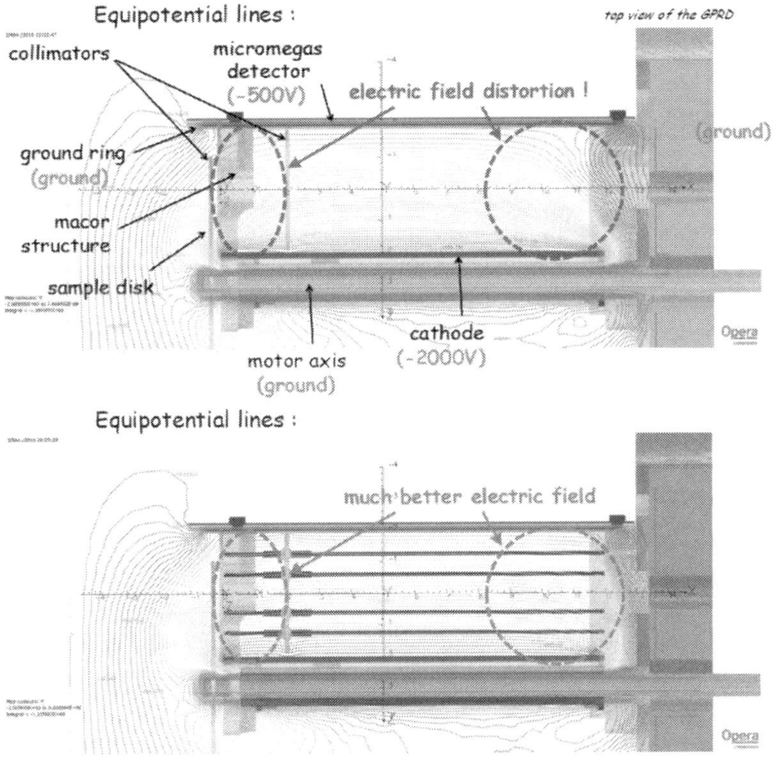

Figure 5. Equipotential lines simulated with the OPERA code. The setup accounts for the two collimators and the field cage (bottom) and no field cage (top).

shown in Fig.6, where the signal amplitude of the first raw of pads is plotted as a function of the irradiation time. The electrostatic charge is partially removed and one third of the signal amplitude recovered by connecting the collimator to the cathode and grounding it.

The right side of Fig.6 shows the OPERA simulation of the electric field when an electrostatic charge of -2 nC is added on the entrance collimator, confirming the field distortion.

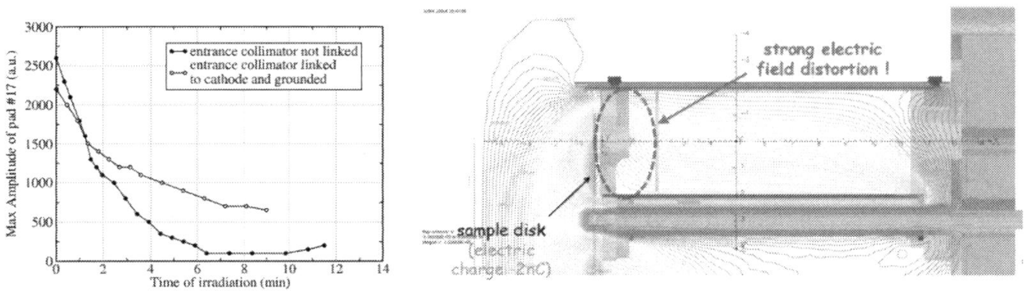

Figure 6. Evolution of the signal amplitude of a pad in the ΔE region with irradiation time (left). (Right) As fig.5, a static charge of −2 nC is added on the entrance collimator.

Tests with the field cage and grounded collimators show no signal amplitude reduction during the irradiation, indicating that the effect is now strongly suppressed.

4 Conclusions and perspectives

Accurate neutron flux measurements via the elastic scattering ^{1}H(n,p) reaction is challenging for neutron energies below 1 MeV. γ-rays produced by the neutron production target generate electrons which induce a strong background in silicon detectors, preventing an accurate determination of the neutron flux. To overcome this issue, a new gaseous proton recoil telescope constituted by a double ionisation chamber was designed and experimentally tested at the CENBG with monoenergetic neutrons. Tests confirmed its low electron sensitivity, making it well suited for neutron energies down to 200 keV in a γs and electrons dense environment. The ΔE-E coincidences, coupled to a 3D proton track analysis, enables the disentanglement between direct and scattered neutrons.

Electrostatic simulations showed that the electric field is slightly distorted by surrounding materials, and heavily distorted by the space charge accumulation on ceramic pieces of the detector when the field cage is removed. The use of a field cage is therefore mandatory. Under the present experimental conditions, it was not possible to measure the proton detection efficiency of the detector, and in particular that every acceptable recoiling proton is actually detected. A quantitative experiment, using a proton microbeam is planned in the short term (middle 2019).

References

[1] M. Chadwick, M. Herman, P. Obložinský, M. Dunn, Y. Danon, A. Kahler, D. Smith, B. Pritychenko, G. Arbanas, R. Arcilla et al., Nuclear Data Sheets **112**, 2887 (2011)

[2] G. Aliberti, G. Palmiotti, M. Salvatores, T. Kim, T. Taiwo, M. Anitescu, I. Kodeli, E. Sartori, J. Bosq, J. Tommasi, Ann. of Nucl. En. **33**, 700 (2006)

[3] M. Salvatores, *Uncertainty and target accuracy assessment for innovative systems using recent covariance data evaluations*, Vol. 26 (Nuclear energy agency, 2008), www.oecd-nea.org

[4] J.C. Hopkins, G. Breit, Nuclear Data Tables A **9**, 137 (1971)

[5] G.M. Hale, A.S. Johnson, in *Proc. 17th Int. IUPAP Conf. on Few-Body Problems in Physics* (W. Gloeckle and W. Tornow (Elsevier, Amsterdam, 2004), 2004), pp. S120–S122

[6] S.J. Bame Jr., E. Haddad, J.E. Perry Jr., R.K. Smith, Rev. of Sc. Instr. **28**, 997 (1957)
[7] T. Ryves, Nucl. Instr. and Meth. **135**, 455 (1976)
[8] NEA, *Nuclear data high priority request list* (2018), http://www.nea.fr/html/dbdata/hprl
[9] P. Marini, L. Mathieu, L. Acosta, M. Aïche, S. Czajkowski, B. Jurado, I. Tsekhanovich, Nucl. Instr. and Meth. A **841**, 56 (2017)
[10] R. Beyer, E. Grosse, K. Heidel, J. Hutsch, A. Junghans, J. Klug, D. Légrády, R. Nolte, S. Röttger, M. Sobiella et al., Nucl. Instr. and Meth. A **575**, 449 (2007)
[11] T.H.R. Skyrme, P.R. Tunnicliffe, A.G. Ward, Rev. of Sc. Instr. **23**, 204 (1952)
[12] E.F. Bennett, Rev. of Sc. Instr. **33**, 1153 (1962)
[13] E.F. Bennett, T.J. Yule, Rad. Res. **50**, 219 (1972)
[14] M.W. Wu, T.C. Guung, C.C. Pei, T.N. Yang, W.S. Hwang, D.J. Thomas, Nucl. Instr. and Meth. A **422**, 79 (1999)
[15] B. Daub, V. Henzl, M. Kovash, J. Matthews, Z. Miller, K. Shoniyozov, H. Yang, Nucl. Instr. and Meth. A **701**, 171 (2013)
[16] M. Kovash, B. Daub, J. French, V. Henzl, K. Shoniyozov, J. Matthews, Z. Miller, H. Yang, in *Advancements in Nuclear Instrumentation Measurement Methods and their Applications (ANIMMA), 2011 2nd International Conference* (2011), pp. 1–3
[17] D. Schmidt, B. Asselineau, R. Böttger, H. Klein, L. Lebreton, S. Neumann, R. Nolte, G. Pichenot, Nucl. Instr. and Meth. A **476**, 186 (2002)
[18] R. Babut, V. Gressier, Jour. of Instr. **2**, P01005 (2007)
[19] R. Nolte, *Detection of neutrons*, in *Ecole Joliot-Curie 2014* (2014), `http://ejc2014.sciencesconf.org`
[20] P. Marini, L. Mathieu, M. Aiche, T. Cheron, P. Hellmuth, J. L. Pedroza, S. Czajkowski, B. Jurado, I. Tsekhanovich, EPJ Web Conf. **146**, 03015 (2017)
[21] P. Marini, et al., arXiv:1811.01005 [physics.ins-det] (2018)
[22] E. Pollacco, S. Anvar, H. Baba, P. Baron, D. Bazin, C. Belkhiria, B. Blank, J. Chavas, P. Chomaz, E. Delagnes et al., Phys. Proc. **37**, 1799 (2012), proceedings of the 2nd International Conference on Technology and Instrumentation in Particle Physics (TIPP 2011)
[23] P. Marini, et al., to be submitted for publication
[24] S. Sorieul, P. Alfaurt, L. Daudin, L. Serani, P. Moretto, Nuclear Instruments and Methods in Physics Research Section B: Beam Interactions with Materials and Atoms **332**, 68 (2014)
[25] N. Škoro, J. of Phys.: Conf. Series **399**, 012017 (2012)
[26] *Vector Fields, OPERA-3d Reference Manual, Vector Fields Limited, England* (2004)

Systematic measurements of prompt fission γ rays – and what they tell us about fission fragment de-excitation

Andreas Oberstedt[1,*], *Angélique* Gatera[2], *Alf* Göök[2], and *Stephan* Oberstedt[2]

[1]Extreme Light Infrastructure – Nuclear Physics (ELI–NP) / Horia Hulubei National Institute for Physics and Nuclear Engineering (IFIN-HH), 077125 Bucharest-Magurele, Romania

[2]European Commission, DG Joint Research Centre, Directorate G - Nuclear Safety and Security, Unit G.2 Standards for Nuclear Safety, Security and Safeguards, 2440 Geel, Belgium

Abstract. In a recent measurement of prompt γ rays from the spontaneous fission of ^{252}Cf an experimental setup was chosen that allowed the study of angular correlations between γ rays and the nuclei from which they were emitted. The observed angular distribution of prompt fission γ rays was then analyzed to estimate the relative contributions from radiation of different multipolarity. Preliminary results are presented and compared to previous angular correlation measurements from different fission processes and recent results from calculations with the Monte Carlo Hauser-Feshbach code FIFRELIN.

1 Introduction

The energy release in nuclear fission is distributed in kinetic and excitation energy of the two fragments. The latter manifests itself in fragment deformation and intrinsic excitation energy. In an early and quite simplified picture, it was assumed that the fragments are first de-excited by the emission of neutrons until the fragments' remaining excitation energy is lower than the neutron binding energy, i.e. on average at about half of it. Only then de-excitation would continue by the emission of γ rays, eventually reaching the ground state [1]. These γ rays may be divided into two categories, commonly called statistical and discrete γ rays, respectively. The first ones originate mainly from electric dipole transitions and carry away most of the remaining excitation energy of the fragments, while the latter stem mainly from electric quadrupole transitions along the Yrast line and carry away most of the angular momentum of the fragments [2]. This so-called prompt γ-ray emission is a fast process, which basically takes place within the first few nanoseconds after scission [3] and is followed by β decay(s) and thereafter by delayed emission of neutrons and γ rays towards the valley of β stability. The average prompt fission γ-ray multiplicity, i.e. the average number of photons per fission, amounts typically to eight [4], as corroborated by our previous measurements [5–10]. An overview on our past and future activities within this field of research is given in Ref. [11].

In this work we have chosen an experimental setup that allowed measuring angular correlations between fission fragments and prompt γ rays, in order to study the de-excitation of the

*e-mail: andreas.oberstedt@eli-np.ro

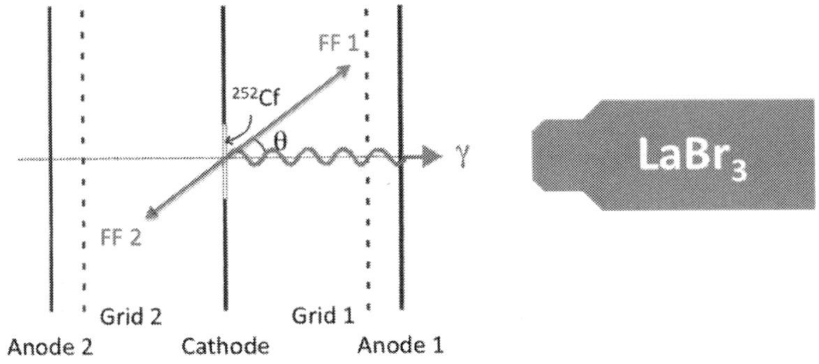

Figure 1. Sketch of the experimental setup used to measure angular correlations between fission fragments (FF 1 or FF 2) and prompt fission γ rays. The fragments were detected with a cylindrical twin Frisch-grid ionization chamber (FGIC) on whose central cathode a ^{252}Cf sample was mounted. For the detection of the γ rays a $LaBr_3$:Ce detector of size 76 mm × 76 mm was employed that was placed along the symmetry axis perpendicular to the sample. The polar angle of the emitted fragments equivalent to the angle relative to the detected photons is denoted by θ.

fragments by prompt γ-ray emission and to deduce information about the relative contributions from radiation of different multipolarity. Below we compare our observations to those from previous angular correlation measurements and give first preliminary results, which are discussed in terms of results from recent calculations with the Monte Carlo Hauser-Feshbach code FIFRELIN [12].

2 Experimental setup and data treatment

The recent measurement of PFGS from ^{252}Cf(sf) was performed with a 76 mm × 76 mm (diameter × length) $LaBr_3$:Ce scintillation detector that was placed perpendicular to the plane of the sample along its symmetry axis. The ^{252}Cf sample was mounted inside a cylindrical twin Frisch-grid ionization chamber (FGIC) [13] in the center of a common cathode. Fission fragments from a thin ^{252}Cf sample are emitted to both sides of the cathode, ionizing the P-10 counting gas, and being detected by means of the electrons moving towards the respective anodes. A Frisch grid is placed in front of each anode in order to shield the latter from the influence of the ions, but it also allows determining the (polar) emission angle of the fragments (see Fig. 1 for a schematic view of the experimental setup). The FGIC provides the fission trigger for the coincident measurement of γ rays, for which energy and time-of-flight (TOF) between sample and detector are recorded. The latter information is used to distinguish prompt γ rays from other photons, emitted e.g. in prompt fission neutron induced reactions, i.e. mainly inelastic neutron scattering. The experimental techniques applied here are in principle the same as described in Refs. [5–10], where more details may be found.

The collected prompt fission γ rays were sorted into energy spectra for $\cos\theta$ bins between 0 and 1 in steps of 0.05. In order to obtain emission spectra, the response function of the detector has to be determined, which is usually done by means of Monte Carlo simulations of the actual setup with computer codes like e.g. Penelope [14] or Geant4 [15]. Thereafter the response function must be unfolded from the raw spectra. However, as a first step, we have chosen here a different, less time-consuming approach. Adding up all spectra gives an angular integrated raw spectrum that may be compared with others, obtained previously for the same fissioning system. Since they all turned out to exhibit a very similar appearance, we may

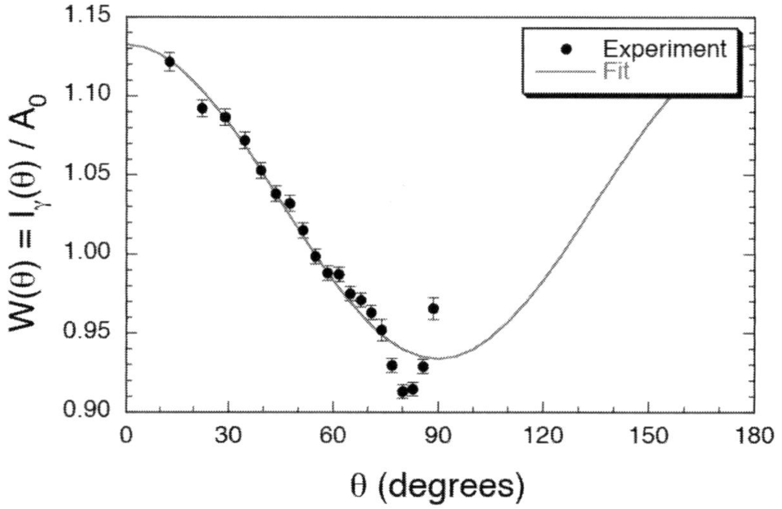

Figure 2. Measured angular distribution of prompt γ rays from ^{252}Cf(sf), depicted as black dots. The error bars correspond to statistical uncertainties only. The line represents the result of a fit according to Eq. 1. The obtained coefficients are $\{A_2/A_0\} = 0.13 \pm 0.03$ and $\{A_4/A_0\} = -0.0092 \pm 0.0086$, with $A_0 = 0.930 \pm 0.005$.

benefit from the fact that similar raw spectra lead to similar emission spectra, provided that detector and setup are comparable [7]. This is the case here. From a previous measurement [5], for which the proper unfolding was carried out, both measured and emission spectra are known, whose ratio gives a transformation function (see Ref. [7] for details). Multiplying this function with the recently measured PFGS gives then the emission spectrum. In the same way angular-dependent emission spectra were created for each bin, from which PFGS characteristics were determined. A detailed inspection of the obtained angular distribution is performed below.

3 Angular distribution of prompt fission γ rays

As described in the previous chapter, emission spectra of prompt fission γ rays were created for $cos\,\theta$ bins, covering the polar angle range $0° < \theta < 90°$. The integrated total multiplicity, i.e. the average number of emitted photons per fission, was determined to $\overline{M}_\gamma = 8.28 \pm 0.51$, where the statistical uncertainty is 0.07 and the systematic one, mainly due to the transformation function, amounts to 0.44. This result is in good agreement with our last published value $\overline{M}_\gamma = 8.29 \pm 0.13$ [6]. The angular distribution of radiation may be expressed according to

$$I_\gamma(\theta) = A_0[1 + \{A_2/A_0\}P_2(cos\,\theta) + \{A_4/A_0\}P_4(cos\,\theta)] \quad , \qquad (1)$$

where $P_k(cos\,\theta)$ denote Legendre polynomials for k = 2, 4 and θ is the emission angle of the fission fragment relative to the γ ray direction, i.e. perpendicular to the cathode plane of the FGIC. Any possible attenuation coefficients are assumed to be 1. The coefficients $\{A_k/A_0\}$ are determined experimentally and may be compared to theory for different types of radiation. For instance, $\{A_2/A_0\} \approx -0.3$ for pure dipole radiation and $\{A_2/A_0\} \approx +0.3$ for pure quadrupole radiation, while $\{A_4/A_0\}$ is close to zero [16]. Figure 2 shows the measured angular distribution as black dots, together with the result of a fit according to the expression given

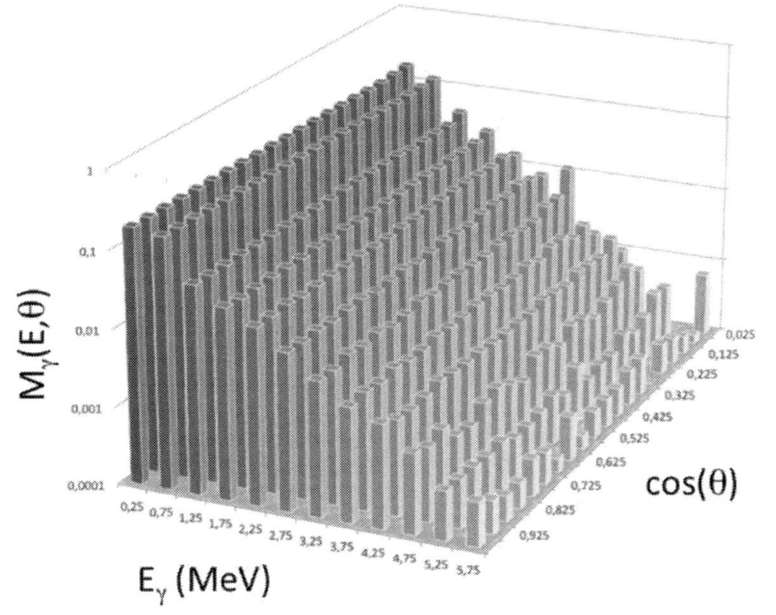

Figure 3. Three-dimensional presentation of measured PFGS from ^{252}Cf(sf) for angular bins of $\Delta\cos\theta$ = 0.05 and energy bins of ΔE = 500 keV.

in Eq. 1 shown as line. It should be noted that the data close to both $\theta = 0°$ and 90° suffers from considerable instrumental errors due to absorption and scattering, but these points were taken into account anyway, since they did not influence much the fit result. The obtained coefficient is $\{A_2/A_0\} = 0.13 \pm 0.03$, which may be considered the result of a superposition of dipole and quadrupole photons (higher multipolarities are less probable and neglected here). Accordingly, it follows that $(+0.3) \times p + (-0.3) \times (1 - p) = 0.13$, with p denoting the probability for quadrupole radiation. This leads to $p = 0.72$, corresponding to 72% quadrupole (L=2) and 28% dipole (L=1) radiation.

In the next step we tried to assess how the ratio between quadrupole and dipole radiation depends on the energy of the prompt fission γ rays. For that reason the PFGS was rebinned into 500 keV intervals (see Fig. 3), for which the corresponding angular distributions were fitted according to Eq. 1 as described above for the entire spectrum. As a consequence, the coefficient $\{A_2/A_0\}$ was determined for each energy bin and the contributions for L=1 and L=2 deduced. The resulting multipolarity-dependent spectra are shown on the left-hand side of Fig. 4. These results are discussed in the next section.

4 Discussion and outlook

Prior to this work, angular distributions of prompt γ rays were measured in thermal-neutron induced fission of the target nuclei ^{233}U, ^{235}U and ^{239}Pu [17], as well as the spontaneous fission of ^{252}Cf [18], however the latter one only for γ rays between 0.1 and 1.5 MeV. A comparison of the coefficients in Eq. 1 from this work and from Ref. [18] for the region $E_\gamma = 0.1 - 1.5$ MeV exhibits some difference, since we obtain $\{A_2/A_0\} = 0.134$ and $\{A_4/A_0\} = 0.002$, while corresponding values of 0.0683(1) and -0.0308(2), respectively, are reported in Ref. [18]. Anyway, a qualitative comparison of all these measurements exhibits a good agree-

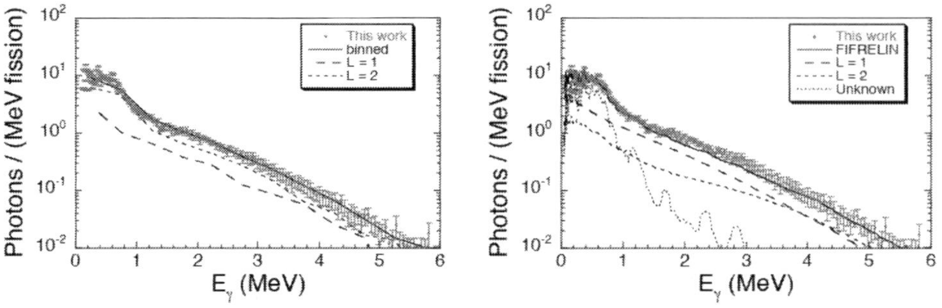

Figure 4. Measured PFGS for ^{252}Cf(sf) from this work, depicted by red dots. In the left part it is compared to the rebinned spectrum of 500 keV intervals (full black line), from which the contributions for dipole (L=1) and quadrupole (L=2) radiation were determined (long and short dashed lines, respectively). The right part shows in addition to the measured spectrum a spectrum and its contributions for L = 1 and L = 2 as well as a contribution for which the multipolarity is unknown (all three as black lines), calculated recently with FIFRELIN (see text for details).

ment, indicating a dominating contribution of quadrupole radiation, of which E2 is much more likely than M2. If this E2 radiation is assigned to γ decays along the Yrast line (cf. Sect. 1), the corresponding multipolarity-dependent multiplicity would be $\overline{M}_{\gamma,L=2} \approx 6.0$, with a total average multiplicity in excellent agreement with $\overline{M}_{\gamma} = 8.29 \pm 0.13$ [6]. Each quadrupole photon will then carry away $2\hbar$, which implies an average angular momentum of $J \approx 12\hbar$, which is in rather good agreement with $J \approx 10\hbar$ reported earlier [19] and results from GEF calculations for different fission fragments [20]. It has to be mentioned though that also lower fission fragment spins of $6\hbar$ to $8\hbar$ have been reported (see [18] and references therein). Hence, further experimental studies are necessary. The remaining dipole multiplicity would then be $\overline{M}_{\gamma,L=1} \approx 2.3$ for statistical γ rays.

Calculations based on the Monte Carlo Hauser-Feshbach code FIFRELIN [12] have previously been rather successful in reproducing experimental results for PFGS characteristics as well as the distinct peak structure of the low-energy part of the energy spectra for ^{235}U(n$_{th}$, f) [8] and ^{252}Cf(sf) [6]. For comparison, the calculated average PFGS multiplicity has been adjusted to the measured one. Even multipolarity-dependent spectra were generated [21], which however seem to show a dominating contribution of L=1 radiation. Apparently, this is in contrast with the experimental results presented in this work. New FIFRELIN calculations were performed recently [22], whose results are shown on the right-hand side of Fig. 4. However, it must be noted here that, apart from γ rays of known multipolarity 1 and 2, there is a considerable amount of photons, whose multipolarity is not assigned (denoted with "unknown" in the figure), since they correspond to transitions between levels taken from the RIPL-3 database [23]. In fact, they constitute about 44% of the calculated total average PFGS multiplicity, as shown in Table 1. Since a dominating contribution of quadrupole radiation has been observed in all PFGS measurements, as referred to above, we tend to conclude that most of these transitions of unknown multipolarity should lead to quadrupole radiation. Therefore, the sum of these "unknown" transitions and the calculated ones with L=2 should be compared to the experimentally obtained contribution for L=2. With this condition comparison of the values given Table 1 shows indeed a rather good agreement.

This agreement would have been even better, if a probably more realistic assumption for the coefficient for pure dipole radiation, $\{A_2/A_0\} \approx -0.2$, had been made, considering the value $\{A_2/A_0\} = -0.212(26)$ from the example given in Ref. [16]. With this value, $\overline{M}_{\gamma,L=1} \approx$

Table 1. Prompt γ-ray characteristics for the spontaneous fission of ^{252}Cf. Experimental results from this work for the average γ-ray multiplicity $\overline{M}_{\gamma}$, the average energy per photon ϵ_{γ}, and the total energy $E_{\gamma,tot}$ released in fission, are given for the integral spectrum as well as for the multipolarity-dependent spectra and compared to corresponding values recently calculated with FIFRELIN [22] (see text for details). Relative multiplicities are also given, under the assumption that higher multipolarities may be neglected.

	Multipolarity L	Exp. transitions (this work)		Calculations (FIFRELIN)	
$\overline{M}_{\gamma}$	all	8.28 ± 0.51		8.28	(adjusted)
$\overline{M}_{\gamma}$	1	2.31	(28%)	3.20	(39%)
$\overline{M}_{\gamma}$	2	5.97	(72%)	1.45	(17%)
$\overline{M}_{\gamma}$	experimental			3.63	(44%)
ϵ_{γ}	all	0.79 ± 0.10 MeV		0.76 MeV	
ϵ_{γ}	1	0.86 MeV		0.94 MeV	
ϵ_{γ}	2	0.76 MeV		1.03 MeV	
ϵ_{γ}	experimental			0.50 MeV	
$E_{\gamma,tot}$	all	6.51 ± 0.76 MeV		6.30 MeV	
$E_{\gamma,tot}$	1	1.99 MeV		3.00 MeV	
$E_{\gamma,tot}$	2	4.52 MeV		1.49 MeV	
$E_{\gamma,tot}$	experimental			1.81 MeV	

2.8 and $\overline{M}_{\gamma,L=2} \approx 5.5$ would have been obtained, corresponding to relative contributions of 34% and 66%, respectively. Of course, these results are still considered preliminary and must be compared with other related information on fission fragments, such as average neutron separation energies and average moments of inertia, in order to verify the deduced energies carried away by dipole and quadrupole photons during the de-excitation process of fission fragments. Hence, these activities are to be continued.

Acknowledgments

One of the authors (A. O.) acknowledges the support from the Extreme Light Infrastructure Nuclear Physics (ELI-NP) Phase II, a project co-financed by the Romanian Government and the European Union through the European Regional Development Fund - the Competitiveness Operational Programme (1/07.07.2016, COP, ID 1334), with which this work had been finalized.

References

[1] C. Wagemans (editor), The Nuclear Fission Process, CRC Press Boca Raton (1991).

[2] A. Hotzel, P. Thirolf, Ch. Ender, D. Schwalm, M. Mutterer, P. Singer, M. Klemens, J. P. Theobald, M. Hesse, F. Gönnenwein, H. v. d. Ploeg, Z. Phys. A **356**, 299 (1996).

[3] P. Talou, T. Kawano, I. Stetcu, J. P. Lestone, E. McKigney, and M. B. Chadwick, Phys. Rev. **C94**, 064613 (2016).

[4] R. Vandenbosch and J.R. Huizenga, Nuclear Fission, Academic Press New York and London (1973).

[5] R. Billnert, F.-J. Hambsch, A. Oberstedt, and S. Oberstedt, Phys. Rev. **C87**, 024601 (2013).

[6] A. Oberstedt, R. Billnert, F.-J. Hambsch, and S. Oberstedt, Phys. Rev. **C92**, 014618 (2015).

[7] S. Oberstedt, A. Oberstedt, A. Gatera, A. Göök, F.-J. Hambsch, A. Moens, G. Sibbens, D. Vanleeuw, and M. Vidali, Phys. Rev. **C93**, 054603 (2016).

[8] A. Oberstedt, T. Belgya, R. Billnert, R. Borcea, T. Bryś, W. Geerts, A. Göök, F.-J. Hambsch, Z. Kis, T. Martinez, S. Oberstedt, L. Szentmiklosi, K. Takács, and M. Vidali, Phys. Rev. **C87**, 051602(R) (2013).

[9] S. Oberstedt, R. Billnert, T. Belgya, T. Bryś, W. Geerts, C. Guerrero, F.-J. Hambsch, Z. Kis, A. Moens, A. Oberstedt, G. Sibbens, L. Szentmiklosi, D. Vanleeuw, and M. Vidali, Phys. Rev. **C90**, 024618 (2014).

[10] A. Gatera, T. Belgya, A. Göök, F.-J. Hambsch, M. Lebois, A. Oberstedt, S. Oberstedt, L. Qi, L. Szentmiklósi, G. Sibbens, and M. Vidali, Phys. Rev. **C95**, 064609 (2017).

[11] S. Oberstedt, R. Billnert, F.-J. Hambsch, M. Lebois, A. Oberstedt, and J.N. Wilson, Eur. Phys. J. A **51** (2015) 178.

[12] O. Litaize and O. Serot, Phys. Rev. **C82**, 054616 (2010).

[13] C. Budtz-Jørgensen, H.-H. Knitter, Ch. Straede, F.-J. Hambsch and R. Vogt, Nucl. Instr. and Meth. A **258**, 209 (1987).

[14] http://www.oecd-nea.org/tools/abstract/detail/nea-1525.

[15] S. Agostinelli and the GEANT4 collaboration, Nucl. Inst. Meth. A **506** (2003) 250.

[16] E.S. Paul, UK Nuclear Physics Summer School, Bristol, 27. August – 6. September 2013, http://ns.ph.liv.ac.uk/ ajb/summerschool/files/ESP-Lecture4.pdf

[17] M.M. Hoffman, Phys. Rev. **133**, B714 (1964).

[18] Yu. N. Kopach, P. Singer, M. Mutterer, M. Klemens, A. Hotzel, D. Schwalm, P. Thirolf, M. Hesse, and F. Gönnenwein, Phys. Rev. Lett. **82**, 303 (1999).

[19] A. Bogachev, L. Krupa, O. Dorvaux, E. Kozulin, M. Itkis, M.-G. Porquet, A. Astier, D. Curien, I. Deloncle, G. Duchene, B.J.P. Gall, F. Hanappe, F. Khalfallah, M. Rousseau, L. Stuttge, N. Redon, and O. Stezowski, Eur. Phys. J. A **34**, 23 (2007).

[20] K.-H. Schmidt, B. Jurado, C. Amouroux and C. Schmitt, Nuclear Data Sheets **131**, 207 (2016).

[21] D. Regnier, "Contribution à l'étude des gammas prompts de fission", PhD thesis, Université de Grenoble, France (2013).

[22] A. Chebboubi, private communication (2017).

[23] R. Capote, M. Herman, P. Oblozinsky, P. Young, S. Goriely, T. Belgya, A. Ignatyuk, A. Koning, S. Hilaire, V. Plujko, M. Avrigeanu, O. Bersillon, M. Chadwick, T. Fukahori, Z. Ge, Y. Han, S. Kailas, J. Kopecky, V. Maslov, G. Reffo, M. Sin, E. Soukhovitskii, P. Talou, RIPL - Reference Input Parameter Library for Calculation of Nuclear Reactions and Nuclear Data Evaluations, Nucl. Data Sheets **110** (12) (2009) 3107.

Performance validation of the first arm of FALSTAFF: ^{252}Cf and ^{235}U fission fragment characterisation

D. Doré[1], E. Berthoumieux[1], Q. Deshayes[1], L. Thulliez[1], P. Legou[1], M. Combet[1], M. Kebbiri[1], A. Marcel[1], J-Ph. Mols[1], M.O. Frégeau[2], S. Herlant[2], X. Ledoux[2], J. Pancin[2)], S. Oberstedt[3]

[1] Irfu, CEA, Université Paris-Saclay, Gif-sur-Yvette, F-91191, France
[2] GANIL, F-14050 Caen, France
[3] European Commission, Joint Research Centre, Directorate G Nuclear Safety and Security, Unit G.2, Retieseweg 111, 2440 Geel, Belgium

Abstract. The renewed interest for the study of nuclear fission is mainly motivated by the development of GEN-IV reactor concepts, mostly foreseen to operate in the fast neutron energy domain. To support this development, new high-quality nuclear data are needed. In this context, a new experimental setup, the FALSTAFF spectrometer, dedicated to the study of nuclear fission is under development. Employing the double-velocity (2V) and energy-velocity (EV) methods, the fission fragment mass before and after neutron evaporation will be deduced and the correlation between prompt neutron multiplicity and fragment mass will be determined. The first arm of the spectrometer is achieved. It is composed of two SED-MWPC detectors (a combination of a foil to produce secondary electrons and a Multi-Wire Proportional Chamber to detect them) and an axial ionization chamber. The SED-MWPC give access to the velocity (V) via time-of-flight and position measurements. The ionization chamber measures the fragment kinetic energy (E) and the energy loss profile. Preliminary results for spontaneous fission of ^{252}Cf and from the thermal-neutron induced fission experiment on ^{235}U, performed at the Orphée reactor (CEA-Saclay, France), are presented.

1 Introduction

Nowadays the fission process still presents a great interest from both theoretical and experimental point of view. New developments on microscopic calculations and the future generation of nuclear reactors are among the main motivations for new experimental programs devoted to the study of nuclear fission. The FALSTAFF project [1-2] aims at providing highly constrained data to improve significantly the description of the fission process. The goal is to determine the neutron multiplicity as a function of the fragment characteristics (pre-neutron mass and kinetic energy) in neutron-induced fission of specific actinides in the MeV range. This experiment is particularly rich as it allows accessing numerous aspects of fission, e.g. the deformation at scission, the influence of single-particle structure in neutron-rich isotopes, the dissipated energy and the energy partition between the two fragments. In order to maintain a fair detection efficiency, neutrons are not measured, but their multiplicity is deduced in average from the measured fragment masses before and after their evaporation.

Based on time-of-flight and residual energy technique, the FALSTAFF setup will consist of two arms allowing the simultaneous measurement of the complementary fragment velocities and energies. The combined measurements of velocities and energies provide information on the mass of the fragments before (pre-) and after (post-) the neutron evaporation. The 2V method (measurement of both fragment velocities in coincidence) gives access to the pre-neutron mass. This method is meaningful only when the mass of the fissioning system is known, namely for incident neutron energies below the threshold of the second chance fission channel opening ($E_n < 5.5$ MeV). The limitation of this technique is linked to the intrinsic velocity dispersion brought by neutron evaporation, which degrades the mass resolution. The post-neutron mass is obtained with the EV (energy-velocity) method. The average prompt neutron multiplicity as a function of the fragment mass distribution may then be deduced from the event-wise difference between pre- and post-neutron masses.

The experimental setup will be described in section 2. Results for the spontaneous fission of ^{252}Cf and the thermal neutron-induced fission of ^{235}U will be presented in section 3 and 4 respectively.

2 Experimental setup

A sketch of the FALSTAFF setup is shown in Fig. 1. Each arm is composed of two ToF detectors (SeD Start and SeD Stop) and an axial ionization chamber. The necessary timing and position resolutions for the ToF detectors are obtained by using a combination of an emissive foil and a MWPC detector. Ionization chambers will be placed behind the stop detectors to measure the fragment residual energies. In addition, the light-fragment nuclear charge could be estimated thanks to the fragment energy-loss profile in the ionization chamber.

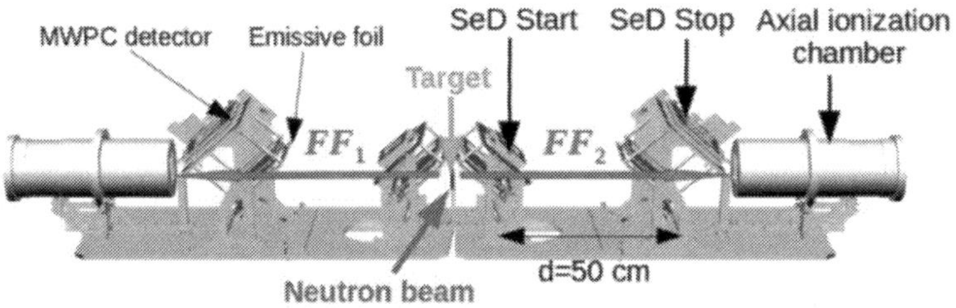

Fig. 1. Sketch of the FALSTAFF spectrometer (see text for details).

2.1 The time-of-flight (ToF) detectors

For the ToF detectors [3-5] an aluminized Mylar foil of 0.5 μm thickness, inclined at 45 deg with respect to the horizontal direction and polarized at -10 kV, is placed along the path of the fission fragments. A grounded grid placed at 1 cm from the emissive foil generates a high electric field close to the foil. The secondary electrons produced by the ion crossing the Mylar foil are then accelerated to 10 keV and fly towards the MWPC detector. The accelerated electrons pass through the MWPC entrance foil (made from 0.9 μm aluminized Mylar) with 70% efficiency [5]. They create ionization electrons inside the gas which are then amplified while drifting towards the cathode. The amplification takes place over the whole drift path due to the low pressure of the gas (0.6-0.8 kPa). The time signal is then read from the wire plane by a fast amplifier. The spatial information is taken from the charge induced on a 2D pixelized cathode for the stop detector while the start detector is made of three perpendicular wire planes from which two planes are used to obtain the position.

Timing and spatial resolutions of each ToF detectors were measured with a dedicated setup. Values of 100-120 ps and 1-2 mm were obtained for the time and spatial resolutions respectively.

2.2 Axial ionization chamber

The choice of an axial field for the ionization chamber was motivated by the possibility to reconstruct the energy loss profile by detecting the number of electrons as a function of the time of arrival on the anode. Moreover, the dead zones are minimized for this type of ionization chamber.

The cylindrical axial ionization chamber [3] has an outside effective diameter of 440 mm and a length of 400 mm. The cathode is the detector entrance window separated from the anode by a distance of 395 mm. A potential of -8 kV is applied to the cathode, while the anode is the readout electrode. The entrance window is a Mylar foil, 0.9 μm thick and with a diameter of 232 mm. To polarize the Mylar foil and to ensure a good electrical conductivity, 50 nm of aluminum are deposited on each side. In order to avoid strong mechanical tension, the entrance window is supported by a grid of 12 nylon wires, 6 vertical and 6 horizontal. These wires have a diameter of 350 μm. The electric field uniformity is assured by 38 aluminum rings, 270 mm in diameter, spaced out by 10 mm and interconnected by a 10 MOhm resistance allowing to maintain a uniform field along the chamber.

A Frisch grid is installed in front of the anode to ensure the shielding from the influence of the charges moving in the ionization region. The main part of the electronic signal on the anode is induced by electrons after they have crossed the grid. The Frisch grid is made up of 68 tungsten wires, 170 μm in diameter and spaced out by 4 mm. The ionizing gas is isobutane at a pressure between 2.7 and 4.0 kPa flowing at 0.5 Ln/ h.

The energy resolution measured with an alpha source (^{241}Am) was estimated to 3.9 %. A better resolution is expected for fragments since the signal amplitude is larger.

2.2 Acquisition systems

Two independent acquisition systems (DAQ) are used to collect the data:

1- the cathode pad/strips signals from the start and stop detectors are digitized using the GET acquisition system [6] on a μTCA crate,

2- all others signals (timing information from start and stop detectors, anode and grid signal from the axial ionization chamber) are recorded by using the standard GANIL acquisition based on a VME crate.

For the start and stop detectors the anode signals are digitized by using a Matacq card [7] having a 2 GHz sampling rate. The anode and grid signals from the ionization chamber are digitized by using a CAEN V1724 card having 100 MHz sampling rate.

3 Results for the spontaneous fission of ^{252}Cf

A ^{252}Cf source was placed behind a collimator at 10 cm in front of the emissive foil of the start detector. The trigger was constructed from the coincidence of start and stop detector signals. The time distribution is calibrated by using an alpha source. Geant4 simulations were used to calibrate the residual fragment energy distribution.

In Fig. 2 the experimental velocity, energy and mass distributions are shown in red. They are compared to simulated distributions. These simulations were performed with GEANT4. Two different event generators were used: FIFRELIN [8] in green dotted line and GEF [9] in blue dashed line. The measured resolutions for the position and the time (σ(t)= 120 ps; σ(pos) = 2 mm) and the expected one for the energy (dE/E= 1%) are taken into account in the simulations. It has previously been shown [1-3] that these resolutions allow to reproduce the correlation between neutron multiplicity and pre-evaporation fragment mass well enough to study the evolution of the correlation as a function of the excitation energy.

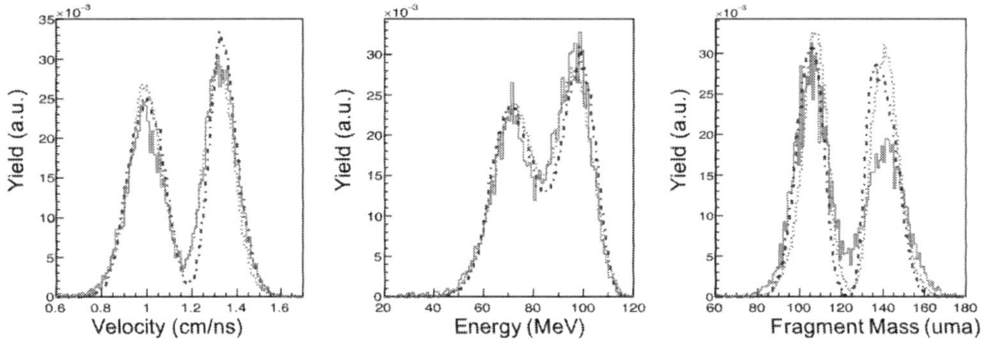

Fig. 2. Velocity (left panel), energy (middle panel) and post-evaporation fragment mass (right panel) distributions for FALSTAFF (red curve) are compared to GEANT4 simulations with two different event generators: FIFRELIN (green dotted curve) and GEF (blue dashedcurve).

The simulated velocity and energy distributions show a rather good agreement with experimental distributions. One observes that the main differences for the fragment mass distribution are the width of the experimental distribution for heavy fragments and the lack of heavy fragments compared to light ones. The discrepancy for the width could be due to a difference in simulated and experimental resolutions although this is neither visible in the velocity nor in the energy distributions. The lack of heavy fragments is not visible for all data taking. This fact comes from the signal-to-noise ratio which varied from one run to another. The signals from the heavy fragments being small, some good events were rejected due to the signal being below the trigger threshold. A comparison (Fig. 3) with the experimental mass distributions from SPIDER [10] and VERDI [11], which are 2V-2E spectrometers, leads to same conclusion and is depicted in Fig. 3.

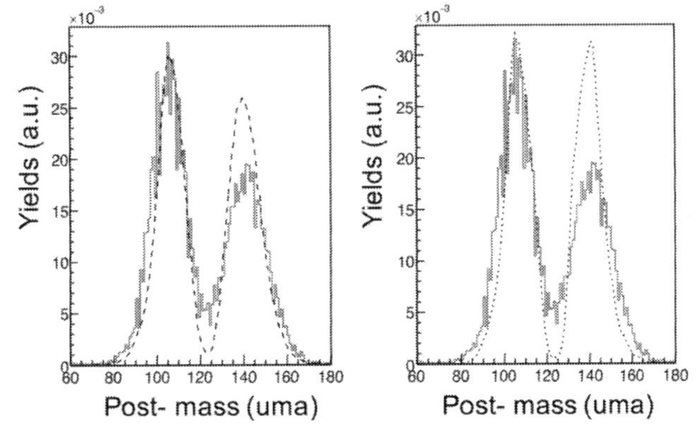

Fig. 3. ^{252}Cf FALSTAFF fragment mass distribution (red curve) is compared to VERDI (left panel) and SPIDER (right panel) results.

4 Results from the thermal neutron-induced fission of ^{235}U

An experiment was performed in 2018 at the Orphée reactor to test one arm of the FALSTAFF spectrometer in real conditions, by measuring the post-neutron emission mass distribution of ^{235}U thermal neutron induced fission. A neutron flux of 7.4E+06 n/cm^2/s was measured at the neutron beam entrance window of FALSTAFF. The experiment was performed with a ^{235}U target of 6.4 µg/cm2 and a restricted FALSTAFF geometrical efficiency of 0.4 %. Velocity and corrected energy distributions obtained for this experiment are presented in Fig. 4. Simulated distributions are also shown.

The velocity distribution is slightly distorted for the light fragment peak compared to simulations. This was also observed for the ^{252}Cf(sf) and has to be investigated. In the distributions, the loss of heavy fragment is seen again. This problem could be solved by amplifying the signal earlier in the electronic chain.

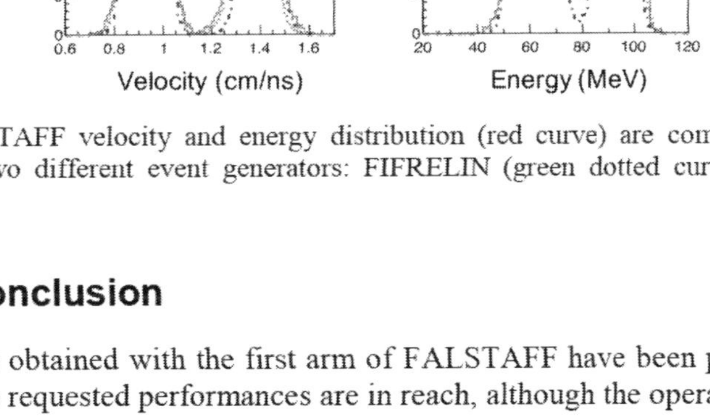

Fig. 4. ^{235}U FALSTAFF velocity and energy distribution (red curve) are compared to GEANT4 simulations with two different event generators: FIFRELIN (green dotted curve) and GEF (blue dashed curve).

5 Outlook/Conclusion

The recent results obtained with the first arm of FALSTAFF have been presented. For the ToF detectors, the requested performances are in reach, although the operation turned out to be unstable. For the ionization chamber, the electronic chain and the Frisch grid have to be improved in order to reach an energy resolution of 1%.

The operation of the spectrometer in experimental conditions using a neutron beam showed no additional performance degradation with respect to the source measurement. After an optimization phase, the next step of the project will be the construction of the second arm of the spectrometer to achieve the full 2V-EV measurement of fission fragments.

Acknowledgments

This work was partially supported by CEA, NEEDS, P2IO and within the Seventh Framework Programme through CHANDA (EURATOM contract no. FP7-Fission-2013-605203). The authors thank the team members of the Orphée reactor for their support and O. Litaize et O. Serot for providing results of FIFRELIN.

References

1. D. Doré et al., EPJ Web of Conference **42** (2013) 01001.
2. D. Doré et al., Nuclear Data Sheets **119** (2014) 346-348.
3. L. Thulliez, Ph.D. Thesis, Université Paris-Saclay, 25 Sept. 2017.
4. L. Thulliez et al., EPJ Web of Conferences **146** 04028 (2017).
5. J. Pancin et al. J. Instrum., **4** (2009) P12012.
6. E. Pollacco et al., NIM A **887** (2018) 81-93.
7. D. Breton, E. Delagnes, IEEE Nuclear Science Symposium, Medical Imaging Conference, and 15th International Room Temperature Semiconductor Detector Workshop, San Diego, United States, Oct. 2006, IEEE, 2, pp.990-995, 2006.
8. K.H. Schmidt et al., Technical report, JEFF Report 24, 2014.
9. A. Litaize et al., Phys. Rev. C **82** (2010) 054616.
10. K. Meierbachtol et al., NIM A**788** (2015) 59-66.
11. M.O Frégeau et al., NIM A**817** (2016) 35-41.
12. K. Jansson et al., Eur. Phys. J. A **54** (2018) 114.

Prompt Fission Neutron Investigation in $^{235}U(n_{th},f)$ and $^{252}Cf(sf)$ Reactions

Shakir Zeynalov, *Pavel* Sedyshev, *Valery* Shvetsov and *Olga* Sidorova

Joint Institute for Nuclear Research, 141980 Dubna Moscow region, Russia

Abstract. The prompt neutron emission in thermal neutron induced fission of ^{235}U and spontaneous fission of ^{252}Cf was investigated by using digital signal electronics. The goal was to check a new revised data analysis software with fission fragment (FF) kinetic energy corrections after prompt fission neutron (PFN) emission. The revised software was used to reanalyze old data measured in EC-JRC-IRMM, where $^{252}Cf(sf)$ reaction was investigated. Both measurements were done using similar twin Frisch grid ionization chamber for fission fragment detection with equivalent NE213 fast neutron detector. About $0.5*10^6$ FF with PFN coincidences have been analyzed in both measurements. The fission fragment kinetic energy, mass and angular distribution were investigated along with prompt neutron time of flight and pulse shape analysis using a six channel synchronous waveform digitizer (WFD) with sampling frequency of 250 MHz and 12 bit resolution in the $^{235}U(n_{th},f)$ reaction. Similar WFD with sampling frequency of 100 MHz was used for PFN investigation in $^{252}Cf(sf)$ reaction. These two experiments were considered as a reference for further investigations with a new setup composed of position sensitive ionization chamber to detect FF and an array of 32 liquid scintillators recently constructed in Dubna to detect neutrons.

1 Introduction

The nuclear fission is considered as the process of a charged drop evolving under the competition between attractive nuclear and repulsing coulomb forces, leading eventually to the split of the nucleus mainly into two parts of comparable masses. The main part of FF excitation energy is released by the prompt fission neutrons, emitted by FF after full acceleration by coulomb forces. The experimental investigations of various characteristics of PFN emission is needed to understand the nuclear fission dynamics from the scission point down to rupture. One of the interesting observation is the increasing $\bar{\nu}(A)$ from the heavy fragment with increase of the excitation energy of the fissioning system [1] which has still no clear explanation. Therefore the systematic study of correlations between fragments and neutrons characteristics is needed. The studies of PFN emission in fission induced by neutrons from energies extending from resonances up to a few MeV could possibly contribute to better understanding the mechanism of PFN emission from the excited FF. The experiments on sub-barrier fission, induced by thermal neutrons are of particular interest because no measurements were done so far on mass and energy distributions for this systems [2]. In this work we report results of PFN investigation in thermal neutron-induced fission of ^{235}U and spontaneous fission of ^{252}Cf. The main goal of the experiments was the feasibility check of the apparatus and the data analysis procedure.

2 Experimental Setup and FF data analysis

A convenient way to study of PFN emission in neutron-induced fission is to use a conventional twin back-to-back ionization chamber, with two chambers sharing a common cathode such as the one of Budtz-Jorgensen and Knitter [3]. The cathode was made from a thin conductive foil and at the same time served as backing for the fissile deposit. For binary fission events two complementary FF are simultaneously detected in the two independent chambers. Free electrons released by FF deceleration were inducing pulses on the chamber anodes and on the common cathode during drift along the applied externally electric field. The pulse height in each chamber was proportional to the corresponding FF kinetic energy and the pulse shape conveys information on the FF angle (Θ) in respect to the electric field applied in the direction of the normal to the cathode plane. From the correlated energies obtained in the above double-energy (2E) experiment, FF masses and velocities could be found in a similar way than exposed in [4]. If the fissile target is located on the common cathode and the fast neutron detector positioned at the certain distance along the normal to the target, the angle between FF and PFN emission would be equal to Θ. The PFN velocity may be determined from the known flight path and the measured time delay between cathode and neutron detector (ND) pulses. Measured FF and PFN velocity vectors then may be used for PFN emission kinematics. The PFN multiplicity distributions in respect to FF kinetic energy release and mass split may be reconstructed comparing of two sets of FF measurements. In the first experiment fission fragment mass and kinetic energy release should be evaluated from the measurement independent from ND. In the second experiment FF mass and kinetic energy release should be evaluated for the FF coincided with ND. The detailed information on PFN emission in fission is available from the measured dependence of the number $\nu(A,TKE)$ of PFN emitted by the FF with mass number A and TKE release of two fission fragments [3-4]. The late function allowed obtaining the averaged characteristics of $\bar{\nu}(A)$ or $\bar{\nu}(TKE)$ by integrating over respective variable, if the mass yield matrix - $Y(A,TKE)$ is known:

$$\bar{\nu}(A)=\frac{\int_0^{\infty}\nu(A,TKE)Y(A,TKE)dTKE}{\int_0^{\infty}Y(A,TKE)dTKE},\quad \bar{\nu}=\int_0^{\infty}\nu(A,TKE)Y(A,TKE)dTKEdA,\ 200=\int_0^{\infty}Y(A,TKE)dTKEdA \qquad (1)$$

Similar relation could be written for averaging over A:

$$\bar{\nu}(TKE)=\frac{\int_0^{\infty}\nu(A,TKE)Y(A,TKE)dA}{\int_0^{\infty}Y(A,TKE)dA},\quad \bar{\nu}=\int_0^{\infty}\nu(A,TKE)Y(A,TKE)dTKEdA,\ \ 200=\int_0^{\infty}Y(A,TKE)dTKEdA \qquad (2)$$

$\nu(A), \nu(TKE)$ can be easily determined if the distributions of $\nu(A,TKE)$ and $Y(A,TKE)$ are known. The experimental method and data analysis procedure implemented in this work was adopted from Ref. [3], where it was described in detail. For each fission event the FF and PFN kinetic energies, FF masses along with the angle between PFN and FF motion should be determined. All this information can be used to reconstruct the PFN emission kinematics both in the laboratory (LF) and in the centre of mass (CMF) frames. The measurements were carried out using the experimental setup presented in Fig. 1 (on the right). Reaction kinematics is sketched in Fig. 1 (on the left). The experimental setup consisted of the twin back-to-back ionization chamber (TIC), which was designated for the measurements of FF kinetic energy release and the cosine of angle between fission axis and the cathode plane normal. The experimental data was collected using a digital pulse processing (DPP) system, consisting of six synchronous waveform digitizers (12 bit, 250 MS/sec). The FF energies and angles were obtained from the chamber signal waveforms using DPP, realized in form of recursive procedures. The neutron energy was derived from time-of-flight (TOF) calculated as the delay between the

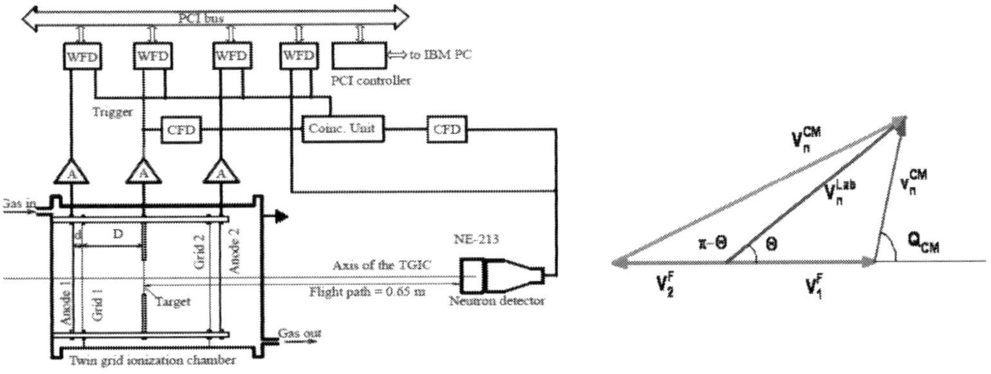

Fig. 1. Experimental setup(left). Vector diagram for PFN emission from FF (right)

cathode and the neutron detector pulses. The measured FF energy release should be corrected for energy losses in the target layer and target backing if the cosine of angle Θ is known. In this work anode current pulse waveform was derived from the output pulse of charge sensitive preamplifier and the pulse "centre of gravity" (T) calculated as described in Ref. [5]. The relation between T and the cos(Θ) was derived in Ref. [6] using Ramo-Shokley theorem and numerically calculated weighting potentials:

$$\cos(\Theta) = (T_{90} - T)/(T_{90} - T_0) \qquad (3),$$

where the used variables have the following meaning: D-is the cathode-anode distance, d-is the grid-anode distance, σ-is the Frisch grid inefficiency value, $T_0 = T(\cos(\Theta) = 1)$, $T_{90} = T(\cos(\Theta) = 0)$. The below formula was used to correct anode signal pulse heights for grid inefficiency:

$$P_A^C = P_A / (1 - \sigma \cdot (1 - \frac{T}{T_{90}}) \cdot (1 + \frac{d}{2D})) \qquad (4)$$

After the emission angle was calculated, the average anode pulse height versus 1/cos(Θ) was plotted for both chambers to find energy loss correction for FF detected in each chamber. The data sets are fitted as a linear function of 1/ cos(Θ), the slopes of which were assigned to energy loss correction in respective chamber. The correction for FF pulse height, caused by momentum transfer to working gas atoms by FF (non ionizing collisions) during its deceleration, is called pulse height defect (PHD). The PHD depends on the FF mass and kinetic energy and was corrected in data analysis using parameterization suggested in Ref. [6]:

$$E = \alpha \cdot PH + \beta + PHD(A_{post}, E_{post}) \qquad (5),$$

where E_{post}, A_{post} - FF kinetic energy and mass respectively after neutron emission. The fitting parameters α and β - are chosen to fit the values of $\overline{TKE}$ and the $<A_H>$ as given in Ref. [4] p. 323.

3 Data analysis procedure

Measurement of PFN time-of-flight in present experiment was done using cathode pulse of TIC as a "T-zero" signal and the ND signal as "Stop" signal. The signals were digitized with 250 MHz sampling rate and stored during experiment for further off-line data analysis.

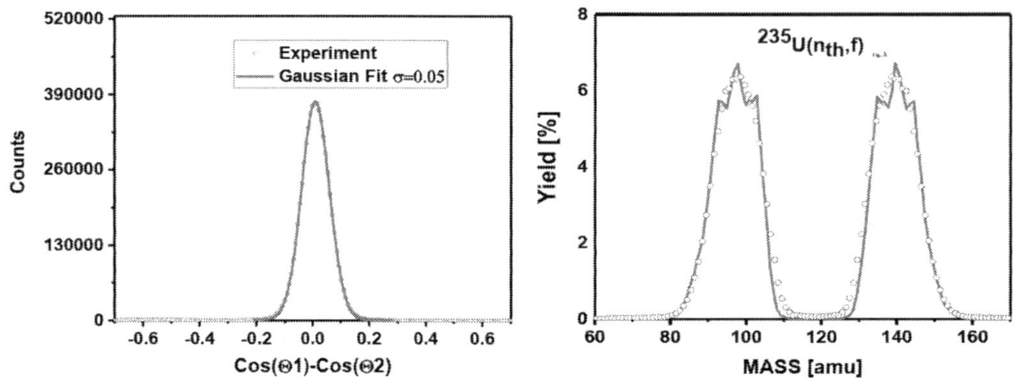

Fig. 2. Demonstration the precision of cos(Θ) measurement in the range 0.5<cos(Θ)<1.0 (on the left). Comparison of measured mass yield (dots) with data (solid line) taken from Ref.[4]. For calculations the successive approximation procedure to obtain pre-PFN emission mass described in Ref. [3] was used (on the right).

Time difference between these two signals was analyzed by implementing standard constant fraction time marking (CFTM) algorithm both to the cathode and to the ND waveforms.
The realization of the algorithm described in Ref. [9] was applied to the cathode waveform. The copy of the original signal is delayed by approximately 400ns of the cathode signal rise time (~1000 ns) and summed with scaled and inverted original signal. The "T-zero" time is assigned to the crossing point of resulting signal with time axis. The crossing point was calculated using parabola interpolation between two successive samples. The time mark for the ND signal was found in the similar way. It should be noted that to achieve the best timing resolution in CFTM realization one should convert sampled waveform to continuous form using Shannon's sampling formula [9]. In this study the PFN kinetic energy spectrum reconstruction was not needed, therefore the resolution ~2.5 ns achieved in this simple implementation was found sufficient for PFN analysis. The neutron multiplicity was estimated by counting the coincidence between cathode pulse and PFN signal of ND. Due to high gamma radiation background both from the target and surrounding materials the PFN counts needed to be separated from the gamma radiation using pulse shape analysis as described in Ref. [9].

The PFN detected by ND was mainly emitted from FF moving towards the ND, but the probability, that it was emitted by FF moving in opposite direction (complementary FF) was not zero and these events were considered as background. The background created by the complementary FF was investigated in Ref. [5] and was slightly modified in our approach. According to the reaction kinematics depicted in Fig. 1, the kinetic energy of the complementary FF in the CMF, must be much higher than the kinetic energy of the FF moving toward the ND. Bearing in mind the exponential drop of the PFN energy spectrum in the CMF, the contribution to the PFN from both FFs could be evaluated using the probabilities defined as:

$$W_x = 1/(1+\exp(E^x_{CM} - E^y_{CM})),$$
$$W_y = \exp(E^x_{CM} - E^y_{CM})/(1+\exp(E^x_{CM} - E^y_{CM})) \qquad (6),$$

where W_x, W_y - are the probabilities of PFN emission with CM kinetic energies $E^{x,y}_{CM}$ of PFN from the forward FF and its complement respectively. A comparison of the mass distributions plotted using measured data and probabilities, defined by Eq. (6) are similar to

Ref. [3], where the background from the complementary fragments for ^{252}Cf(sf) was found to be small.

The angular distribution of PFN emitted in FF CM reference frame was homogeneous, proving that almost all of the PFN are emitted from fully accelerated FFs. The transformation from the LAB to CM reference frame was done using the following formula:

$$\cos(\Theta_{CM}) = (V_n^{Lab} \cdot \cos(\Theta_{Lab}) - V_1^F) / V_n^{CM} \quad (7)$$

4 Results of PFN analysis in ^{235}U(n_{th},f) reaction.

The PFN distribution was evaluated considering neutron emission from the fully accelerated FF, using the reference frame moving along with the FF towards the ND. We used the Jacobean factor and conversion formulae from CM to LAB reference frame as was described in Ref. [10]:

$$\bar{\nu}(A,TKE) == \int_0^\infty \frac{Yc(A,TKE,V_{LAB}) \cdot V_{CM} \cdot (V_{LAB} - V_F \cdot \cos(\Theta))}{\varepsilon(V_{LAB}) \cdot V_{LAB}^2} dV_{LAB} / Y(A,TKE) \quad (8),$$

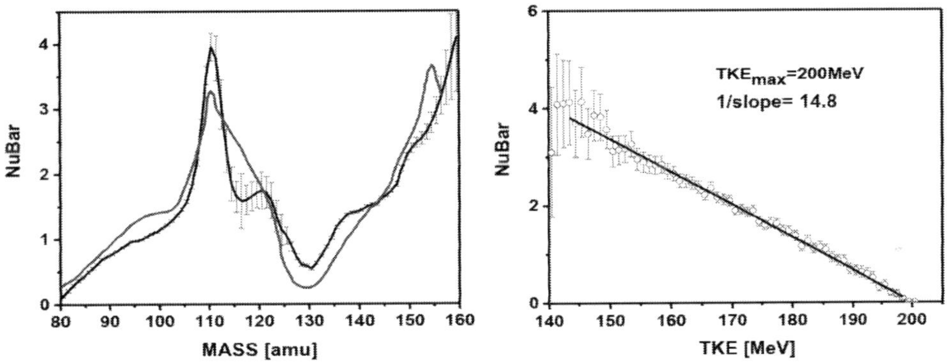

Fig. 3. The average PFN emission distribution mass dependence (black line), measured in ^{235}U(n_{th},f) reaction in comparison with data (blue line) from Ref. [7] (on the left). The average PFN emission distribution dependence on FF TKE, measured in ^{235}U(n_{th},f) reaction with TKE$_{max}$= ~200 MeV (on the right).

where $Y_C(A,TKE,V_{LAB})$ - the number of FF coincidences with ND, $\varepsilon(V_{LAB})$ - is the ND efficiency dependence on PFN velocity in LAB frame, V_{LAB} - the PFN velocity measured in LAB frame, V_F - FF fragment velocity in LAB frame, V_{CM} - is the PFN velocity in CM frame. The distribution $Y(A,TKE)$ was calculated without coincidence with ND as described above, but distribution $Yc(A,TKE,V_{LAB})$ was calculated with FF energy corrected due to FF recoil after PFN emission as described in Ref. [11]. Dependence of average PFN multiplicities on mass and TKE were evaluated using formulae (1,2) and plotted in Fig. 3.

It should be noticed that measurement of $\bar{\nu}(A)$ in ^{235}U(n_{th},f) was more complicated in comparison with ^{252}Cf(sf) due to very high peak to valley ratio in FF mass distribution for ^{235}U. Low statistics in the valley area of mass distribution makes data analysis very sensitive to background. This is why in this report we reanalyzed the data with the same data analysis software that was used in the analysis of the ^{235}U(n_{th},f) reaction. The test of our data analysis procedure in $^{252}Cf(sf)$ reaction was necessary to have confidence in our results for $^{235}U(n_{th},f)$ which more precisely reproduced the mass distribution of the reaction. Our target was made on the polyamide backing with thickness of 35 μg/cm^2 and the sample layer thickness was ~50 μg/cm^2. Experiment was carried out on the IBR2 pulsed reactor with thermal neutron flux ~2*10^6 n/cm^2. The sample diameter was 60 mm.

5 Results of PFN analysis in ^{252}Cf(sf) reaction.

Measurement of PFN in spontaneous fission of ^{252}Cf was done in 2007-2009 at IRMM using an experimental setup similar to the one presented in Fig.1. The ionization chamber was absolutely identical to the chamber used in above reported study. The PFN detectors were the same size but in case of experiment with Cf-target the distance between target and PFN detector

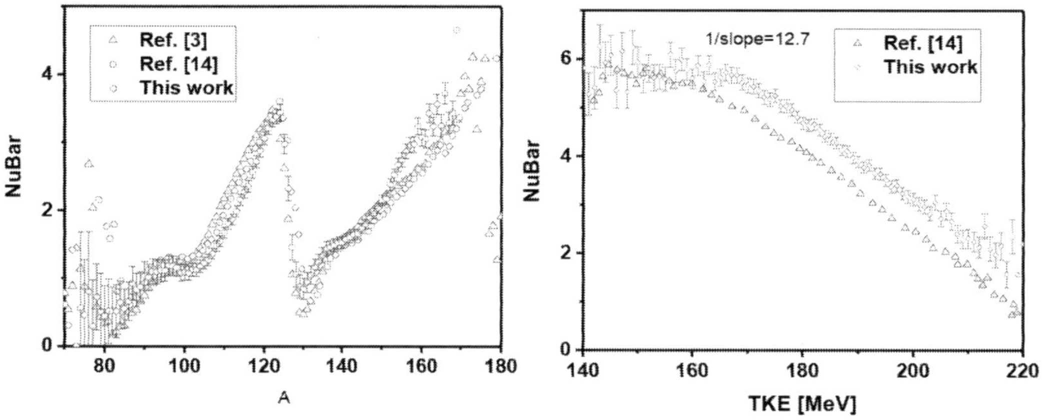

Fig. 4. The average PFN emission distribution mass dependence, measured in $^{252}Cf(sf)$ reaction in comparison with data from Refs. [3],[14 (on the left). The average PFN emission distribution dependence on FF TKE, measured in $^{252}Cf(sf)$ reaction with Q_{max}= ~239 MeV in comparison with data from Ref. [14] (on the right) Q_{max}= ~230.5 MeV.

was 0.725 m. Measurement was done with the data acquisition hardware and the software identical to that used in $^{235}U(n_{th},f)$ reaction investigation (except of WFD sampling frequency, which was 100 MHz). Data analysis was done using the same procedure as described above.

6 Conclusions and outlook

The average PFN emission was evaluated using the same data analysis software in the analysis of $^{235}U(n_{th},f)$ and $^{252}Cf(sf)$ reactions. The average PFN emission dependence on FF mass and TKE are presented in Fig. 3 and Fig. 4 for $^{235}U(n_{th},f)$ and $^{252}Cf(sf)$ respectively. The distributions for $^{252}Cf(sf)$ presented for FF recoil take into account effects of corrections according to Ref. [11]. It should be noted that the FF mass dependence of average neutron number also should be corrected according to formula (11) from Ref. [11], but this correction is very small for both in $^{235}U(n_{th},f)$ and $^{252}Cf(sf)$ reactions.

As it was pointed out in the review paper [12], use of a large neutron detector (with higher PFN detection efficiency), in many aspects is more preferable in PFN investigations. Bearing this in mind we developed new setup for PFN investigation, which would combine the properties of both detectors with high and low efficiency. New experimental setup consists of a position sensitive twin fission chamber with capability to reconstruct the orientation of the fission axis in 3D [13] and to provide sufficient accuracy on the FF kinetic energy measurement. The twin chamber consists of a common cathode, where the fissile target can be located. Two segmented anodes are located at 40 mm distance symmetrically from the both sides of the cathode plane plain. Segmented anodes consist of triangular strips, electrically isolated from each other. Each strip is connected to the nodes of the resistive chain filter. Detailed operational principle of position sensitive ionization chamber, formulae for FF "charge centre-of -gravity" and FF orientation evaluation was provided in Refs. [8,13]. Fast neutron detector consist of 32 type VS-0499-100 scintillation detector modules of diameter 76 mm and 51 mm from SCIONIX HOLLAND BV (see Fig. 5). The setup is planned to be used in experiments in the end of this year at IREN which is a resonance neutron source facility in Dubna.

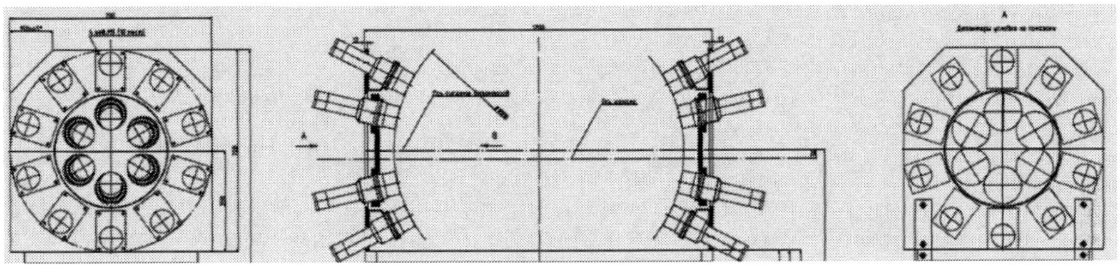

Fig. 5. Fast neutron detector composed of 32 liquid scintillation modules with position sensitive twin ionization chamber in centre (not shown in figure)

References

[1] A.A. Naqvi, F. Kappeler, and F. Dickmann, Phys. Rev. **C34**,218 (1986)
[2] A. Al-Adili, D. Tarrio, F.-J. Hambsch, A. Gook, K. Jansson, A. Solders, V. Rakopoulos, C. Gustavson, M. Lantz, A. Materrs, S. Oberstedt, A.V. Prokofiev, M. Viladi, M. Osterlund, and S. Pomp, *EPJ Web of Conferences* **122**, 01007 (2016)
[3] C. Budtz-Jorgensen and H.-H. Knitter, Nucl. Phys., **A490**, 307 (1988)
[4] C. Wagemans, *The Nuclear Fission Process*, CRC Press, Boca Raton , FL, 1991
[5] S. Zeynalov, O. Zeynalova, F.-J. Hambsch and S. Oberstedt, Bull. Russ. Acad. Sci.: Phys. **73**,506 (2009)
[6] Zeynalov, S., Hambsch, F.-J., Oberstedt, S., 2011. Jour. Korean Phys. Soc. **59**, 1396
[7] V.F. Apalin, Yu. N. Gtitsuk, I.E. Kutikov, V.I. Lebedev, and L.A. Mikaelyan, Nucl. Phys., **55**, 249 (1964).
[8] O. Zeynalova, Sh. Zeynalov, M. Nazarenko, F.-J. Hambsch, and S. Oberstedt, *AIP Conf. Proc.* **1404**, 325(2011)
[9] G. Knoll, *Radiation Detection and Measurement*, John Willey & Sons, Inc, Third edition, 2001
[10] H.R. Bowman, J.C.D. Milton, S.G. Thompson, and W.J. Swiatecki, Phys. Rev. **129** (1963) 2133
[11] A. Gavron, Nucl. Instrum. and Meth., **115** (1974) 99
[12] H. Nifenecker, C. Signarbieux, R. Babinet, J. Poitou, *Neutron and gamma emission in fission*, IAEA-SM-174/207
[13] Sh. Zeynalov, P. Sedyshev, O. Sidorova, V. Shvetsov, Applications of Nuclear Techniques (CRETE17), *International Journal of Modern Physics: Conference Series*, Vol. **48** (2018) 1860123.
[14] A. Gook, F.-J. Hambsch, and M. Vidali, Phys. Rev. C **90**, 064611 (2014)

Status of fission fragment observables measured with the LOHENGRIN spectrometer

S. Julien-Laferrière [1,2], *L. Thombansen*[1], *G. Kessedjian*[1], *A. Chebboubi*[2], *O. Serot*[2], *C. Sage*[1], *O. Méplan*[1], *M. Ramdhane*[1], *O. Litaize*[2], *D. Bernard*[2], *J. Nicholson*[2], *A. Blanc*[3], *H. Faust*[3], *P. Mutti*[3], *U. Köster*[3], *A. Letourneau*[4], *T. Materna*[4], *M. Rapala*[4]

[1] LPSC, Université Grenoble-Alpes, CNRS/IN2P3, F-38026 Grenoble Cedex, France
[2] CEA, DEN, DER, SPRC, LEPh, Cadarache center, F-13108 Saint Paul lez Durance, France
[3] Institut Laue-Langevin, F-38042 Grenoble Cedex 9, France
[4] CEA, DSM, IRFU, SPhN, LEARN, Saclay center, F-91191 Gif-sur-Yvette, France

Abstract. Nuclear fission yields are key parameters to evaluate reactor physics observables, such as fuel inventory, decay heat, spent fuel radiotoxicity, criticality but also for understanding the fission process. Despite a significant effort allocated to measure fission yields during the last decades, the recent evaluated libraries still need improvements in particular in the description of the uncertainties with the associated correlations. Additional kinds of measurements provide complementary information in order to test the models used in the nuclear data evaluation. Moreover, some discrepancies between these libraries must be explained. A common effort by the CEA, the LPSC and the ILL aims at tackling these issues by providing precise measurement of isotopic and isobaric fission yields with the related variance-covariance matrices. Nevertheless, the experimental program represents itself a large range of observables requested by the evaluations: isotopic yields, nuclear charge polarization, odd-even effect, isomeric ratio and their dependency with fission fragment kinetic energy as a probe of the nuclear de-excitation path in the (E^*, J^π) representation. Measurements for thermal neutron induced fission of ^{241}Pu have been carried out at the Institut Laue Langevin using the LOHENGRIN mass spectrometer. Experimental program, observables reachable, results and comparison to model calculations are shown.

1 Fission studies for applications

The collaboration is involved in an experimental program using thermal neutrons to study the fission process and to answer the specific requests concerning the fission yields for the current and innovative fuel cycle studies. Innovative cycles cover the multi-recycling of the plutonium in the GEN-III reactors or the studies involving the thorium cycle. Nevertheless, for the actual generation of reactors, improvements on the knowledge of fission yields are expected for different goals:

- Fission fragments as probes of the fuel burn-up or the neutron fluency using cumulative yields (e.g. ^{140}Ba, ^{147}Nd);

- Fission yields are input data for the fuel inventory, the decay heat calculations and the Beta/neutrino spectra calculations (using independent yields associated with reaction and decay data);
- Probes of impact of the dynamics in the fission process, in particular the descent from the saddle point to the scission point;
- Fission yields are used for the determination of the prompt particles spectra emitted by the fission fragments. For instance, the calculation of the total prompt gamma spectra for all actinides present in the fuel are required for the determination of the reactor vessel heating and is relying on models calculations.

The program on fission yields with the LOHENGRIN spectrometer at ILL has been initiated for several years [1-2] and we developed different methodologies to obtain absolute fission yields with the description of the Probability Density Function (PDF) and the covariance matrices associated to the measurements [3]. This information on measurements is required in order to consider statistical tests as the cornerstone of the evaluation program [4-5].

2. Experimental setup

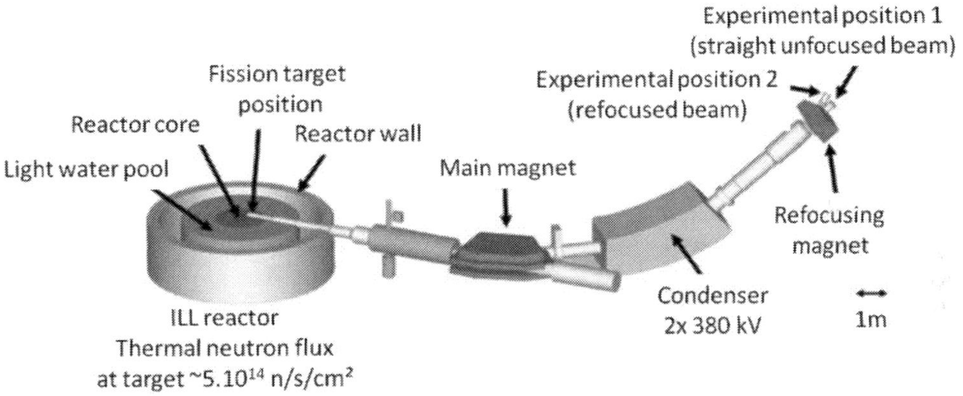

Fig. 1. The LOHENGRIN mass spectrometer at the ILL, Grenoble. The ILL high neutron flux compensates the small geometrical acceptance of the instrument requested for the mass resolution.

The LOHENGRIN recoil mass spectrometer [6, 7] of which a descriptive sketch is shown on Fig. 1, has been built in the 70's at the ILL and is since a major instrument for nuclear data measurements in particular due to its high mass resolution ($\Delta A/A \sim 1/400$). A purify ^{241}Pu target (see Table 1) is placed close to the reactor core under a high thermal neutron flux ($\sim 5 \cdot 10^{14}$ n/s/cm2 at target position) and undergoes a significant amount of fission fragments from $^{241}Pu(n,f)$ (more than 98.85% during 15 days irradiation). Fission products which enter in the apparatus (solid angle $\leq 3.2{\cdot}10^{-5}$ sr) are deflected through a magnetic and an electric fields. The ion beam is selected according to the A/q and E_k/q ratios, where A is the fission product mass, E_k its kinetic energy and q its ionic charge. The ratios A/q and E_k/q can be achieved with different triplet (A, E_k, q) leading to a possible degeneracy. Two experimental positions exist, at position 1, the ion beam directly enters in an ionization chamber. This position is used to measure isobaric yields, Y(A) with a dual anode Frisch grid ionization chamber. At position 2, the ion beam is refocused and implanted on a moving tape. Two clovers of four high purity germanium crystals each are placed in the vicinity of the tape and detect γ-rays resulting from the fission products β− disintegration, allowing to measure isotopic yields Y(A, Z).

Table 1. Initial target composition and limits on fission rates after 6 day irradiation for the July 2016 experiment with the LOHENGRIN spectrometer.

Actinide	**Composition (%)**	**Relative (n,f) rates (%)**
^{238}Pu	< 0.0003	$< 1\ 10^{-3}$
^{239}Pu	0.02	$< 3.1\ 10^{-2}$
^{240}Pu	5.30	$< 7\ 10^{-4}$
^{241}Pu	71.20	> 98.85
^{242}Pu	17.40	$< 1.5\ 10^{-2}$
^{241}Am	6.06	$< 0.04\ 10^{-2}$
$^{242GS}Am + ^{242m}Am$	-	< 1.039

3. Experimental program: isotopic and isomeric yields

The first goal of the experiments consists in the precise measurements of mass, isotopic and isomeric yields with a control of the systematic effects and the determination of the covariance matrices associated to the analyses. For these observables, their dependency with fission fragment kinetic energy increases substantially the information on fission process. Theoretically, the full characterization of fission events is provided by the determination of the yield as a function of the mass A, the nuclear charge Z, the excitation energy E* and the spin-parity (J^{π}) for each fragment: Y(A, Z, E*, J^{π}). Unfortunately, this quantity cannot be determined experimentally. The experimental observables correspond to the fission yields as a function of the kinetic energy E_k and the ratio of the isomeric over ground states populations (IR): Y(A, Z, E_k, IR). Usually the kinetic energy dependency of these populations is described using ratio measurements of isotopes or isomers in order to limit the systematic corrections and to increase the precision (1-5 % achievable). However, the interpretation of these ratios is not straightforward. Indeed, the comparison of experimental and theoretical IR implies the use of several models with different degree of impact. For instance, the sharing of the excitation energy between both fragments (for instance [8-10]) as well as the mechanism by which these fragments acquire their angular momenta, the level density models, the gamma strength function and the spin parity distribution before and after prompt neutron emission [11] must be considered. Thus, isotopic and isomeric ratios are the crucial observables to assess the dynamical aspects of the fission process.

In references [12] [13], the authors described two experiments performed at LOHENGRIN to illustrate the expected dependence of the isomeric ratio as a function of the kinetic energy. Different experimental methodologies are being developed today by the collaboration to measure and interpret isotopic and isomeric yields. The goal is the construction of a library of measurements in order to test the decay cascade on a large range of isotopes.

3.1 Isotopic yields and charge polarization

Isotopic and long-life isomeric (few μs to ms) yields are now currently measured with the LOHENGRIN spectrometer using the gamma spectrometry. This consists in a cumulative measurement with duration of about 30 min per point over the kinetic energy or ionic charge distributions. After corrections for the radioactive decays of the fragments, we obtain the isotopic yields (see Fig. 2). Particular effort is oriented on the determination of the systematic uncertainties and the covariance associated to the measurement process [3]. From this data, combined with mass yields measurements, we are able to extract the nuclear charge polarization (see Fig. 3). If we observed a good agreement with the JEFF3.1.1 and JEFF3.3 libraries for the mass 130 and around mass 140, a strong structure appears for the masses 132, 136 and 138. This region corresponds to the already observed structures at Z=50 and Z=54. Complementary measurements on the neighboring masses are planned at ILL to better understand this phenomenon.

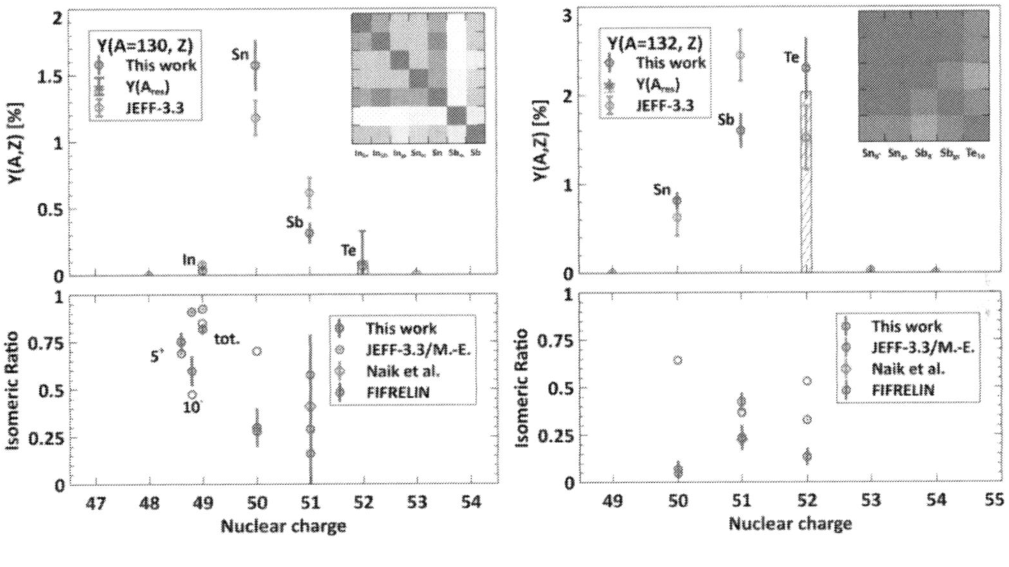

Fig. 2. Results for the masses 130 (left) and 132 (right). (Top plots) Isotopic yields with their correlations have been compared with the library JEFF-3.3. The hatched bars correspond to the residual isotopic yields which are not measured with gamma spectrometry but deduced from the mass yields measurements. (Bottom plots) Isomeric ratios for different isotopes compared to the FIFRELIN calculations, the JEFF-3.3 evaluations and the H. Naik et al. interpretations [14].

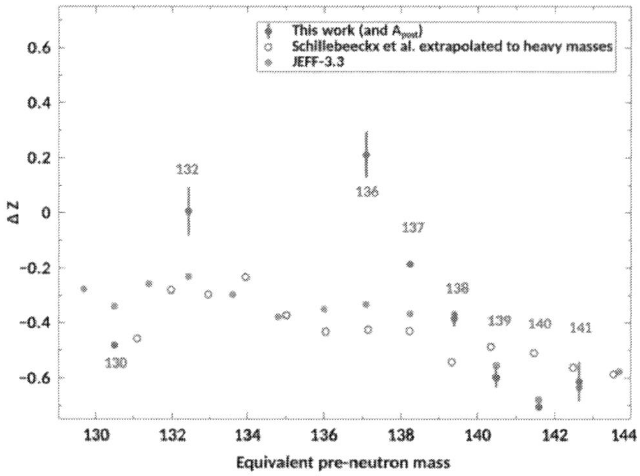

Fig. 3: Our measurements (red points labeled with the measured post-neutron emission mass A_{post})) of the nuclear charge polarization as a function of the pre-neutron mass. Results are compared to the Jeff libraries and the extrapolated points from light fragment measurement (P. Schillebeckx et al. [22])

3.2 Isotopic and isomeric yields and their dependency on the kinetic energy

Previous studies made by the collaboration show that under specific conditions, isomeric ratios (in that case μs isomeric states) are dependent on fission fragment kinetic energy [15]. Measurements are obtained via a coincidence between an ionization chamber and the array of γ-rays detector. By using a γ de-excitation code such as FIFRELIN, the angular momentum distribution of the fission product can be assessed. Recently the collaboration has developed similar measurements with the evolution of isotopic yields as a function of fission fragment kinetic energy. These results on the mass 139 are presented on Fig. 4 (left) in comparison to the FIFRELIN calculations. It represents a test for the models allowing the validation of the neutron and gamma cascade. Moreover we can extract the odd-even effect and neutron multiplicity as a function of excitation energy of nuclei (see Fig. 4 right).

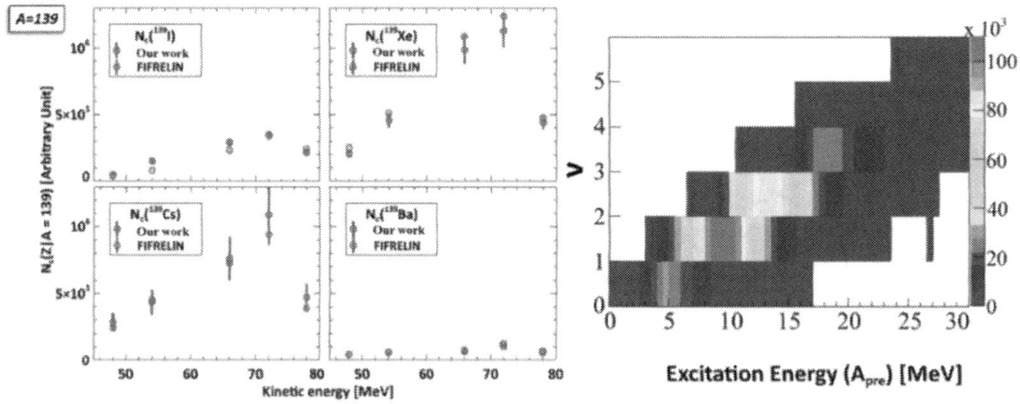

Fig. 4: (left) Kinetic energy dependence of cumulative isotopic yields for the mass 139 and the comparison to the FIFRELIN calculations. (Right) neutron multiplicity as a function of excitation energy deduced from FIFRELIN calculations.

3.3 Nanosecond converted Isomeric Ratio

Indirect measurement of nanosecond isomeric ratios can be determined from the ionic charge distribution [16-18]. The method consists in the deconvolution of the ionic charge distribution per isotope obtained by gamma spectrometry. The converted isomeric ratio (CIR) is defined as the converted isomer population over the total ionic population (converted and unconverted). According to the statistical models form H. Betz [19-20], we assume that the ionic charge distribution associated to the unconverted population follows a Gaussian distribution due to the charge equilibrium in the cover of the target (here Ni foils). A deviation from this Gaussian distribution indicates a charge changing due to the conversion from ps and ns isomers (see Fig. 5).CIR measurements are compared to FIFRELIN calculations (Fig. 6). We note a good tendency even if some differences have to be explored in details given the assumption used of the analysis or the models considered for the decay cascade calculations. It is interesting to note that this observable is not dedicated to one isomer but for all transitions at low energy. For certain nuclei, many isomers and bands are taking into account at the limit of the knowledge of the nuclear structure. Then it corresponds to integral measurements used to test the overall decay cascade.

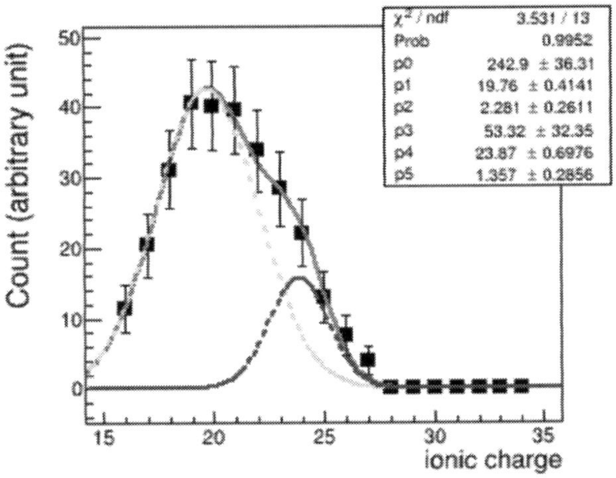

Fig. 5: Ionic charge distribution de-convoluted using the Gaussian assumption for the unconverted state according to the Betz model [19-20]. The blue curve corresponds to the ns-isomer contribution.

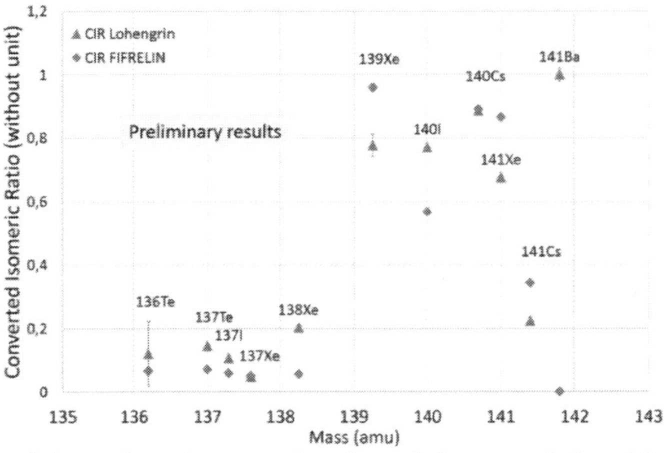

Fig. 6: Converted isomeric ratios as a function of the mass induced by the fission of $^{241}Pu(n_{th},f)$ in comparison to FIFRELIN calculations [18].

4. Conclusion and perspectives

A general status of the experimental program using the gamma spectrometry with the LOHENGRIN mass spectrometer at the ILL has been presented. Classical measurements of fission yields are now expanded with their kinetic energy dependency in order to test the limit of fission models. New measurement and validated models are central in order to progress in the evaluation project. Recent results for thermal neutron induced fission of ^{241}Pu have been presented. The gamma spectrometry method, in particular for the Converted Isomeric Ratios, considers the nuclear structure as the basic data allowing the study of a large timescale of isomeric states as probe of the fission process. In this topic, the development of the FIPPS project [21] at ILL is dedicated to the fission product spectroscopy and the fission process studies.

Acknowledgements

This work was supported by CEA, IN2P3 and "le défi NEEDS". The authors are grateful for the support of the ILL and all the staff involved from CEA Cadarache and LPSC.

References

1. O. Serot et al., Nuclear Data Sheets 119, 320–323 (2014)
2. F. Martin et al., Nucl. Data Sheets 119, 328 (2014)
3. S. Julien-Laferrière, A. Chebboubi, G. Kessedjian, O. Serot, EPJ N - Nuclear Sciences & Technologies, EDP Sciences, 2018, 4, pp.25
4. N. Terranova, O. Serot, P. Archier, C. de Saint Jean and M. Sumini, Annals of Nuclear Energy 109, 469–489 (2017)
5. B. Voirin, G. Kessedjian, A. Chebboubi, O. Sérot, S. Julien-Laferriere et al., EPJ Nuclear Sci. Technol. 4, 26 (2018)

6. H.R. Faust, P. Geltenbort, F. Gönnenwein and A. Oed, ILL Internal Scientific Report 81FA45S (1981)
7. G. Fioni, H. R. Faust, M. Gross, M. Hesse, P. Armbruster, F. Gönnenwein and G. Müzenberg, Nucl. Instrum. Meth. A 332, 175-180 (1993)
8. O. Litaize and O. Serot, Phys. Rev. C 82, 054616 (2010)
9. K. H. Schmidt and B. Jurado, Phys. Rev. C 83, 014607 (2011).
10. H. Faust et al., J. Korean Phys. Soc. 59, 879 (2011).
11. D. Regnier, PhD thesis, Grenoble University (2013)
12. J. P. Bocquet et al., Physics and chemistry of fission, IAEA, Proceedings of symposium, Jülich 1979.
13. F. Gönnenwein, I. Tsekhanovich, V. Rubchenya, International Journal of Modern Physics E, Vol. 16, No. 2 (2007) 410-424
14. H. Naik et al. Systematics of fragment angular momentum in low-energy fission of actinides. Nucl. Phys. A, 587 :273–290, 1995.
15. A. Chebboubi, G. Kessedjian, O. Litaize, O. Serot, H. Faust et al. Kinetic energy dependence of fission fragment isomeric ratios for spherical nuclei 132Sn, Phys. Lett. B, 2017, 775, pp.190-195.
16. T. Rzaca-Urban, J. Genevey, T. Materna et al., Phys. Rev. C 80, 064317 (2009)
17. A. Chebboubi et al., EPJ Web of Conferences, 146, (2017) 04021.
18. L. Thombansen, Master 1 report, Phelma engineering school, Grenoble, supervision G. Kessedjian, LPSC, 2018
19. H. D. Betz, Rev. Mod. Phys., 44 :465–539, Jul 1972.
20. H. D. Betz. Heavy Ion Charge States. In S. Datz, editor, Applied Atomic Collision Physics, Volume 4 : Condensed Matter, volume 4, page 2, 1983.
21. C. Michelagnoli et al., EPJ Web of Conferences,193, 04009 (2018)
22. P. Schillebeeckx et al., Nucl. Phys. A, 580, 15–32, 1994.

Prompt and Delayed Neutron Emissions and Fission Product Yield Calculations with Hauser-Feshbach Statistical Decay Theory and Summation Calculation Method

Shin Okumura[1,3,*], *Toshihiko* Kawano[2,**], and *Satoshi* Chiba[1,***]

[1]Laboratory for Advanced Nuclear Energy, Tokyo Institute of Technology, 2-12-1 Ookayama, Meguro-ku, Tokyo, 152-8550, Japan
[2]Theoretical Division, Los Alamos National Laboratory, Los Alamos, NM 87545, USA
[3]NAPC-Nuclear Data Section, International Atomic Energy Agency, Vienna A-1400, Austria

Abstract. We demonstrate the neutron emission and fission product yield calculations using the Hauser–Feshbach Fission Fragment Decay (HF^3D) model and β decay. The HF^3D model calculates the statistical decay of more than 500 primary fission fragment pairs formed by the neutron induced fission of ^{235}U. In order to calculate the prompt neutron and photon emissions, the primary fission fragment distributions, i.e. mass, charge, excitation energy, spin and parity are deterministically generated and numerically integrated for all fission fragments. The calculated prompt neutron multiplicities, independent fission product yield are fully consistent each other. We combine the β-decay and the summation calculations with the HF^3D model calculation to obtain the cumulative fission product yield, decay heat and delayed neutron yield. The calculated fission observables are compared with available experimental data.

1 Introduction

The nuclear data of fission yields are essential ingredients for numerous nuclear applications such as the conception of new generation of reactors and nuclear fuel cycle developments. With increase in requirements by those applications, fission yields from wide variety of actinides and wide ranges of energy have become of great interest.

Numerous theoretical efforts have been made with both microscopic and macroscopic methods to understand the fission dynamics. These approaches provide a quantum description of the fission dynamics and mass, kinetic energy distributions of excited fission fragments. Dynamical description of fission by Langevin equations is one of the examples, which can treat the deformation of each fragment independently and allows us to analyse key information about fission fragments such as mass distribution and total kinetic energy (TKE) [1].

These fission fragments formed just after fission will evaporate prompt neutrons and photons to reach their ground state or meta-stable state. Many models and codes have been developed for the study of prompt neutron and photon emissions, and yield distribution to reproduce experimental results. One of the reliable approaches is the Hauser-Feshbach stastical decay of fission fragments.

*e-mail: s.okumura@iaea.org
**e-mail: kawano@lanl.gov
***e-mail: chiba.satoshi@nr.titech.ac.jp

The post-neutron emitted fission fragments, so called fission products, will β decay that leads the β-delayed neutron and photon emissions, to reach the stable isotopes. Many efforts have been done to understand beta decay properties of nuclei far away from beta stability. Theoretical approach such as the gross theory treats the sum of strengths of the transition from initial to final states in quantum mechanics [2]. For reactor application purpose, summation calculation has been used for the prediction of the decay heat from fission products. The summation calculation requires information about half-lives, γ- and β-energies released in the decay. Since the decay can proceed via a very large number of excited levels, the decay data have been evaluated and provided as an evaluated library such as ENDF/B-VII.1 decay data library.

However, an accurate prediction of fission observable through out the fission (pre prompt emissions), de-excite and the beta decay processes by theoretical calculations in a consistent manner still remains difficult. Therefore the evaluations of nuclear data have been done inconsistently [3] and some ingredients have been provided based on some empirical models.

We demonstrate the calculations of the fission observables such as independent/cumulative fission product yields (FPY), prompt/delayed neutron and photon emissions, and decay heats in a consistent manner starting from a unique fission fragment distribution until all fission fragment pairs de-excite and β decay until they reach to their stable isotopes. We employ the Hauser-Feshbach Fission Fragment Decay (HF^{3}D) model to calculate the prompt neutron multiplicity and the independent FPY [4]. The β-decay chains of each nuclide in the independent FPY are tracked to obtain the cumulative FPY by using ENDF/B-VII.1 decay data library. The decay heat and delayed neutron yield are calculated by the summation calculation method using a Oyak computer code [5] that also incorporates ENDF/B-VII.1 decay data library. The calculated results are compared with experimental data and the evaluated nuclear data libraries.

2 Method

For the de-excitation process of the fission fragments, we employed the HF^{3}D model [4]. The main features of the HF^{3}D model are (1) generation of the distributions of mass and charge yields, excitation energy, level density, spin, and parity $Y(A, Z, E_{ex}, J, \Pi)$ for all primary fission fragments, and (2) integration over all these distributions. More than 500 fission fragment pairs are generated by the HF^{3}D model for ^{235}U. The concept of the generation of the fission fragment distributions by the HF^{3}D model is discussed in Ref [4]. The mass distribution of primary fission fragment yield $Y(A)$ is generated from the five Gaussians fitted to the experimental data. The charge distributions of every mass number are generated using the Wahl systematics (Z_P model) [6] to fill the fission fragment distribution. The experimentally available total kinetic energy TKE(A) distribution is fitted by analytical functions in order to generate total excitation energy (TXE) distributions for fragment pairs. The TXE can be calculated from TKE by taking into account the energy balance of the reaction. The HF^{3}D model calculates the neutron multiplicities $\overline{\nu}_{l,h}^{(k)}$ for the k-th pair of fission fragments by integrating the neutron evaporation spectrum $\phi_{l,h}^{(k)}$ from the light or heavy fragment in the center-of-mass system,

$$\overline{\nu}_{l,h}^{(k)} = \int dE_x \sum_{J\Pi} \int d\epsilon\, R(J,\Pi) G(E_x) \phi_{l,h}^{(k)}(J,\Pi,E_x,\epsilon)\,, \tag{1}$$

where $R(J,\Pi)$ is the probability of nucleus having the state of spin J and parity Π, and $G(E_x)$ is the distribution of excitation energy. By performing this calculation for each fission fragment pair, one can obtain the prompt neutron spectrum $\chi(E)$ and multiplicities $\overline{\nu}$, and the independent FPY $Y_I(A, Z)$ from ^{235}U(n,f).

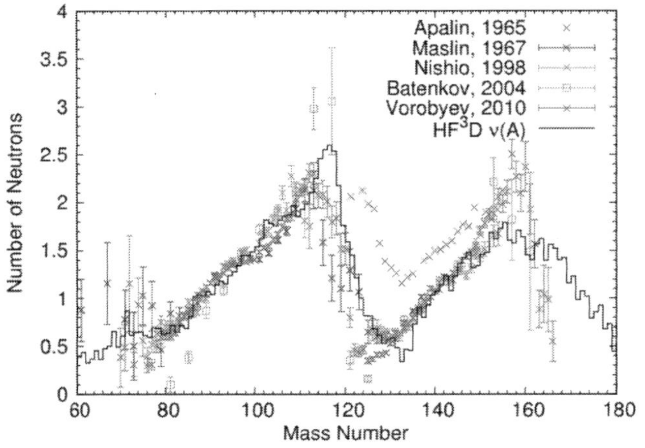

Figure 1. The mass dependent neutron multiplicity for ^{235}U(n_{th},f) calculated with the HF^{3}D model compared with experimental data.

For the Hauser-Feshbach calculations, the neutron optical potential by Koning-Delaroche, level density systematics based on the KTUY05 mass model, and the Kopecky and Uhl γ-ray strength function for the E1 transition, and the standard Lorentzian model for higher multipolarities are used.

For the β-decay process, the cumulative FPY was obtained by performing the β-decay calculation of the nuclides for the calculated independent FPY by tracing the β-decay path using the ENDF/B-VII.1 decay data library. The β- and γ-energy components in the decay heats, and delayed neutron yield were calculated by the summation calculations method [5].

3 Results and Discussion

3.1 Prompt neutron multiplicity

The calculated average neutron multiplicity $\overline{\nu}$ for ^{235}U(n_{th},f) was 2.38 which is 2.5% smaller than the evaluated value in JENDL-4.0 ($\overline{\nu}$ = 2.44). The mass dependent neutron multiplicity $\overline{\nu}(A)$ is shown in Figure 1 together with available experimental data [7–11]. Our calculation well follows the experimental data in the range with the mass number 60<A<150. Generally, simulations rarely calculate $\overline{\nu}$ for mass ranges A<60 and 150<A when the Monte Carlo technique is used for generations of fission fragment distributions because the fission fragment yields are too small to sample in these ranges, whilst our calculation gives $\overline{\nu}$. This is because our calculation is based on the deterministic method. However, such small yields in this mass range also makes experiment difficult and as a concequence experimental data has a large uncertainty.

3.2 Fission product yield

Figure 2 illustrates the mass dependence of the independent FPY for ^{235}U(n_{th},f). Our calculation well reproduces the distribution of independent FPY in JENDL/FPY-2011 for small peaks at A = 99 and 134, and the tendency of independent FPY in the wide range of mass numbers which varies from 10^{-11} to 10^{-2}.

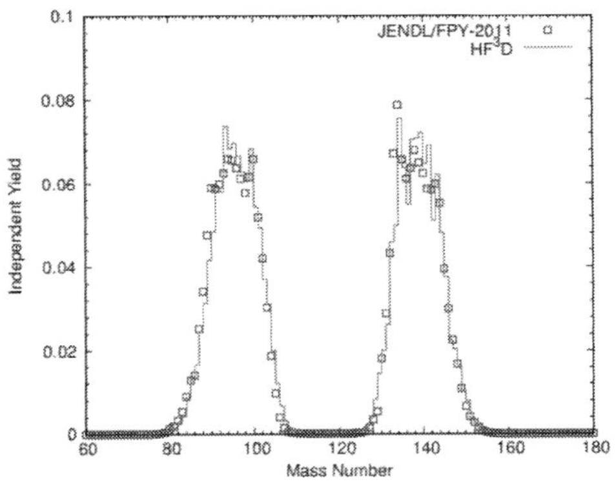

Figure 2. The mass total independent FPY of $^{235}U(n_{th},f)$ calculated by the HF³D model and the evaluated data in JENDL/FPY-2011

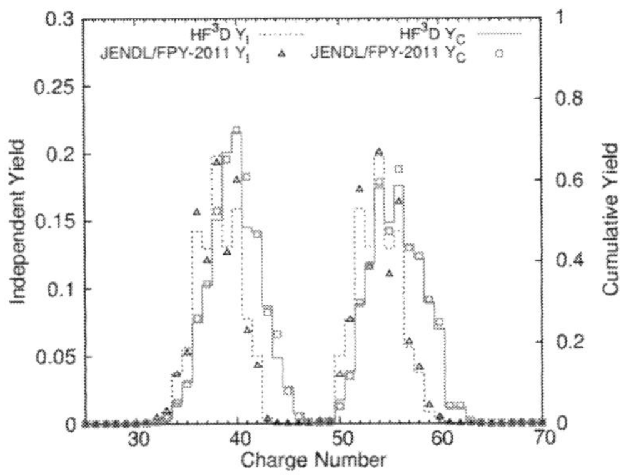

Figure 3. The charge total independent and cumulative FPY of $^{235}U(n_{th},f)$ calculated by the HF³D model and the evaluated data in JENDL/FPY-2011.

The cumulative FPY was obtained by performing the β-decay calculation using the calculated independent FPY. Since the post-neutron emitted (independent) fission product decays by β-decay path in the same mass number and β-delayed neutron emission yield is quite small to give an obvious change in the mass dependent fission yield, we compare the charge dependent fission yield before and after the β-decay. Figure 3 shows the charge dependence of the independent and cumulative FPYs for $^{235}U(n_{th},f)$. The calculated cumulative FPY well reproduces the evaluated data in JENDL/FPY-2011. By comparison between the independent and cumulative FPYs, the β-decay calculation successfully tracks the β-decay chains for each nuclide in the independent FPY data and reproduces the cumulative FPY.

3.3 Summation calculations

We have performed a set of preliminary calculations of the summation method. The calculated decay heats by β and γ emissions are shown in Figure 4 together with the experimental data of Lowell and Oak Ridge National Laboratory for $^{235}U(n_{th},f)$. The major differences in the decay heat from the γ emissions below 10 second after the fission burst is mainly due to an over-estimation of the independent FPY of ^{97g}Y and some other nuclides. In addition, the calculation strongly depend both on the independent FPY and the decay data library that is used to track the β-decay chain and sum-up the energy release from both the β and γ emissions.

The calculated total average number of delayed neutrons per fission ($\overline{\nu}_d$) for $^{235}U(n_{th},f)$ is 0.01659. The calculated result is 4.7% higher than evaluated value in JENDL-4.0 ($\overline{\nu}_d$=0.01585). Figure 5 shows the cooling time dependence of calculated $\overline{\nu}_d$ compared with the experimental data. There are some nuclides that are over-estimated in the independent FPY, e.g. $^{94,95}Rb$ and ^{96}Y, compared to the evaluated nuclear data libraries. These nuclides strongly affect $\overline{\nu}_d$.

It is worth mentioning that we did not introduce any adjustments at each calculation step in this study. In order to reproduce experimental data more accurately, one can optimize the starting distributions and can introduce some constrains. The deterministic method to generate the distributions loses correlation information among the distributions. However, this method is quite effective to examine the fission fragment distributions which are generated by the theoretical simulation of fission process such as the Langevin method, because many of comparable fission observables are after the β decay. Such the connection also makes pos-

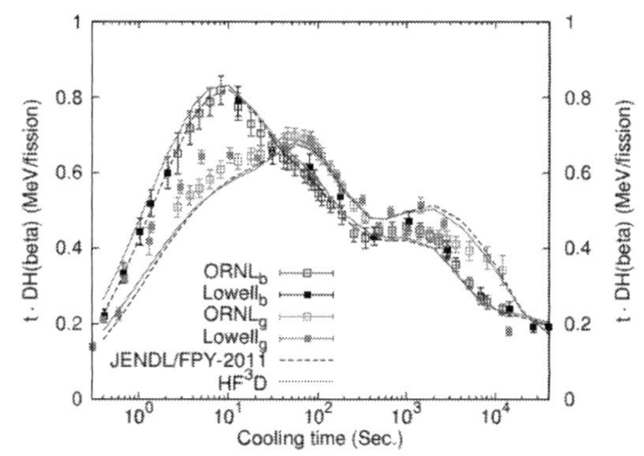

Figure 4. Cooling time dependence of the decay heats by β (b) and γ (g) emissions comparing with the experimental data for ^{235}U(n_{th},f).

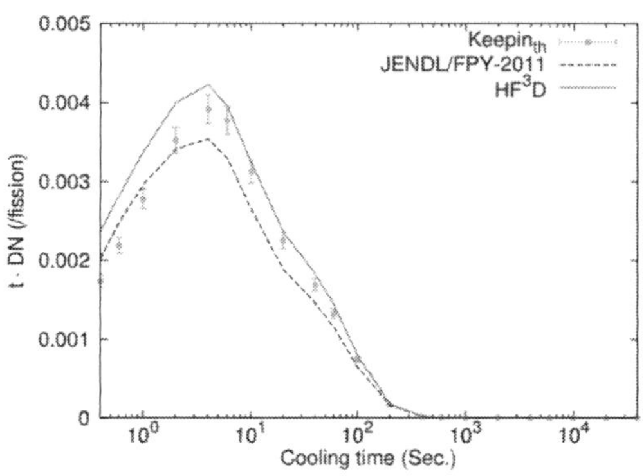

Figure 5. Cooling time dependence of the delayed neutron yield comparing with the experimental data for ^{235}U(n_{th},f).

sible to expand the nuclear data evaluation for variety of actinides and wide range of incident neutron energies.

4 Conclusion

In this work we demonstrated a fully consistent approach to calculate some fission observables starting from a unique fission fragment distribution. We employed the recently developed Hauser-Feshbach fragment decay model, HF^{3}D, to calculate the prompt neutron multiplicity and the independent FPY. The β-decay path of each fission product is tracked using the ENDF/B-VII.1 decay data library to obtain the cumulative FPY. The decay heat and delayed neutron yield are calculated by the summation calculation method. All obtained fission observables were compared with the experimental data and showed fairly good agreement with experimental tendency without any parameter adjustments. The scalability of the combination of HF^{3}D model, β-decay, and summation calculation enables us to perform series of consistent calculations for the other actinides and energy ranges. We have performed a set of preliminary calculations of the incident energy dependent prompt and delayed neutron emissions [12].

Acknowledgements

This work was partially supported by the grant 'Development of prompt-neutron measurement in fission by surrogate reaction method and evaluation of neutron- energy spectra,' entrusted to JAEA by Ministry of Education, Culture, Sports, Science and Technology. This work was partly carried out under the auspices of the National Nuclear Security Administration of the U.S. Department of Energy at Los Alamos National Laboratory under Contract No. DE-AC52-06NA25396.

References

[1] M.D. Usang, F.A. Ivanyuk, C. Ishizuka, S. Chiba, Phys. Rev. C **96**, 064617 (2017)
[2] K. Takahashi, M. Yamada, Kondoh, Atomic Data and Nuclear Data Tables **12**, 101 (1973)

[3] P. Jaffke, Nuclear Science and Engineering **190**, 258 (2018)
[4] S. Okumura, T. Kawano, P. Jaffke, P. Talou, S. Chiba, Journal of Nuclear Science and Technology **55**, 1009 (2018)
[5] M.S. K. Oyamatsu, H. Takeuchi, J. Katakura, Journal of Nuclear Science and Technology **38**, 477 (2001)
[6] A.C. Wahl, Tech. Rep. LA-13928, Los Alamos National Laboratory (2002)
[7] V. Apalin, Y. Gritsyuk, I. Kutikov, V. Lebedev, L. Mikaelian, Nuclear Physics **71**, 553 (1965)
[8] E.E. Maslin, A.L. Rodgers, W.G.F. Core, Phys. Rev. **164**, 1520 (1967)
[9] K. Nishio, Y. Nakagome, H. Yamamoto, I. Kimura, Nuclear Physics A **632**, 540 (1998)
[10] O.A. Batenkov, G.A. Boykov, F. Hambsch, J.H. Hamilton, V.A. Jakovlev, V.A. Kalinin, A.B. Laptev, V.E. Sokolov, A.S. Vorobyev, AIP Conference Proceedings **769**, 1003 (2005)
[11] A. Vorobyev, O. Shcherbakov, A. Gagarski, G. Valski, G. Petrov, EPJ Web of Conferences **8**, 03004 (2010)
[12] S. Okumura, T. Kawano, P. Jaffke, P. Talou, T. Yoshida, S. Chiba, CERN Proc. **1**, 173 (2019)

Constraining Fission Yields Using Machine Learning

Amy Lovell[1,2,*], *Arvind* Mohan[1,3,**], *Patrick* Talou[4,***], and *Michael* Chertkov[3,****]

[1]Center for Nonlinear Studies, Los Alamos National Laboratory, Los Alamos, NM 87545, USA
[2]Nuclear Physics Group, Theoretical Division, Los Alamos National Laboratory, Los Alamos, NM 87545, USA
[3]Condensed Matter Physics and Complex Systems Group, Theoretical Division, Los Alamos National Laboratory, Los Alamos, NM 87545, USA
[4]Materials and Physical Data Group, X Computational Physics Division, Los Alamos National Laboratory, Los Alamos, NM 87545, USA

Abstract. Having accurate measurements of fission observables is important for a variety of applications, ranging from energy to non-proliferation, defense to astrophysics. Because not all of these data can be measured, it is necessary to be able to accurately calculate these observables as well. In this work, we exploit Monte Carlo and machine learning techniques to reproduce mass and kinetic energy yields, for phenomenological models and in a model-free way. We begin with the spontaneous fission of ^{252}Cf, where there is abundant experimental data, to validate our approach, with the ultimate goal of creating a global yield model in order to predict quantities where data are not currently available.

1 Introduction

It is important to reliably and consistently calculate fission observables for applications such as energy, non-proliferation, defense, and astrophysics. Often, fission observables are calculated independently of one another, leading to inconsistencies within evaluations (recently shown, for example, in [1]). However, now with tools such as CGMF [2], we can form a consistent picture of fission from scission to the emission of prompt neutrons and γ rays.

There are many models available to describe single prompt fission observables (e.g. emitted neutron energy spectra or neutron average multiplicity). Some of these models calculate fission yields from shape parameterizations of nuclei using fundamental interactions, i.e. [3–7]; this is typically computationally expensive. However, most are phenomenological and must be tuned individually to reproduce data. While there are many optimization schemes available (including Monte Carlo techniques), machine learning algorithms can, in principle, learn the complex mapping between observables of the same system and different systems, enhancing the calculation of correlated observables and giving more predictive power. Currently, machine learning is still in its infancy in nuclear physics, and only a few examples are available (i.e. [8, 9]).

*e-mail: lovell@lanl.gov
**e-mail: arvindm@lanl.gov
***e-mail: talou@lanl.gov
****e-mail: chertkov@lanl.gov

In this work, we begin with a study of ^{252}Cf observables, focusing first on mass and total kinetic energy yields. Because it spontaneously fissions, a large body of data have been collected, ranging from yields, to multiplicity distributions and correlated observables. It is therefore an ideal case to study a variety of optimization methods and understand the strengths of each before tackling systems where less experimental data is available. The standard optimization method that we discuss has also been studied for ^{252}Cf in a previous work [10], giving us a baseline with which to benchmark our results before exploring the more novel machine learning techniques.

This proceedings is organized as follows. In Section 2, we discussed the physical models that are used in this work, followed by the numerical methods in Section 3. We then summarize our results in Section 4 and conclude in Section 5.

2 Theory

Fission is a rich and complex process that can be described using a range of physical models. Here, we only focus on prompt fission, in particular, the emission of prompt neutrons and γ rays from the fission fragments.

2.1 CGMF

To model the decay of the fission fragments and resulting correlations, the Monte Carlo code CGMF, has been developed [2]. The Hauser-Feschbach statistical model is used to follow the decay of the two daughter fission fragments on an event-by-event basis. At each step of the decay, probabilities for emitting neutrons or γ rays are calculated. For each decay, the energy and angle of the fission fragments and emitted particles are recorded, enabling the calculation of correlations between various observables.

In order to initialize these decays, yields in mass, charge, total kinetic energy, spin, and parity are required, $Y(A, Z, TKE, J, P)$. Currently, the sum of three Gaussians is used for the mass distribution, the TKE distribution is described by a single Gaussian for each fragment mass, the charge is determined from Wahl systematics [11], the spin distribution is proportional to $(2J+1)\exp(-0.5J(J+1)\hbar^2/(\alpha\mathcal{I}_o(A,Z)T))$ ($\mathcal{I}_o(A,Z)$ being the ground-state moment of inertia of nucleus (A,Z) and α the spin cut-off parameter), and the parity is chosen to be positive or negative with equal probability. The total kinetic energy is shared between the two daughter fragments based on an effective temperature ratio between the light and heavy fragments, which can either be constant or mass-dependent.

2.2 Brosa Yield Model

Currently, the models for $Y(A)$ and $Y(TKE)$ in CGMF are uncorrelated. In 1990, Brosa [12] published a model for $Y(A, TKE)$ based on separate fission modes or channels which correspond to different families of shapes of the fissioning system: standard (shown to be comprised of three modes, S1, S2, and S3), superlong (SL), and superasymmetric (SX).

For each mode, the yields are described by

$$Y(A, TKE) = \sum_m Y_m(A) Y_m(TKE|A), \tag{1}$$

where

$$Y_m(A) = \frac{w_m}{\sqrt{8\pi\sigma_m^2}}\left[\exp\left(-\frac{(A-\bar{A}_m)^2}{2\sigma_m^2}\right) + \exp\left(-\frac{(A-A_{cn}+\bar{A}_m)^2}{2\sigma_m^2}\right)\right], \tag{2}$$

and

$$Y_m(TKE|A) = \left(\frac{200}{TKE}\right)^2 \exp\left[2\frac{d_m^{max} - d_m^{min}}{d_m^{dec}} - \frac{T_m(A)}{d_m^{dec}} - \frac{(d_m^{max} - d_m^{min})^2}{T_m(A)d_m^{dec}}\right]. \tag{3}$$

Here,

$$T_m(A) = \frac{(Z_{cn}/A_{cn})^2(A_{cn} - A)Ae^2}{TKE} - d_m^{min}, \tag{4}$$

where Z_{cn} and A_{cn} are the charge and mass of the fissioning system. Each mode, m, has six free parameters, w_m, $\bar{A}_m$, σ_m, d_m^{max}, d_m^{min}, d_m^{dec}, which are, respectively, the weight of the mode, the mean mass of the heavy fragment, the width of the mass distribution, the most favorable semilength for fission, the minimum semilength below which fission will not occur, and the the length scale for the decay of the exponential. Since the weights describe the relative probabilities of each mode being populated, they must sum to one.

3 Methods

We have been exploring two paths to construct these yields. The first uses more traditional optimization techniques, in particular a Markov Chain Monte Carlo (MCMC). The second uses a machine learning algorithm. These are described below.

3.1 Monte Carlo Methods

First, a Markov Chain Monte Carlo [13] is used to find a best fit for the free parameters in Equations 2 and 3. This is done by exploring parameter space through randomly generated sets of parameters. Each randomly generated parameter set, i, depends on the previous one, $i-1$, through a Gaussian distribution $x_i \sim \mathcal{N}(x_{i-1}, \epsilon x_0)$, where ϵ is a scaling factor, and x_0 is a fixed parameter set. Here, x_0 is the initial parameter set as given by Brosa in [12]. This combination allows flexibility in the step size while keeping the step size appropriate for the scale of each parameter.

Each new parameter set is accepted or rejected based on the criteria

$$R < \frac{\exp(-\chi_i^2/2)}{\exp(-\chi_{i-1}^2/2)}, \tag{5}$$

where R is a random number sampled uniformly in $[0, 1]$, and χ_i^2 is the χ^2-value of the ith parameter set, defined as

$$\chi^2 = \sum_j \left(\frac{Y_j^{\text{th}} - Y_j^{\text{exp}}}{\sigma_{Y_j}^{\text{exp}}}\right)^2. \tag{6}$$

Here, Y_j^{th} is a calculated yield for a given value of mass and kinetic energy as in Eq. 1, Y_j^{exp} is the experimental yield for the same mass and kinetic energy, and $\sigma_{Y_j}^{\text{exp}}$ is the experimental error on Y_j^{exp}.

In this work, we used three values of ϵ to ensure that we were not stuck in a local minimum due to the choice of x_0 and to find the minimum χ^2 value (or equivalently, the maximum likelihood). These values were 20ϵ, 5ϵ, and ϵ (with $\epsilon = 0.001$), in sequence from largest to smallest, to find the parameter set that minimized Eq. 6.

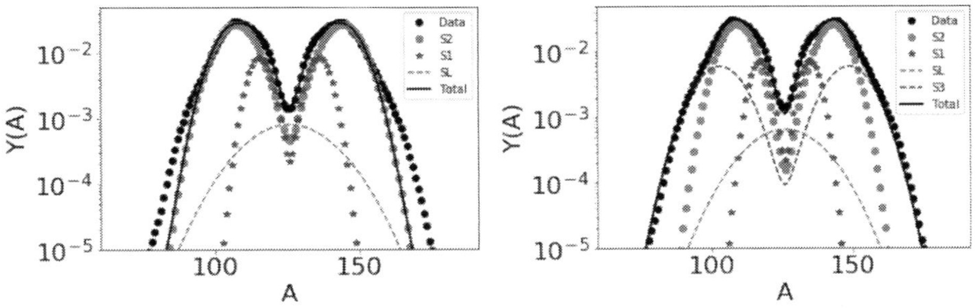

Figure 1. Mass yields when (left) three Brosa modes were fit using the MCMC and when (right) four Brosa modes were fit. Black solid lines show the total fit compared to the experimental data from [15].

3.2 Probabilistic Neural Networks

Neural networks (NN) are a machine learning algorithm that try to learn a complex relationship between input and output using a large-scale, data-driven optimization over hundreds of thousands of parameters. The base unit of the network is a neuron, several of which can be arranged in layers (the input and output each comprise a layer), and each neuron has a weight and a bias that are learned. The combination of these layers form the NN. The structure of the layers is driven by the specific application.

In a typical NN, the weights and biases are optimized based on a maximum likelihood estimation (MLE), which is a deterministic approach and cannot take into account uncertainties in the training data. Inherently, nuclear data contain uncertainties which should be included. In addition, a standard MLE will perform an average over discrepant data sets which may not reflect the confidence in each individual set.

To take this into account, we use a probabilistic machine learning approach - the Mixture Density Network (MDN) [14]. This approach predicts the complex mapping between input and output, $y = f(x)$, as a mixture of Gaussians,

$$f(x) = \alpha_1 \mathcal{N}(\mu_1, \sigma_1) + ... + \alpha_n \mathcal{N}(\mu_n, \sigma_n). \tag{7}$$

The values of α_i, μ_i, and σ_i are learned by the NN, and the user has control over the number of Gaussians that are included (as well as the underlying architecture of the NN).

4 Results and Discussion

We began with a MCMC to constrain the Brosa mode parameters for ^{252}Cf spontaneous fission yields. First, data from [15] were used to constrain, simultaneously, the Brosa parameters for the S2, S1, and SL modes. We then introduced the S3 mode, rerunning the MCMC for the four modes, and finally included the SX mode as well. The results for $Y(A)$ using three and four modes are shown in Fig. 1, along with the contributions of from each of the modes. (The contribution from the SX mode was effectively negligible.)

The $Y(A, TKE)$ distributions from these three parameterizations were then used as input for CGMF, from which we compared the resulting prompt fission observables to those that were calculated from the current models for $Y(A)$ and $Y(TKE)$. This is shown in Table 1. It is clear that the newly calculated values of $\bar{\nu}$ are significantly higher than the precise experimental value. This is due to the fitted value of $\langle TKE \rangle$ - as $\langle TKE \rangle$ and $\bar{\nu}$ have been shown to be highly anti-correlated, e.g. [16]. To remedy this, we instead fit the five Brosa

N_{modes}	$\langle TKE\rangle$ (MeV)	σ_{TKE} (MeV)	$\bar{\nu}$
O	185.77	8.97	8.758
3	184.46	10.20	3.956
4	184.16	10.74	3.946
5	184.27	10.71	3.948
5_O	185.85	8.85	3.767

Table 1. Observables calculated from CGMF from the default parameterization (O), three, four, and five modes, as well as when five modes were fit to the default yields of CGMF (5_O).

Mode	w	$\overline{A}$	σ	d^{min}	d^{max}	d^{dec}
S2	0.638	142.8	5.04	14.6	18.0	0.226
S1	0.111	135.3	3.34	9.31	17.6	0.095
SL	0.021	126.0	13.2	13.9	17.8	0.376
S3	0.229	148.8	7.02	16.7	18.7	0.299
SX	0.001	152.1	0.032	14.7	21.0	0.172

Table 2. Brosa parameters corresponding to the CGMF calculation in the last row of Table 1.

modes to $Y(A, TKE)$ sampled from the current Gaussian implementation within CGMF. This gives much more accurate results for $\langle TKE\rangle$ and $\bar{\nu}$ (last row of Table 1). The corresponding Brosa parameters are listed in Table 2.

Once $\langle TKE\rangle$ and $\bar{\nu}$ were better reproduced, a sensitivity study was performed to determine the linear response of several observables calculated with CGMF to the parameters of the Brosa modes. This response is calculated as

$$S_{ij} = \frac{x_0^i}{R_0^j}\left(\frac{\partial R^j}{\partial x^i}\right)_{x_0}, \tag{8}$$

where x_0^i are the best-fit parameters, and R_0^j are the calculated observables at x_0. The derivative in Eq. 8 is calculated at x_0 using the three-point midpoint formula. The ratio x_0^i/R_0^j is included for normalization. Responses were calculated for $\langle TKE\rangle$, σ_{TKE}, $\bar{\nu}$, $\langle\nu(\nu-1)\rangle$ (second factorial moment), $\langle\nu(\nu-1)(\nu-2)\rangle$ (third factorial moment), $\bar{\nu}_\gamma$, $\langle\nu_\gamma(\nu_\gamma-1)\rangle$, and $\langle\nu_\gamma(\nu_\gamma-1)(\nu_\gamma-2)\rangle$; x_0 were the parameters from the five mode fit, excluding $\bar{A}_{SL}$, fixed at 126, and w_{SL}, used to enforce the normalization condition described in Sec. 2.2.

Figure 2 shows the results of these calculations. The most influential parameters are d_{S2}^{max}, d_{S3}^{max}, $\bar{A}_{S2}$, and $\bar{A}_{S3}$. The two d^{max} parameters strongly influence the distributions of TKE and number of neutrons, while $\bar{A}$ has a reduced - but noticeable - effect on the distribution of the number of γ rays. This is consistent with what was seen in [10] where $\bar{A}$ and d^{max} of the most prominent mode (S2) had the largest response. When all five modes were included, the S2 and S3 modes contributed to approximately 80% of the yields; these are also the two modes that cause the most impact on the observables calculated here.

Using a MCMC works well when we have an adequate amount of data with which to constrain the model. However, many of the systems that we are interested in describing do not have measured $Y(A, TKE)$ distributions, and it is not clear that we will find the same level of agreement between theory and experiment when this is not available. In addition, to extrapolate to systems where even less is known, trends in energy and mass have to be

https://doi.org/10.1051/epjconf/201921104006

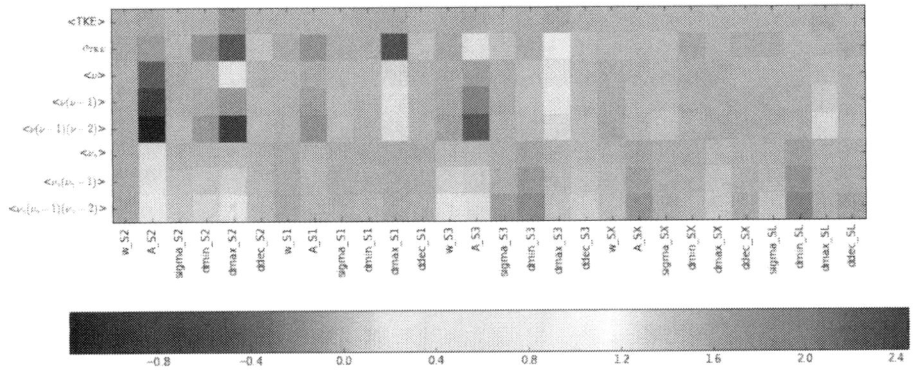

Figure 2. Responses of various observables calculated with CGMF to small changes in the Brosa mode parameters, as defined by Eq. 8.

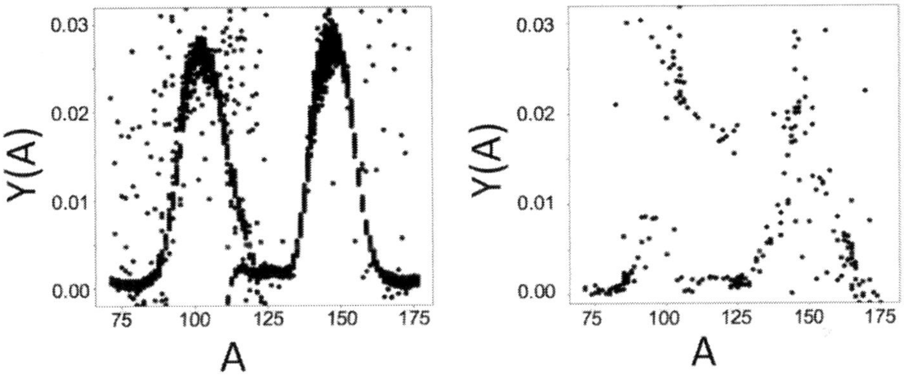

Figure 3. Results of the MDN attempt to learn $Y(A)$ for a sparse data set of ^{238}Pu(sf) (left) using transfer learning with ^{252}Cf(sf) (right) without transfer learning.

explicitly determined for these parameters. For these reasons, we have also begun to explore machine learning techniques.

In this work, we also explore the suitability of a MDN to estimate fission observables. Although the ultimate goal is to calculate all of the prompt fission observables, we began with mass yields. First, we used a three-layer MDN and train 20 Gaussians on CGMF calculations for $Y(A)$ of ^{252}Cf spontaneous fission. This method was able to learn the weights and Gaussian parameters to reproduce the inputted simulations.

We then used simulations of $Y(A)$ from ^{252}Cf to learn the mass yields for spontaneous fission of ^{238}Pu through transfer learning. To do this, we first trained the MDN on ^{252}Cf yields to learn the weights and biases. We transfered these weights and biases to another similar MDN network, which was then trained on a sparse set of ^{238}Pu mass yields containing only 3% of the generated yields. Figure 3 shows the results of this transfer learning from ^{252}Cf(sf) to ^{238}Pu(sf). Even though only a sparse data set was used for ^{238}Pu, the resulting yields show good agreement with the actual yields. However, if the sparse data set is used without the pretraining on ^{252}Cf, the ^{238}Pu yields are nonsensical, as shown in the right panel of Fig. 3.

5 Conclusions

In conclusion, we explored two ways of constraining fission fragment yields, using both parametric models as well as machine learning techniques. Using a Markov Chain Monte Carlo, we were able to reproduce fission fragment yields for the spontaneous fission of ^{252}Cf as a function of mass and total kinetic energy. These yields were implemented into the Hauser-Feschbach Monte Carlo code, CGMF, in order to calculate correlated fission observables. Although the experimental value of $\bar{\nu}$ was not reproduced when fitting to the experimental data of [15], this value was reproduced when $Y(A, TKE)$ sampled from the default version CGMF was used to constrain the Brosa mode parameters. Using not only the yields, but also neutron and γ-ray observables, it should be possible to optimize the Brosa parameters across all of these observables. We took a step towards this global optimization by performing a sensitivity study of the parameters to some of these observables calculated with CGMF.

In parallel, we have been using Mixture Density Networks, a type of probabilistic neural network, to construct fission yields in a non-parametric way. Using a network of three layers and 20 Gaussians, we were able to reproduce simulated $Y(A)$ from CGMF for ^{252}Cf(sf). Furthermore, we showed that by pretraining on ^{252}Cf mass yields, we were able to reproduce sparse ^{238}Pu(sf) mass yields through transfer learning. Although this small demonstration is just a proof-of-principle calculation, it gives an idea of the power of machine learning, not only for predicting the yields of unmeasured isotopes, but also ultimately to make predictions for fissioning systems where there is sparse or non-existent data.

The authors would like to thank Samuel Jones, Ionel Stetcu, and Harsha Nagarajan for useful discussions. This work was supported by the Office of Defense Nuclear Nonproliferation Research & Development (DNN R&D), National Nuclear Security Administration, US Department of Energy. It was performed under the auspices of the National Nuclear Security Administration of the U.S. Department of Energy at Los Alamos National Laboratory under Contract DEAC52-06NA25396. We gratefully acknowledge the support of the U.S. Department of Energy through the LANL/LDRD Program and the Center for Non Linear Studies.

References

[1] P. Jaffke, Nucl. Sci. Eng. **190** 258 (2018)

[2] P. Talou, T. Kawano, I. Stetcu, P. Jaffke, M.E. Rising, and A.E. Lovell, Comp. Phys. Comm. *in preparation*

[3] P. Möller and T. Ichikawa, Eur. Phys. J. A **51** 173 (2015)

[4] A. Sierk, Phys. Rev. C **96** 034603 (2017)

[5] M. Verriere, N. Schunck, and T. Kawano, arXiv:1811.05568v1 [nucl-th] 13 Nov 2018

[6] N. Schunck, D. Duke, H. Carr, and A. Knoll, Phys. Rev. C, **90** 054305 (2014)

[7] D. Regnier, N. Dubray, N. Schunck, and M. Verriere, Phys. Rev. C **93** 054611 (2016)

[8] L. Neufcourt, Y. Cao, W. Nazarewicz, and F. Viens, Phys. Rev. C, **98** 034318 (2018)

[9] L. Neufcourt, Y. Cao, W. Nazarewicz, E. Olsen, and F. Viens, arXiv:1901.07632v1 [nucl-th] 22 Jan 2019

[10] A. Carter, *Sensitivity Analysis and Optimization of Cf-252(sf) Observables in CGMF*, LA-UR-17-31022, Master's Thesis, University of Michigan (2018)

[11] A.C. Wahl, Technical Report LA-13928 (2002)

[12] U. Brosa, S. Grossmann, and A. Müller, Phys. Rep. **197**, 167-262 (1990)

[13] N. Metropolis, A. W. Rosenbluth, M. N. Rosenbluth, A. H. Teller, and E. Teller, J. Chem. Phys. **21**, 1087-1092 (1953)

[14] C.M. Bishop, Neural Computing Research Group Report NCRG/94/004 (1994)

[15] A. Göök, F. Hambsch, and M Vidali, Phys. Rev. C **90**, 064611 (2014)
[16] P. Talou, B. Becker, T. Kawano, M.B. Chadwick, and Y. Danon, Phys. Rev. C **83**, 064612 (2011)

Detailed validation of two deterministic models of prompt emission. Systematic behaviours of residual quantities from the sequential emission

Anabella Tudora [1,*]

[1] University of Bucharest, Faculty of Physics, Ro-077125 Bucharest-Magurele, Romania

Abstract. The results of the PbP and sequential emission modellings describe very well the recent experimental data of $^{235}U(n,f)$. The application of the sequential emission treatment to 49 fission cases has emphasized systematic behaviours of different quantities characterizing the fragments and the prompt emission. These allowed the determination of a general form of the residual temperature distribution for each emission sequence and the inclusion of sequential emission into the Los Alamos model.

1 Basic features of the Point-by-Point and sequential emission modellings

Detailed descriptions of both models were already reported, see Ref.[1] and references therein for the Point-by-Point (PbP) model and Ref.[2] for the sequential emission treatment. Consequently only a few basic features, showing the similarities and differences between these modellings, are briefly mentioned in the following.

Both models work with the same fragmentation range, which is deterministically constructed as following. The initial fragment mass range is going from symmetric fission up to a very asymmetric split (with a step of 1 mass unit). For each mass number A, three or five charge numbers Z are taken into account, as the nearest integer values above and below the most probable charge $Zp(A) = Z_{UCD}(A) + \Delta Z(A)$. The isobaric charge distribution is taken as a narrow Gaussian centred on $Zp(A)$. For each fragmentation the calculations are done at TKE values covering a large range (e.g. from 100 to 200 MeV) with a step size of 5 MeV or even less (e.g. 1 or 2 MeV). Both models use the same TXE partition based on modelling at scission ([1, 3] and references therein), which consists of the calculation of the extra-deformation energy of fragments at scission compared to full acceleration, and the partition of the available excitation energy at scission under the assumptions of statistical equilibrium at scission and level density of nascent fragments in the Fermi-gas regime.

In the PbP model the sequential emission is globally taken into account by a residual temperature distribution $P(T)$ on which the neutron evaporation spectrum in the centre-of-mass frame at a given residual temperature is integrated. In the PbP model the compound nucleus cross-section of the inverse process $\sigma_c(\varepsilon)$ is provided by optical model calculations

* Corresponding author: anabellatudora@hotmail.com

with phenomenological parameterisations adequate for nuclei appearing as fragments (e.g. Becchetti-Greenlees, Koning-Delaroche). Other prescriptions for $\sigma_c(\varepsilon)$ are possible, too, e.g. analytical expressions of $\sigma_c(\varepsilon)$ or even a constant $\sigma_c(\varepsilon)$. The PbP model can work with different prescriptions for the level density parameter of fragments, but usually energy-dependent level density parameters provided by the super-fluid model are employed.

The sequential emission treatment is based on the successive equations of residual temperature, which are solved for each emission sequence k corresponding to each initial fragment A, Z at each TKE value [2]. These recursive equations are solved under the approximations of non-energy dependent level density parameters of fragments, e.g. provided by the Egidy-Bucurescu systematic for the back-shift Fermi-gas (BSFG) model, and an analytical expression of $\sigma_c(\varepsilon)$ (see Ref.[2], for details).

The primary results of the PbP model are the multi-parametric matrices of different quantities, generically labelled $q(A,Z,TKE)$ (e.g. prompt neutron multiplicity $\nu(A,Z,TKE)$, prompt γ-ray energy $E\gamma(A,Z,TKE)$, average prompt neutron energy in the centre-of-mass frame $<\varepsilon>(A,Z,TKE)$ etc.). In the sequential emission treatment the matrix $q(A,Z,TKE)$ is obtained by averaging the quantities corresponding to each sequence $q_k(A,Z,TKE)$ over the number of sequences $k_{max}(A,Z,TKE)$ associated to each initial fragment at each TKE value.

2 Detailed validation of these modellings

The first and most relevant validation of a prompt emission model consists of the comparison of the multi-parametric matrices for different quantities (e.g. $\nu(A,TKE)$, $E\gamma(A,TKE)$, etc.) with the experimental data. This comparison *validates the prompt emission model itself* because the fragment distribution $Y(A,TKE)$ is not involved.

The comparison of different single distributions of prompt emission quantities (e.g. $\nu(A)$, $\nu(TKE)$, $<\varepsilon>(A)$ etc.) and of total average quantities (e.g. $<\nu_p>_{tot}$, $<E\gamma>$, prompt neutron spectrum in the centre-of-mass and laboratory frames, etc.) with the experimental data *validates the prompt emission model together with the Y(A,TKE) distribution* (on which the primary results, i.e. $q(A,TKE)$, are averaged).

The primary model results, i.e. multi-parametric matrices of different quantities, can be compared in detail with the existing experimental data, using the following 2D representations (a) the quantity as a function of TKE for a given fragment mass and (b) the quantity as a function of A for a given TKE value.

The recent experimental data of $\nu(A,TKE)$ for $^{235}U(n,f)$ measured by Göök et al. [4] offer the possibility to validate the PbP and sequential emission models themselves.

Fig.1 shows these $\nu(A,TKE)$ data in the 2D representation denoted by (a) (exemplified for nine fragment mass pairs), which are compared in the upper part with the PbP calculation performed in 2017 [1] and in the lower part with the sequential emission calculation also previously reported [2]. A comparison of the $\nu(A,TKE)$ data in the 2D representation denoted by (b) (open squares) with the results of PbP (red circles) and sequential emission (blue stars) is illustrated in **Fig.2** for four values of TKE. The excellent description of the $\nu(A,TKE)$ data of Ref.[4] by the PbP model result is easily seen. In the case of sequential emission treatment, the agreement with the data is also good but not so remarkable as in the case of PbP. This fact, as well as the staggering observed in a part of ν(A,TKE) results of sequential emission (in both 2D representations) is due to the limited number of initial fragments, which are taken into account in the deterministic construction of the initial fragmentation range. In the case of PbP even if the fragmentation range is the same, this situation is avoided by the global treatment of the sequential emission using the residual temperature distribution $P(T)$ which covers the entire process of successive neutron emission corresponding to each initial fragment.

Consequently the *ν(A,TKE)* results for ^{235}U(n_{th},f) provided by the PbP and sequential emission modellings reported in Refs.[1, 2] are confirmed by the subsequent measurements of Göök et al. [4]. This fact can be considered as a valuable validation of both models.

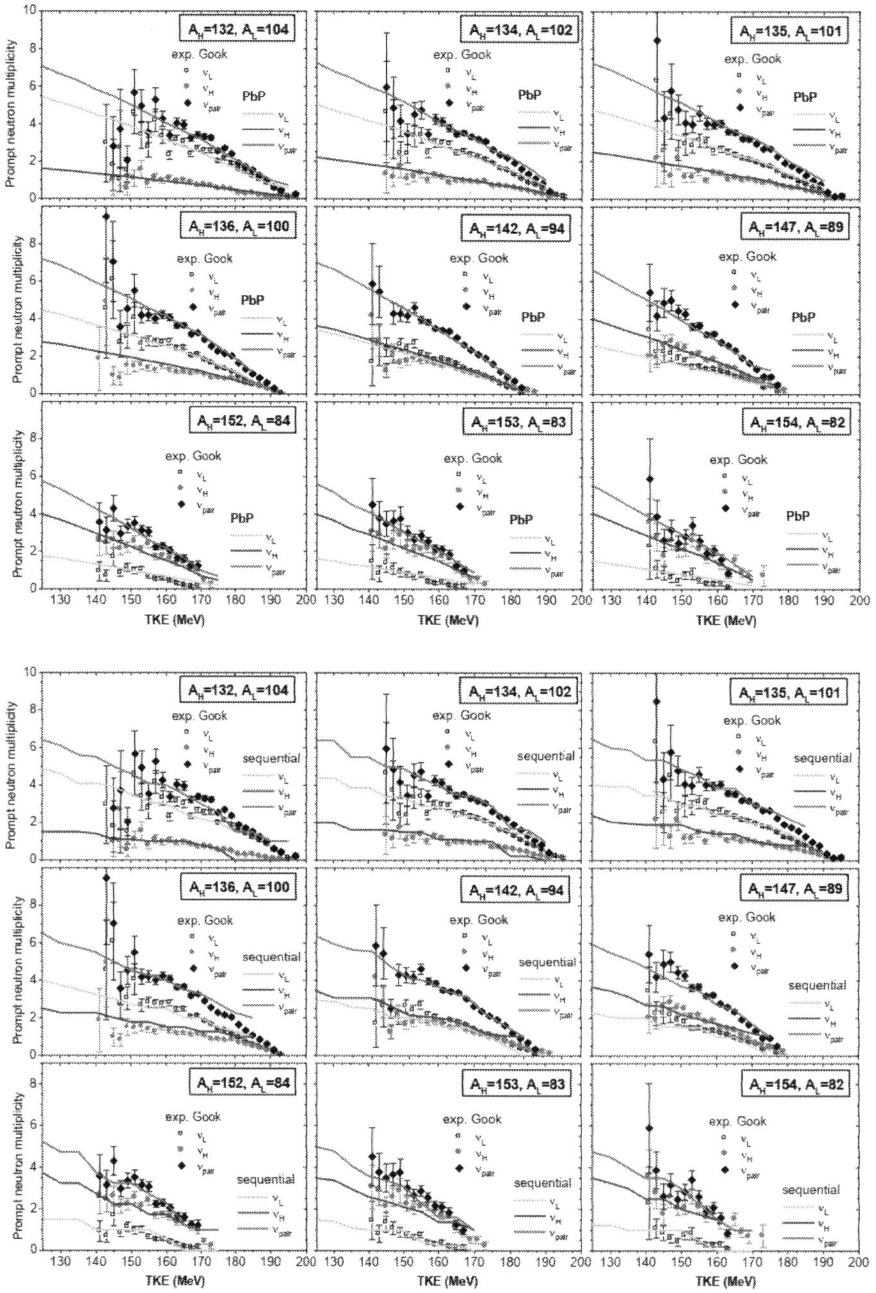

Fig.1. Comparison of the *ν(A,TKE)* results of PbP (upper part) and sequential emission (lower part) with the recent data of Göök et al. in the representation as *ν(TKE)* for a given fragment mass.

PbP and sequential emission results of single distributions of different quantities, describing well the existing experimental data, were already reported, see Refs. [2, 3] and references therein.

The comparison of model results for the prompt neutron spectrum in the centre-of-mass frame $\Phi(\varepsilon)$ with experimental data deserves a special mention because the data for this quantity are almost inexistent. Fortunately the recent experimental $\Phi(\varepsilon)$ data reported in Ref.[4] offer this possibility.

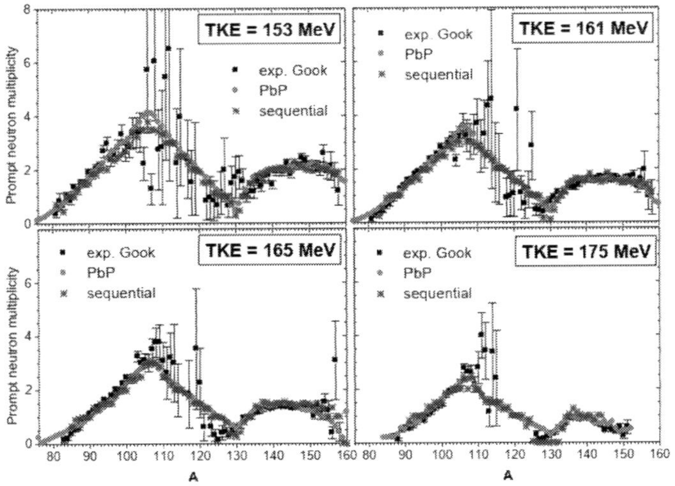

Fig.2. Comparison of the $\nu(A,TKE)$ results of PbP (red circles) and sequential emission (blue stars) with the recent data of Göök et al. (black squares) in the representation as $\nu(A)$ at a given TKE value (which is indicated in each frame).

As an example, **Fig.3** shows the very good description of experimental $\Phi(\varepsilon)$ data for selected fragment mass ranges around the most probable fragmentation [4] (light fragments in the upper part and heavy fragments in the lower part) by the PbP results [1] (representing the spectrum obtained by averaging over the light and heavy fragment groups, respectively).

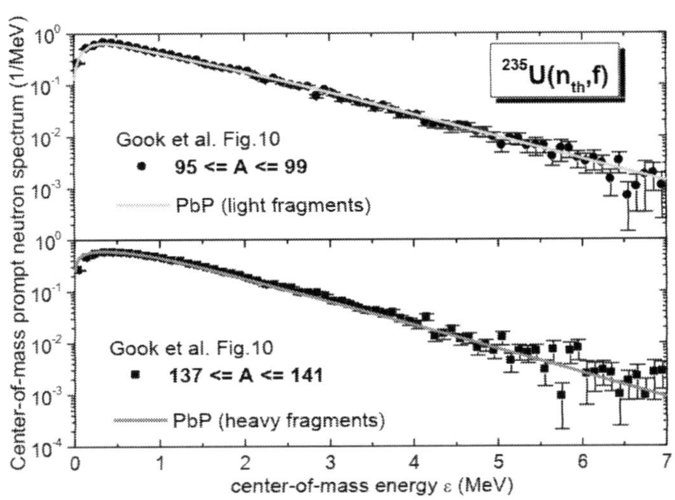

Fig.3. Experimental prompt neutron spectrum in the centre-of-mass frame for selected fragment mass ranges around the most probable fragmentation of Göök et al. in comparison with the PbP result.

3 Systematic behaviours of residual quantities resulting from the sequential emission treatment and possible applications (preliminary results)

As it was mentioned in Ref.[2] the development of a sequential emission modelling has had as initial goal the determination of a general form for *P(T)* needed in prompt emission models with a global treatment of the sequential emission, like PbP and Los Alamos (LA). For this reason the sequential emission modelling was applied to many fissioning systems benefiting of reliable experimental *Y(A,TKE)* data, i.e. the spontaneous fission of ^{252}Cf and of even-even isotopes $^{236\text{-}244}$Pu, the thermal neutron induced fission of ^{235}U, ^{233}U and ^{239}Pu and the fast neutron induced fission of ^{234}U, ^{238}U and ^{237}Np, at incident energies below the threshold of the second chance fission. This means a total number of 49 fission cases covering a large range of fissioning nuclei and TXE values. A first finding, reported in Ref.[2], refers to the total average residual temperatures $<T>$ (corresponding to the sum of residual temperature distributions following the successive emission of all neutrons from the light and heavy fragments and from all fragments) which can be related only to the average temperature of initial fragments $<T_i>$. I.e. the ratios $<T>/<T_i>$ are the same, of about 0.6, for all studied cases irrespective of the prescriptions used for $\sigma_c(\varepsilon)$ and the level density parameters of initial and residual fragments.

The study of such systematic behaviours is in progress. It was ascertained that for each emission sequence (indexed k) the ratios $<T_k>/<T_i>$ are almost the same for all studied fission cases irrespective of the prescriptions used for $\sigma_c(\varepsilon)$ and the level density parameters. This fact is illustrated in the left part of **Fig.4** where the ratios $<T_k>/<T_i>$ of the studied cases are plotted as a function of TXE with different symbols (corresponding to the emission sequences k from 1 to 6). The constant values approximating these calculated ratios for each emission sequence are represented by horizontal lines plotted with the same colours as the respective symbol.

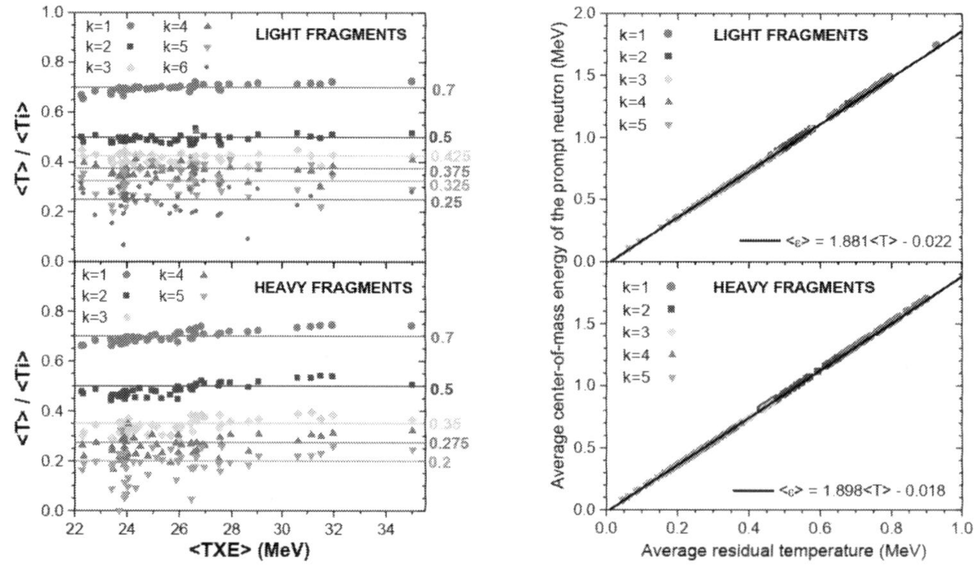

Fig.4. Left part: the ratios $<T_k>/<T_i>$ as a function of TXE resulting from the sequential emission calculations (different symbols corresponding to each sequence) and the constant values approximating these ratios (lines plotted with the same colour as the respective symbol). Right part the average centre-of-mass energy of each emitted neutron $<\varepsilon>_k$ as a function of $<T_k>$ (using the same symbols and colours as in the left part) and its linear fit (black line).

The ratios of the average residual energy to the initial excitation energy $<E_r^{(k)}>/<E^*>$ corresponding to each emission sequence are also almost the same for all investigated fission cases, irrespective of the prescriptions used for the level density parameters and $\sigma_c(\varepsilon)$. Similar to the total average temperature ratio $<T>/<T_i>$ which is of about 0.6 in all cases [2], the ratios of the total average residual energy $<E_r>$ to $<E^*>$ is also the same, of about 0.43, for all studied fissioning systems.

Linear correlations between the average centre-of-mass energy of each emitted prompt neutron $<\varepsilon>_k$ and the corresponding average residual temperature $<T_k>$ and between $<\varepsilon>_k$ and the root-square of $<E_r^{(k)}>$ are also observed. The slopes of these linear dependences allow the determination of global values for the total average level density parameter of the light and heavy fragment groups, which are in good agreement with the general trend of the average level density parameters of the studied cases. Correlations of the average energy carried away per neutron ($<\eta_k> = <\varepsilon>_k + <S_n>_{k-1}$) with different quantities characterizing the fragments (e.g. the average neutron separation energy, the average residual temperature etc.) are also obtained.

The constant values of the ratios $<T_k>/<T_i>$ and $<E_r^{(k)}>/<E^*>$ together with the linear dependence of $<\varepsilon>_k$ on $<T_k>$ can provide indicative values of $<T>$, $<Er>$, $<\varepsilon>$, $<\eta>$ without to use prompt emission models. E.g. if the average initial temperatures of the light and heavy fragment groups $<T_i>_{L,H}$ or of all fragments $<T_i>$ are known for a given fissioning system then $<\varepsilon>_{L,H}$ or $<\varepsilon>$ can be obtained from the linear dependence of $<\varepsilon>$ on $<T>$ (right part of Fig.4) by using the ratio $<T>/<T_i>\approx 0.6$ [2]. As an example for ^{252}Cf(SF): $<TXE>=35$ MeV and the equivalent initial temperature $<T_i>=(<TXE>/<a>)^{1/2}$ based on the average level density parameter which is simply taken as $<a>=252/11\ MeV^{-1}$ lead to $<\varepsilon> = 1.382$ MeV which is in a reasonable agreement with the experimental data.

1.2 Inclusion of the sequential emission into the Los Alamos model

The constant values of the temperature ratios referring to each emission sequence (plotted in the left part of Fig.4), i.e. $r_k = <T_k>/<T_i>$, allow to define a residual temperature distribution $P_k(T)$ for each emission sequence, having a triangular form with a sharp cut-off at high temperatures [2] and the maximum temperature given by $T_{max}^{(k)} = (3/2) r_k < T_i >$. This fact makes possible the inclusion of sequential emission into the LA model. In this case the spectrum in the centre-of-mass frame corresponding to each prompt neutron emitted successively from the light or heavy fragment of the most probable fragmentation is calculated as

$$\Phi_k(\varepsilon) = \int_0^{T\max^{(k)}} \varphi(\varepsilon, T) P_k(T)\, dT \qquad (1)$$

In order to determine the input parameters of the LA model, as average values corresponding to the most probable fragmentation, different prescriptions can be used, concerning: *i)* $\sigma_c(\varepsilon)$ (provided by optical model calculation, analytical expression or constant), *ii)* different methods of TXE partition, *iii)* the level density parameters of fragments (e.g. energy-dependent provided by the super-fluid model or non-energy dependent provided by different systematics, etc.).

An example of prompt neutron spectra in the centre-of-mass frame provided by the LA model including the sequential emission is given for ^{235}U(n_{th},f) in **Fig.5** where the centre-of mass spectrum of the first emitted neutron is plotted with a blue dashed line and of the second emitted neutron with a green dash-dotted line. The total centre-of-mass spectrum (obtained by averaging the spectra of each emitted neutron over the probability for emission of each neutron Pn_k [2]) is plotted with a red solid line. As it can be seen it describes very well the experimental data of Göök et al. [4]. Note, in this example the following prescriptions were used: $\sigma_c(\varepsilon)$ from optical model calculations with the Becchetti-Greenlees parameterisation, the TXE partition from modelling at scission ([1, 3] and references therein) and level density parameters provided by the super-fluid model.

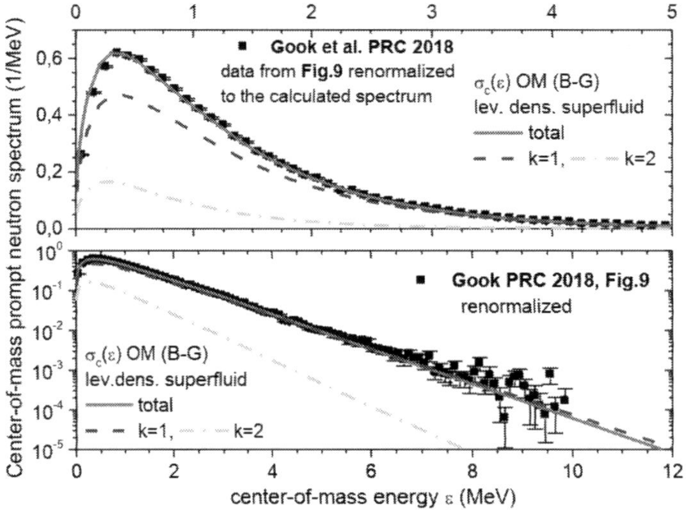

Fig.5: LA model calculation of the centre-of-mass spectrum for the first emitted neutron (dashed blue line), the second emitted neutron (green dash-dotted line) and the total centre-of-mass spectrum (solid red line) in comparison with the experimental data of Göök et al.[4] (black squares).

4 Conclusions

The very good description of the recent prompt emission data of Göök et al. [4] (especially those of prompt neutron multiplicity as a function of A and TKE) by the results of the PbP and sequential emission modellings previously reported [1, 2], constitutes a valuable validation of these models.

The application of the deterministic treatment of sequential emission [2] to 49 fission cases (covering a large range of fissioning nuclei and excitation energies) emphasized systematic behaviours of different average quantities characterizing the initial and residual fragments and the prompt emission. The systematics concerning the average residual temperatures allowed to define a general form of the residual temperature distribution associated to each emission sequence $P_k(T)$. This fact makes possible the inclusion of sequential emission into the Los Alamos model of prompt emission.

Acknowledgements: this work was done in the frame of the Romanian research project PN-III-P4-PCE-2016-0014 (contract 7/2017).

References

1. A. Tudora, F.-J. Hambsch, Eur. Phys. J. A **53**, 159 (2017)
2. A. Tudora, F.-J. Hambsch, V. Tobosaru, Eur. Phys. J. A **54**, 87 (2018)
3. R. Capote et al., Nucl. Data Sheets **131**, 1-106 (2016)
4. A. Göök, F.-J. Hambsch, S. Oberstedt, M. Vidali, Phys. Rev. C (2018) (to be published)

Advanced Modelling of $^{238}U(n,f)$ in a Fast Reactor Application

Cristiana Oprea[1,2], *Alexandru* Mihul[2,3] and *Alexandru* Oprea[1]

[1]Joint Institute for Nuclear Research (JINR), Frank Laboratory of Neutron Physics (FLNP), 6 Joliot Curie, 141980 Dubna, Moscow Region, Russia
[2]Romanian National Authority for Scientific Research, Mendeleev 21-25, 010362 1 Bucharest, RO
[3]CERN, CMS, CH-1211 Genève 23, Switzerland

Abstract. Fast neutron reactors, as a possible future solution on energy demand of human society, based on fission process of ^{238}U, request new and reliable nuclear data necessary for new generation reactors design. Fission process induced by fast neutrons on ^{238}U was investigated. Fission observables like cross sections and their uncertainties, fission fragment mass distribution, prompt neutrons emission, isomer ratios and other parameters were obtained by using Talys computer code or programs realized by authors. Then the production of isotopes like $^{135,133}Xe$, ^{99}Mo, ^{131}I, ^{89}Y as well as yields of fissile nuclei were evaluated. Obtained theoretical evaluations are compared with existing experimental data.

1 Introduction

Forecast for world energy demand predicts an increasing with about order of magnitude for the year 2100 in comparison with 2015. In the '80 years of the last century the scientists demonstrated that hydrocarbs based energy will end in a few decades. Almost the same situation is for nuclear energy produced by ^{235}U fission. Alternative new sources using solar or wind energy are still expensive and not effective [1]. The most appropriate solution for future energy challenge can be considered the fission of ^{238}U (^{8}U) nucleus [2].

^{8}U fission process in power generation IV reactors is important for a wide range of fundamental and applied researches issues [3,4]. For basic investigations, ^{8}U fission offers new information about configuration of fissionable systems near scission point, anisotropy, emitted gamma, ground states of fission product. ^{8}U fission represents also a source of new isotopes for specific applications in medicine, electronics, nuclear technologies etc.

In this paper fission of ^{8}U nucleus induced by fast neutrons starting from 7 up to 25 MeV was detailed investigated. Cross sections, fragment mass distributions, yields of some nuclides of interest, average prompt neutrons multiplicity, prompt neutrons multiplicity distributions and isomer ratios were obtained with the extensively - used reaction code Talys [5], version 1.9. The isomer ratios were calculated with an author's computer program by using cross sections from Talys. The present evaluations of a large set of fission variables are necessary in the planning and analysis of future experiments to be realized at FLNP - JINR basic facilities.

2 Codes and elements of theory

Fast neutron induced fission process on [8]U can be described in the frame of statistical model; then cross sections can be evaluated by using of Hauser - Feshbach approach [5,6]. For fission, transmissions are calculated using Hill-Wheeler formulas [5, 7].

Fragment mass distribution, isotopes relative yields and production cross sections, average prompt neutrons multiplicity and prompt neutrons distributions were modeled by Brosa model [5,8].

[8]UIsomer ratios knowledge allows extracting the spin distribution of fission fragments, levels density dependence on angular momentum and other important parameters. The loss of charged particles in simple case of thin target can be neglected, than the experimental isomer ratios are defined as:

$$R = Y_m \cdot [Y_g]^{-1} = \int_{Ethr}^{E\max} N_0 \phi_n(E) \sigma_{nf}^m(E) dE \cdot \left[\int_{Ethr}^{E\max} N_0 \phi_n(E) \sigma_{nf}^g(E) dE \right]^{-1} \quad (1)$$

where: $Y_{m,g}$ = yields of isotope in isomer (m) and ground (g) states, respectively; N_0 = number of target nuclei; ϕ_n = flux of incident beam; $\sigma^{m,g}$ = cross section production of m and g states, respectively; E_{thr} = threshold energy of emergent particle; E_{max} = maximum energy of incident beam.

For default Talys run isomer and ground states production cross sections are not given for [8]U fission as in the case of other nuclear reactions. In this case, according to Huizenga approach, it is supposed that the relative yields of isomer and ground states are proportional with spin distribution [9 - 12]. It has the form [9]:

$$P(J) = (2J+1)(2\sigma^2)^{-1} Exp\left[-(J+0.5)^2 (2\sigma^2)^{-1} \right] \quad (2)$$

where: J = spin; σ = spin cut - off parameter.

3 Results and discussions

Results for fast neutron induced fission of [8]U were obtained by varying a large number of Talys input data like parameters of Wood-Saxon potential (volume and surface types) and spin-orbit potential, both with real and imaginary part. A double humped potential barrier was considered. Coupled channels calculations were enabled considering rotational band and deformations available in Talys database. For residual nuclei 10 discrete levels were considered in evaluations and 40 levels for target nucleus and formed fissile nuclei. Experimental fission barriers suggested by Talys were considered [5]. The most important parameters of n + [8]U channel are given in Tables 1 and 2.

Table 1. First and second barriers properties

	Height [MeV]	Width [MeV]	Type of axiality
First barrier	6.35	0.7	Triaxial + left - right asymmetry
Second barrier	5.80	0.5	Left - right asymmetry

Table 2. Optical Wood-Saxon potential parameters. Central. Surface. Spin-orbit

	Volume			Surface			Spin - orbit		
	U_V [MeV]	r_V [fm]	a_V [fm]	U_d [MeV]	r_d [fm]	a_d [fm]	U_{so} [MeV]	r_{so} [fm]	a_{so} [fm]
Real	46.56	1.245	0.660	0	0	0	5.716	1.121	0.590
Im	0.05	1.284	0.594	1.690	1.208	0.614	0.003	1.121	0.590

Total fission cross section of ^{8}U nucleus induced by neutrons with energy from 7 MeV up to 25 MeV is shown in Fig. 1. Results with Talys default run are in Fig. 1.a). Due to (n,xn), (n,xp) and other reactions the following excited fissile nuclei are formed: ^{236}U, ^{237}U, ^{239}U, ^{236}Pa, ^{237}Pa, ^{238}Pa, ^{232}Th, ^{233}Th, ^{234Th}, etc. The contributions of ^{236}U, ^{237}U, ^{238}U, ^{239}U fissile nuclei to the total fission cross section are shown in Fig.1.a). Fission cross sections of ^{236}Pa, ^{237}Pa, ^{238}Pa, ^{232}Th, ^{233}Th, ^{234}Th have low values and therefore are not shown in Fig 1.a). In Fig. 1.b) the total fission calculated cross section is compared with experimental data [13]. Knowledge of the contribution of the above fissile nuclei to total fission cross section is obvious important in nuclear technology. Agreement between theoretical calculations and experimental data was obtained by varying parameters of Wood-Saxon and spin-orbit optical potentials together with levels density. The experimental statistical uncertainties reduction was done by Talys using a Bayesian Monte Carlo procedure based on the EXFOR database and they were in fair agreement with the standards [14,15]. For incident energies higher than 20 MeV, the height of first and second fission barriers of ^{236,237,238,239}U nuclei was modified from 10% up to 30%.

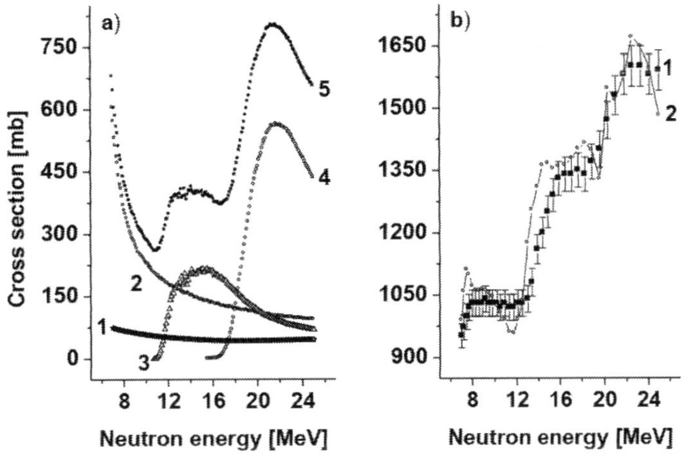

Fig. 1. Total cross section (TXS). a) TXS and contribution of fissile nuclei: 1- ^{239}U; 2- ^{238}U; 3- ^{237}U; 4- ^{236}U; 5- 1+2+3+4. b) Comparison between: 1) Experiment [13]; 2) Talys (Theory)

Relative yields of fission fragment mass distribution were obtained for a large incident neutron energy up to 25 MeV applying Brosa model [5,8]. In Fig. 2 mass distribution results are shown for neutron energy, E_n= 10, 18, 25 MeV, respectively. In Fig. 2.a) the pre-neutrons emission case is represented and in Fig. 2.b) the post-neutrons ones. With the increasing of neutron energy the asymmetry of mass distribution is reducing. Furthermore, with neutron energy, the mass distribution is enlarging, including new light and heavy fragments (see Fig 2.a) and 2.b)). In Fig. 2.c) calculations were verified for 5.5 MeV neutron energy for which the experimental data are well described [16].

Results related to prompt neutron emission are presented in Fig. 3. In this figure are selected 10, 18 and 25 MeV neutrons energy cases also. In Figs. 3a) and 3b) the dependence of average neutron multiplicity on fission fragment mass are represented before and after neutron emission. As it is expected the number of emitted neutrons by fission fragments is increasing with incident energy. As seen in Fig. 2 with the increasing of neutron energy the mass distribution is enlarging with the fragment mass A.

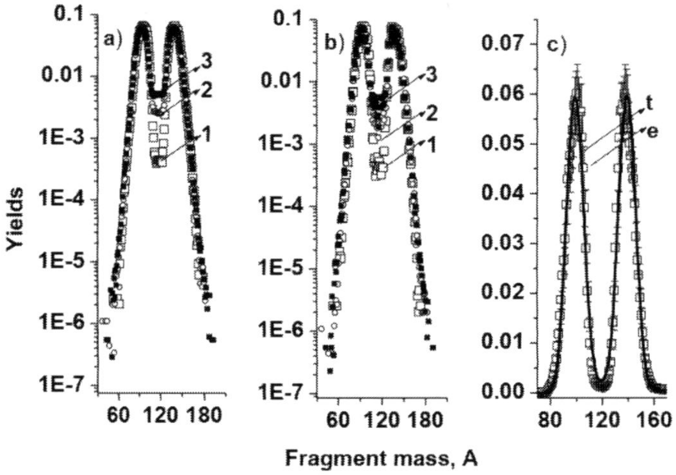

Fig. 2. Mass distribution (MD). Neutron emission: a) Pre; b) Post; c) MD for 5.5 MeV; 1 - 10 MeV; 2 - 18 MeV; 3 - 25 MeV; t – Talys evaluation; e – experimental data

In Fig. 3c) prompt neutron multiplicity distributions are shown for the same energy as in Figs. 3a) and 3b). Multiplicity is rising with incident neutron energy and is limited by excitation energy. Further parameter extracted is the average prompt neutrons (nubar-prompt), $\bar{\nu}_{pr}$. For 10, 18 and 25 MeV, the neutron induced fission cross section is σ_{nf} = 1020, 1406, 1482 mb, and average prompt neutrons, $\bar{\nu}_{pr}$ = 4.15, 5.11, 5.86 neutrons, respectively.

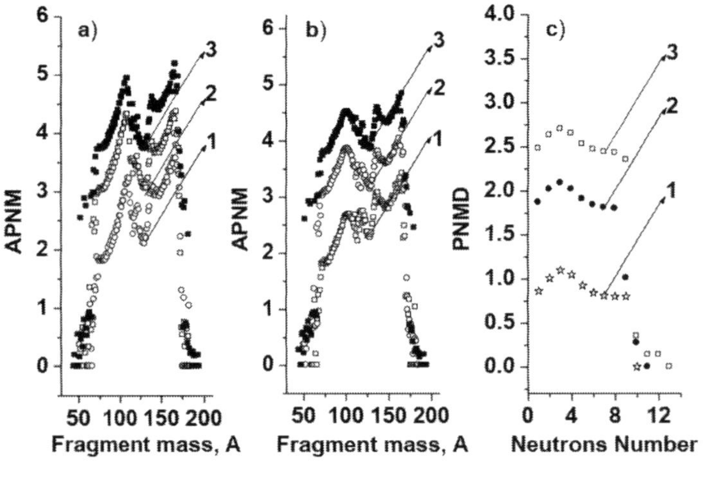

Fig. 3. Prompt neutrons. a) + b) Average prompt neutron multiplicity (APNM); a) Pre b) Post; c) Prompt neutron multiplicity distribution (PNMD). 1 - 10 MeV; 2 - 18- MeV; 3 - 25 MeV

From the large number of nuclides obtained in fission process, ^{99}Mo, ^{131}I, ^{133}Xe and ^{135}Xe isotopes were selected. The ^{99}Mo nucleus has applications of great interest in cancer treatment. The ^{131}I isotope is important in radiological protection, nuclear medicine and also can be used as industrial tracer. The last nuclide, ^{133}Xe, has many applications in medicine.

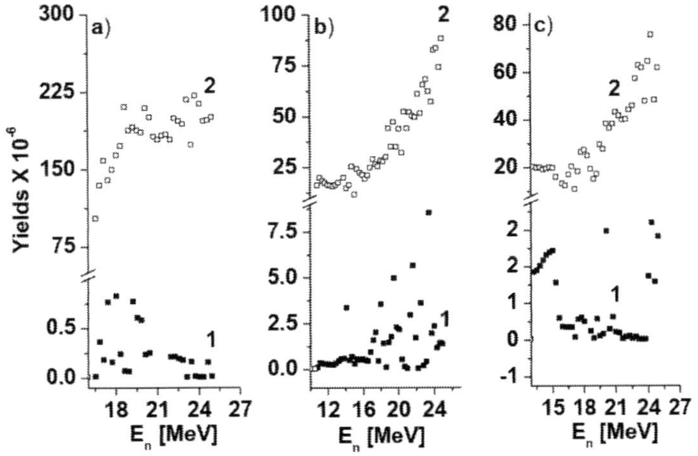

Fig. 4. Relative yields isotope production as function of incident neutron energy: a) ^{99}Mo; b) ^{131}I; c) ^{133}Xe; Neutron emission: 1- Pre; 2- Post

Production of ^{99}Mo, ^{131}I and ^{133}Xe isotopes in fast neutron induced fission of ^{8}U is very low and then not calculated with a default Talys run. In a default Talys run, for reducing computing time, cross sections lower than 10^{-7} mb are not considered. For obtaining results from Fig. 4 it is necessary to extend the limit for cross section calculations up to 10^{-10} mb. In Figs. 4.a - 4.c) are represented the energy dependences of relative yields for production of ^{99}Mo, ^{131}I and ^{133}Xe isotopes before and after neutron emission.

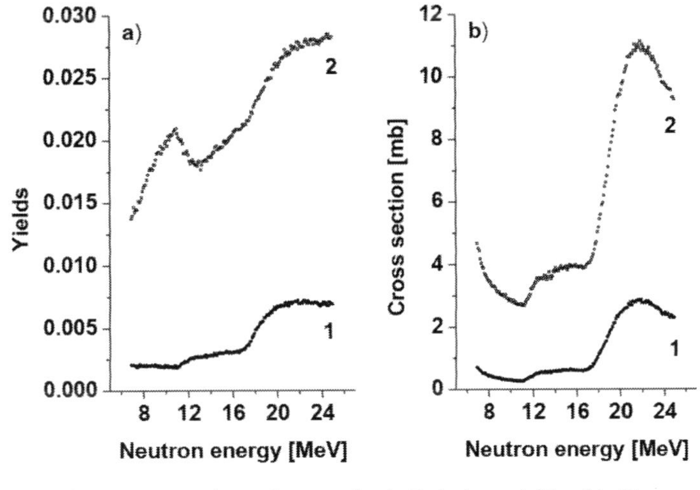

Fig. 5. ^{135}Xe production. Energy dependence of: a) Relative yields; b) Cross section; Neutron emission: 1 - Pre; 2 - Post.

Another isotope of interest for nuclear technology is ^{135}Xe nucleus, which is a major fission product and an important neutron absorber. In Fig. 5 the energy dependence of relative yields and cross section production are represented before and after neutron emission. The results from Fig. 5 are obtained with default Talys precision. Relative yields for ^{99}Mo, ^{131}I, ^{133}Xe nuclei (Figs. 4.a - c) are much lower than in the case of ^{135}Xe (Fig. 5.a). The same situation is in the case of cross section production. Cross section for ^{135}Xe is of order of milibarns (Fig. 5.b) and for ^{99}Mo, ^{131}I and ^{133}Xe isotopes is about 10^{-3}mb or lower.

The relative yields and cross sections evaluated with Talys for a large number of isotopes were further used to calculate isomer ratios.

Isomer ratios in $^{8}U(n,f)^{133m,g}Xe$ process with neutrons from 7 up to 20 MeV were evaluated. Fission cross sections are taken from Talys. Spin distributions are evaluated using expression (2) with parameters from Talys database. The flux of incident neutrons is considered proportional with $(1/E_n)^{0.9}$, characteristic for pulsed neutron source available at FLNP. Spin, parity, time of life of $^{133m,g}Xe$ states are taken from [18]. Using the procedure from [11,12], the calculated isomer ratio is:

$$R = 3.95 \pm 0.42 \tag{3}$$

The error from (3) comes from the fact that integrals are sums with a 0.2 MeV step. The obtained isomer ratio is in agreement with similar experimental data [18].

Another nucleus of interest is Yttrium because it has no natural isotopes [19,20]. Artificial Yttrium isotopes with masses from 87 up to 111 can be produced in ^{8}U fission with fast neutrons. Their production depends on the incident neutron flux and therefore natural ^{89}Y nucleus can be used in fast neutron threshold detectors. For the calculation of relative yields and cross sections of Yttrium isotopes, the cross section lower limit was extended from 10^{-7} mb to 10^{-10} mb. For example, production cross section of ^{89}Y after neutron emission is σ_{nf}= 5.36 × 10^{-4} mb for E_n=25 MeV. Relative yields and cross section production of ^{98}Y are shown in Fig 6.a) and 6.b). Analysis of $^{89}Y(n,xn)$ processes with fast neutrons was done in [21] and $^{89}Y(n,xn)$ cross section results are represented in Figs. 6.c).

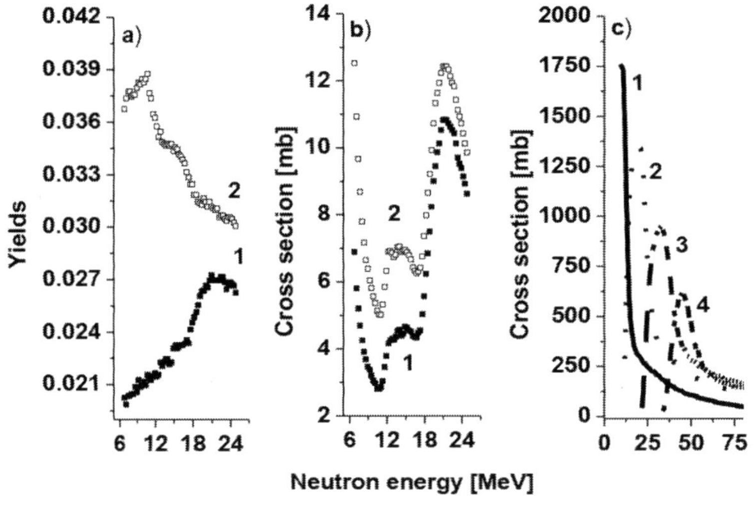

Fig.6. Yttrium isotopes production in $^{8}U(n,f)$. a) Relative yields of ^{98}Y; b) Production cross section of ^{98}Y. Neutron emission: 1 - pre; 2 - post. $^{89}Y(n,xn)$, x=1,2,3,4. Production of: 1 - ^{89}Y; 2 - ^{88}Y; 1 - ^{87}Y; c) Cross section production of Yttrium isotopes in (n,xn) reactions [21]. 1 -^{89}Y; 2 -^{88}Y; 3 -^{87}Y; 4 -^{86}Y

4. Conclusions

Fast neutron induced fission observables on ^{8}U were investigated. The theoretical modeling and calculations were performed for cross sections, mass distributions, average prompt neutron multiplicity, prompt neutron multiplicity distributions, and isotopes production necessary in applications for neutrons with energy from 7 MeV up to 25 MeV were obtained. The isomer ratios using fission cross sections evaluated with Talys and the statistical approach from [11] were calculated. Yttrium isotopes production in neutron induced fission of ^{8}U nucleus as well as $^{89}Y(n,xn)$ reactions necessary for fast neutron threshold detectors were analyzed.

This paper stated the need for new fission experimental data, improvement of the analyses of existing ones and computer modeling of fission measured observables. In this respect future measurements are planned to be realized at JINR Dubna basic facilities at the new neutron source IREN from FLNP and Microtron MT-25 from Flerov Laboratory for Nuclear Reactions.

Acknowledgements. The present work was realized in the frame of the Annual Program of Cooperation between Romanian research institutes and Joint Institute for Nuclear Research leaded by Plenipotentiary Representative of Romanian Government to JINR Dubna and 1128 Theme Plan of FLNP at 2017-2018 years, Research Project 323-107-21-05-2018.

References

1 P. Moriarty, S. J. Wang, Energy Procedia **75,** p. 2523 (2015)
2 A. Horvath, E. Rachlew, Ambio **45**, Supplement **1**, p. 38 (2016)
3 G. H. Marcus, Progress in Nuclear Energy, **37**, **1 - 4**, 19 (2000)
4 V. Nian, Energy Procedia, **75**, 2852 (2015)
5 A. J. Koning, S. Hilaire and M. C. Duijvestijn, TALYS-1.0., *Proceedings of the International Conference on Nuclear Data for Science and Technology*, April 22-27, 2007, Nice, France, editors O. Bersillon, F. Gunsing, E. Bauge, R. Jacqmin, S. Leray, EDP Sciences, 211 (2008)
6 W. Hauser, H. Feshbach, Phys. Rev. **87**, **2**, 366 (1952)
7 D. L. Hill, J. A. Wheeler, Phys. Rev. **89**, 1102 (1953)
8 U. Brosa, S. Grossmann, A. Muller, Phys. Rep **197**, 167 (1990)
9 A. Gilbert, A. G. W. Cameron, Can. J. Phys. **43**, 1446 (1965)
10 A. J. Koning, S. Hilaire, S. Goriely, Nucl. Phys. **A**, **810**, 13 (2008)
11 R. Huizenga, R. Vandenbosh, Phys. Rev. **120,** 1305 (1960)
12 R. Vandenbosh, J. R. Huizenga, Phys. Rev. **120,** 1313 (1960)
13 F. Toversson, A. Laptev, T. S. Hill, Nucl. Sci. and Eng. **178**, **1**, 57 (2014)
14 R. Rochman, O. Leray, A. Vasiliev, H. Ferroukhi, A.J. Koning, M. Fleming, J.C. Sublet, Annals of Nucl. Energy **95**, 125 (2016)
15 N. Otuka et al., Nucl. Data Sheets **120**, 272 (2014)
16 F. Vives, F. - J. Hambsch, H. Bax, S. Oberstedt, Nucl. Phys. **A 662**, 63 (2000)
17 R. B. Firestone, V.S. Shirley, S. Y. F. Chu, C. M. Baglin, J. Zipkin, *Tables of Isotopes*. CD ROM Edition, Wiley-Interscience (1996)
18 G. P. Ford, K. Wolfsberg, B. R. Erdal, Phys. Rev. C **30**, **1**, 195 (1984)
19 G. Audi, A.H. Wapstra, Nucl. Phys. **A, 565**, 1 (1993)
20 G. Audi, A.H. Wapstra, Nucl. Phys. **A, 595**, 408 (1993)
21 C. Oprea, A. Oprea, European Physical Journal, Web Conference EDP Sciences **111, 11002**, 6 (2016)

Current status of the verification and processing system GALILÉE-1 for Evaluated Data

Mireille Coste-Delclaux[1], *Cédric* Jouanne[1,*], *Frédéric* Moreau[1], *Claude* Mounier[1], *Thierry* Visonneau[1], and *Fadhel* Malouch[1]

[1] DEN, Service d'Études des Réacteurs et de Mathématiques Appliquées (SERMA), CEA, Université Paris-Saclay, F-91191, Gif-sur-Yvette, France

Abstract. This paper describes the current status of GALILÉE-1 that is the new verification and processing system for evaluated data, developed at CEA. It consists of various components respectively dedicated to read/write the evaluated data whatever the format is, to diagnose inconsistencies in the evaluated data and to provide continuous-energy and multigroup data as well as probability tables for transport and depletion codes. All these components are written in C++ language and share the same objects. Cross-comparisons with other processing systems (NJOY, CALENDF or PREPRO) are systematically carried out at each step in order to fully master possible discrepancies. Some results of such comparisons are provided.

1 Introduction

GALILÉE-1 system, written in C++ language is a new verification and processing system for evaluated data. It is part of a CEA global development program dedicated to fine modelling of nuclear systems. At the present time, three main components are under development:

- GALION (GALilée Input Output for Nuclear data): dedicated to read evaluated data and write produced data.
- GALVANE (GALilée Verification of the Accuracy of Nuclear Evaluations): dedicated to verify nuclear evaluations that are GALILÉE-1 input data.
- GTREND (Galilée TReatment of Evaluated Nuclear Data): dedicated to provide continuous-energy (CE) and multigroup (MG) data as well as probability tables (PT).

Additional components, such as interface modules creating consistent libraries for application codes or a convivial and automatic chain for creating these libraries, will be developed later.

GALILÉE-1 system originality lays in its two complementary running ways: an integrated one and an open one for providing application codes with processing tools.

* Corresponding author: cedric.jouanne@cea.fr

2 GALILÉE-1 system description

GALILÉE-1 system is built upon GBASE component that defines and implements a set of common objects, shared by all other GALILÉE-1 components. GBASE objects are completely independent from the input and output data formats.

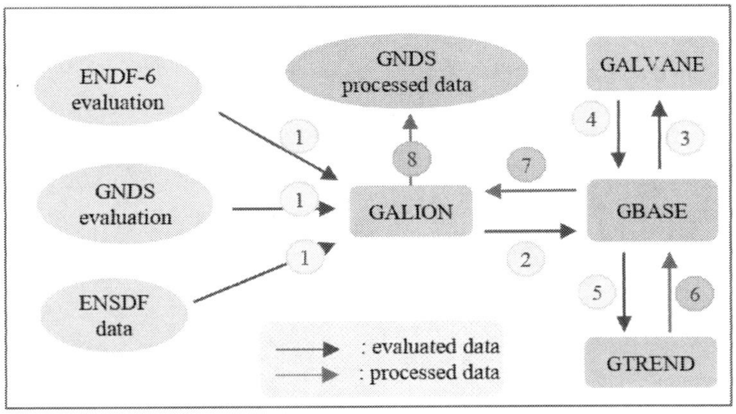

Fig. 1. GALILÉE-1 processing modules.

As shown in Figure 1, GBASE objects are initialized thanks to GALION that reads the evaluation or the structure data. These objects are checked and possibly corrected by GALVANE and then processed data are created by GTREND. One has to note that GALVANE and GTREND only work on GBASE objects, which allows the same verification and processing stages, whatever the evaluation format is. The objects storing processed data are kept in GBASE and can be written on binary or ASCII files by GALION.

3 GBASE Objects

The GBASE object hierarchy is very close to the GNDS object hierarchy. For each nucleus or element, we create a database allowing us to store, in the same object, structure data and interaction data for a given projectile. "GBASE structure data" contain all the information needed to verify and optionally correct the evaluated data: masses, level scheme, spins, energy, half-life, decay modes, etc. "GBASE interaction data" contain:

- the list of products that can be created by the interaction,
- all the information given in an evaluation (JEFF-3.3, ENDF/B-VIII, JENDL-4.0,…) but organized in such a way that processing is easier,
- data processed using GALILÉE-1 (CE data, Probability Tables, MG data, …).

Several GBASE structure data or several GBASE interaction data may exist in the same database.

4 GALION Module

GALION can read evaluated data in ENDF-6 or GNDS (under progress) format as well as structure data in ENSDF or NUBASE format. It supplies tools for creating the GBASE objects corresponding to structure data or evaluated data. It also provides tools for writing continuous energy data in PENDF format.

5 GALVANE Module

One of the goals of GALILÉE-1 system is to test the consistency and the validity of nuclear data evaluations. We plan to perform a complete assessment of evaluated files before any treatment. Currently, GALVANE can diagnose inconsistencies in general information, resonance parameters, Q reaction values, thresholds, excited level schemes, kinematic data of emitted particles and thermal scattering laws. Some checks can be performed by comparing data with the ones contained in structure databases, e.g. NUBASE or ENSDF. This is the case for:

- masses of nuclides, given in terms of neutron mass,
- energies of excited states reached in the inelastic scattering,
- gamma decay schemes of the excited states.

Some additional tests are designed to check the consistency between the data given in an evaluation:

- consistency between thresholds considered for various data of the same reaction,
- energy balance for reaction products,
- spin/parity of resonance parameters,
- normalization of distributions.

6 GTREND Module

GTREND code aims at replacing NJOY [1] and CALENDF [2] codes in CEA application library production. It consists of three main parts, GTREND_CE corresponding to NJOY/RECONR, /BROADR, /UNRESR, /THERMR and /HEATR, GTREND_PT corresponding to CALENDF and GTREND_MG corresponding to NJOY/GROUPR.

Today, GTREND can reconstruct continuous energy cross-sections in the resolved resonance range, averaged cross-sections in the unresolved resonance range, generate a linearization grid, broaden linearized cross-sections, and calculate moment based probability tables.

6.1 Reconstruction in the resolved resonance range

Formalism

Following references [3], [4] and [5], in scattering theory, a channel c is characterized by the pair α of two particles making up the channel, the orbital angular momentum of the pair ℓ, the channel spin s (including associated parity) that is the sum of the spins of the two particles of the pair and the total angular momentum J (including associated parity). A spin group of channels is defined as a set of channels with the same total angular momentum J. The angle-integrated cross-section from entrance channel c to exit channel c', with total angular momemtum J , is given, in terms of the scattering matrix $U_{cc'}$, by:

$$\sigma_{c,c'} = \frac{\pi}{k_\alpha^2} g_{J\alpha} \left| e^{2iw_c}\delta_{cc'} - U_{cc'} \right| \delta_{JJ'} \quad (1)$$

where k_α is the wave number, $g_{J\alpha}$ the spin statistical factor and w_c is the difference between the Coulomb phase shift for a given ℓ and for $\ell = 0$ (this difference is equal to zero for non-Coulomb channels).

The scattering matrix U, that describes the transition between entrance and exit channels, can be written:

$$U = \Omega[1 + 2iP^{\frac{1}{2}}(1 - RL)^{-1}RP^{\frac{1}{2}}]\Omega \quad (2)$$

- Ω is the diagonal matrix which diagonal term is given by $\Omega_{c,c} = e^{-i(w_c - \varphi_c)}$, φ_c being the potential-scattering phase shift,
- L is the diagonal matrix which diagonal term is given by $L_{c,c} = S_c + iP_c - B_c$, S_c, P_c, B_c being respectively the shift factor, the penetrability and a boundary condition (real functions),
- P is the diagonal matrix which diagonal term is equal to P_c,
- R is the channel matrix which terms are defined by:

$$R_{c,c'} = \sum_\lambda \frac{\gamma_{\lambda,c}\gamma_{\lambda,c'}}{E_\lambda - E}\delta_{JJ'} \quad (3)$$

$\gamma_{\lambda,c}$ standing for the channel amplitude, E_λ for the energy of the level, E for the neutron kinetic energy and J and J' for respectively the total angular momentum (with associated parity) of the channel c and c'.

The scattering matrix U can be written in terms of X matrix as:

$$U = \Omega[1 + 2iX]\Omega \quad (4)$$

$$X = P^{\frac{1}{2}}(1 - RL)^{-1}RP^{\frac{1}{2}} \quad (5)$$

In GTREND, the angle integrated cross-section for the interaction leading from particle pair α, for which one particle is a neutron, to particle pair α', is deduced from X matrix in the following way:

$$\sigma_{\alpha\alpha'} = \frac{4\pi}{k_\alpha^2}\sum_J g_{J\alpha} \sum_c \left[\left(sin^2(\varphi_c)(1 - 2Im(X_{cc})) - \sin(2\varphi_c)\, Re(X_{cc})\right)\delta_{cc'} + \sum_{c'}|X_{cc'}|^2\right] \quad (6)$$

The summations are over channels c and c' belonging to the spin group J and such that the particle pair is α for c and α' for c'.

The nuclear formalisms currently supported in GTREND are Single and Multi-level Breit-Wigner, Reich-Moore, and R-Matrix-Limited formalisms. The classical approximations are implemented.

Results

In order to validate GTREND reconstruction at 0 Kelvin in the resolved resonance range, cross-comparisons with NJOY2016 were carried out for all JEFF-3.2 nuclei (~ 470). About 25 nuclei show relative reconstruction discrepancies larger than 10^{-5}, between NJOY2016 and GTREND, for MT1, MT2, MT102 and MT18 when it exists. All observed discrepancies can be explained. Figure 2 shows a comparison of the cross-section reconstructions with NJOY2016 and GTREND.

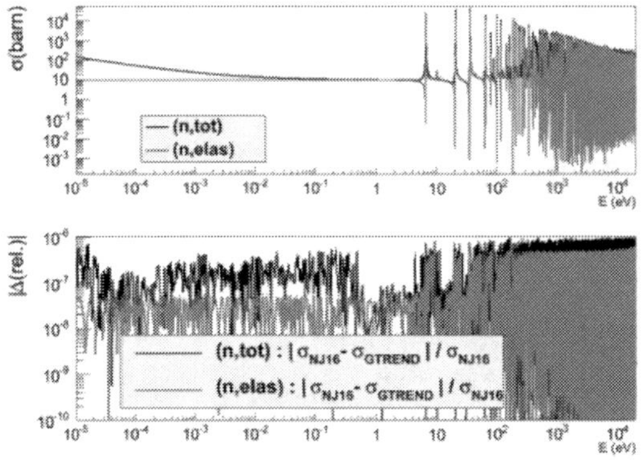

Fig. 2. Comparison between NJOY2016 and GTREND on U238 reconstruction in RRR

6.2 Linear piecewise reconstruction

A tool building a piecewise linear approximation of the cross-sections derived from a nuclear formalism (accurate but time consuming) is very convenient and efficient for the processing itself and for Monte Carlo transport codes. In the frame of GALILÉE-1, we developed a generic linearization tool. Various concrete criteria are implemented such as Punctual, Integral or Mixed ones. When an integral criterion is used a weighting function W is mandatory.

The function F to be represented in a piecewise linear form has to be a "functor" with a specific signature. This allows us to use the same algorithm for various types of functions: resonance formalism, thermal scattering model, Legendre expansion of an angular distribution or tabulated function with various interpolation schemes. The user must give, as input data, well-chosen points in the linearization range.

The basis of the GALILÉE-1 algorithm is very similar to the one implemented in NJOY: halving interval and checking with some kind of criteria. The first two basic criteria are the maximum number of subdivisions of an initial interval and the minimum width of an interval expressed in term of a multiple of the machine epsilon. All the specialized criteria use these two parameters.

A convergence diagnostic is available to check if the "convergence" is reached using these criteria. The diagnostic can be used to trigger a restart of the linearization process with the same or with modified parameters of the criterion. This restart mode is useful if the user cannot set properly the initial points of the linearization.

6.3 Doppler broadening

Description

The Doppler broadening and the thermal modules are designed consistently in GTREND whatever the thermal motion is (free gas or chemical binding model). The Doppler broadening and the calculation of thermal scattering cross-sections can be done starting from a nuclear cross-section given by the true formalism or from a linearized one. At present time, only the "SIGMA1" method designed in PREPRO [5] system that provides an exact Doppler broadening for a piecewise linear representation of a cross-section, is implemented in GTREND.

Results

For validating GTREND, we performed, on U238 JEFF-3.2, a SIGMA1 Doppler broadening at 300K on the NJOY2016/BROADR energy grid and we compared the cross-sections to the NJOY ones. The relative discrepancies are given in Figure 3. The general trend is satisfactory but at high energies, NJOY2016 curve may display unphysical behaviors leading to higher discrepancies.

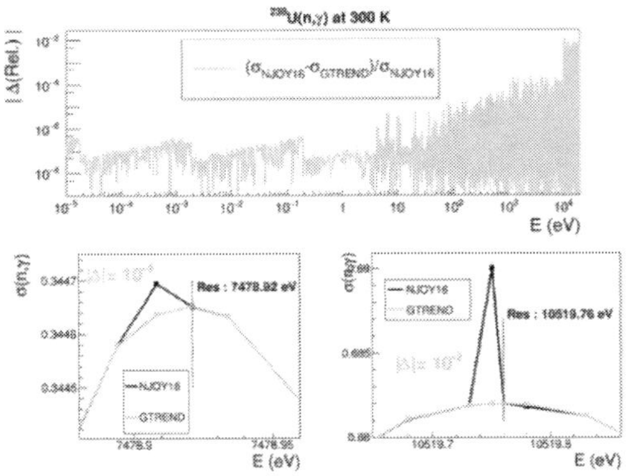

Fig. 3. Relative discrepancies on radiative capture cross-sections of U238 at 300 K

6.4 Probability Table calculation in the Unresolved Resonance Range (URR) on an evaluator-defined energy grid

In this section, we present the first results on probability table calculations in URR. The use of the average data available in the evaluation files requires the sampling of quantities using several distribution laws. Spacing between the resonances is distributed according to the Wigner's law. Partial widths are represented by χ-2's laws with various degrees of freedom.

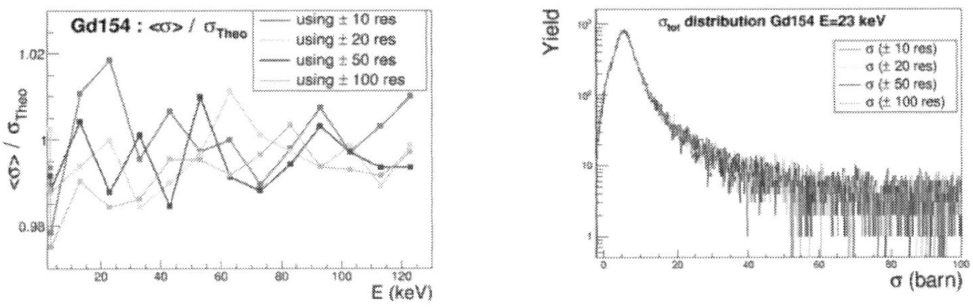

Fig. 4. 4a (left): Ratio between average total cross-section from 30,000 sets of resonances and theoretical cross-section using different number of resonances at 300 K. 4b (right): Histogram of 30,000 total cross-sections using different numbers of resonances.

We have undertaken various tests to validate the calculation of these probability tables. In a first time, we sampled 30,000 sets of random resonances to define the average cross-sections on an energy grid in the unresolved domain. The nucleus selected is Gd154 from JEFF-3.2 library that contains URR from 2.76 keV to 123 keV. We were concerned with the influence of the number of resonances taken into account around the energetic position for the calculations. We sequentially considered 20, 40, 100, and 200 resonances around this computational energy. Figure 4a shows the ratio between the average values calculated for these 30,000 sets of resonances and the theoretical average value as a function of the number of resonances. Figure 4b shows the histograms of the 30,000 total cross-section values calculated at 23 keV for this nucleus as a function of the number of resonances taken into account. These histograms are the input for the different probability table calculation.

Figure 5a presents these calculated probability tables for an incident energy of 23 keV and compares them to the one processed by NJOY's PURR module. We observe a very good agreement for the various probability tables produced with GTREND independently of the number of resonances taken into account. The comparison with the NJOY2016/PURR probability table shows a slightly different distribution, in particular an extension to higher cross-sections in the case of calculations provided by GTREND.

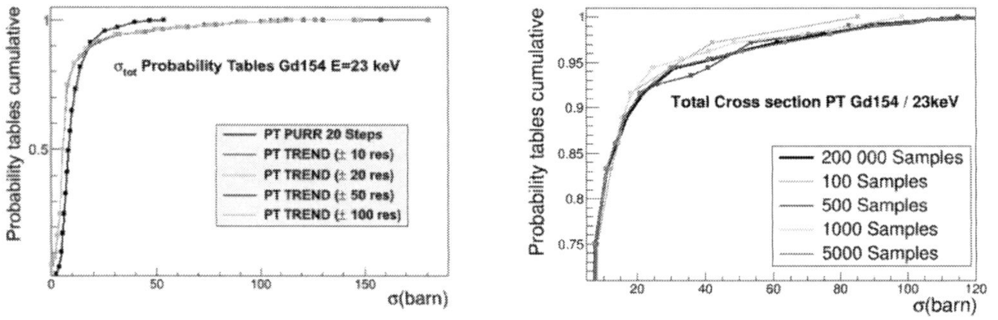

Fig. 5. 5a (left): Comparison of probability table cumulative distributions from PURR and GTREND using various numbers of resonances. 5b (right): Comparison of probability table cumulative distributions calculated with GTREND as a function of the number of resonance sets.

An important question is to define the number of resonance sets to be taken into account in order to establish a converged probability table. To do this, we sampled several series of 100 resonances around 23 keV. Figure 5b shows the evolution of probability tables as a function of the number of resonance sets. The reference is a calculation performed using 200,000 sets of resonances. We observe that from 5,000 sets of resonances the agreement becomes accurate with the reference.

These various tests must be continued and improved in order to obtain a systematic treatment for the calculation of probability tables for all nuclei containing an unresolved resonance domain.

7 Conclusion

GALILÉE-1 system will have soon the capability of producing Monte Carlo transport libraries from validated evaluations, which is a first important step for our renewed library creation chain. It allows R&D activity on new processing methods and is also a powerful tool for nuclear data analysis.

References

1. R.E. MacFarlane et al., LANL, USA, No. LA-UR-17-20093, (2017)
2. J.C. Sublet, P. Ribon, and M. Coste-Delclaux, CEA-R-6277, (2011)
3. A. Lane and R. Thomas, *Rev. Mod. Phys.*, ***30***, 257, (1958)
4. A. Trkov, M. Herman and D. A. Brown, Report BNL-90365-2009 Rev.2, (2012)
5. N. Larson, ORNL/TM-9179/R8, (2008)
6. D. E. Cullen, ORNL, No. PREPRO2012, 004357WKSTN00, USA (2015).

Nuclear data analyses for improving the safety of advanced lead-cooled reactors

Pablo Romojaro[1,*], *Francisco* Álvarez-Velarde[1], and *Nuria* García-Herranz[2]

[1]CIEMAT, Energy Department, Avenida Complutense 40, 28040, Spain
[2]UPM, Energy Engineering Department, Calle José Gutiérrez Abascal 2, 28006, Spain

Abstract. A target accuracy assessment of the effective neutron multiplication factor, k_{eff}, for MYRRHA (Multi-purpose hYbrid Research Reactor for High-tech Applications) lead-bismuth cooled fast reactor has been performed with JEFF-3.3 and ENDF/B-VIII.0 state-of-the-art nuclear data libraries and the SUMMON system. Uncertainties in k_{eff} due to uncertainties in nuclear data have been assessed against the target accuracies provided by SG-26 of the WPEC of OECD/NEA in 2008 for LFR. Results show that k_{eff} target accuracy is still exceeded by more than a factor of two using the latest nuclear data evaluations released in 2018. Consequently, nuclear data assimilation has been carried out using criticality experiments from the International Criticality Safety Benchmark Evaluation Project that are representative of MYRRHA. The results from this work show that the level of accuracy needed in nuclear data cannot be obtained using only differential experiments, but the combination of experimental covariance data and integral experiments together with Generalised Least Squares technique can provide adjusted nuclear data capable of predicting reactor properties with lower uncertainty and consistent with differential data.

1 Introduction

The Lead-cooled Fast Reactor (LFR) is one of the three technologies selected by the Sustainable Nuclear Energy Technology Platform (SNETP) [1] that can meet future European energy needs. Significant efforts are being made by researchers and industry to overcome the main drawbacks for the industrial deployment of LFR, which are the lack of operational experience and the impact of uncertainties in the reactor design, operation and safety assessment.

In nuclear reactor design the uncertainties mainly come from material properties, fabrication tolerances, operative conditions, simulation tools and nuclear data. However, even though the uncertainty in nuclear data is one of the most important sources of uncertainty in reactor design and reactor physics simulations [2], significant gaps between the uncertainties and the target accuracies have been systematically shown in the past [3]. Meeting the target accuracy is required not only to achieve the requested level of safety for this technology, but also to minimize the increase in the costs due to additional security measures.

* Corresponding author: pablo.romojaro@ciemat.es

In this paper, a target accuracy assessment of the effective neutron multiplication factor, k_{eff}, for MYRRHA (Multi-purpose hYbrid Research Reactor for High-tech Applications) [4] lead-bismuth cooled fast reactor is performed with the JEFF-3.3 [5] and ENDF/B-VIII.0 [6] state-of-the-art nuclear data libraries and the SUMMON system [7]. Uncertainty quantification (UQ) analyses have already been carried out for previous MYRRHA designs [8-10] or with older nuclear data evaluations [11]; nevertheless, since JEFF-3.3 and ENDF/B.VIII.0 were released, an updated analysis was required in order to take into account the changes in the new nuclear data libraries and their impact in the criticality safety coefficients.

Furthermore, data assimilation for the main isotopes contributing to the uncertainty has been performed using critical mass experiments from the International Criticality Safety Benchmark Evaluation Project (ICSBEP) [12] that are representative of MYRRHA, in order to obtain adjusted nuclear data with constrained uncertainties and to reduce the uncertainty in the criticality parameter.

2 SUMMON system

The Sensitivity and Uncertainty Methodology for MONtecarlo codes (SUMMON) system [7, 13] has been conceived by CIEMAT as a tool to perform complete automated sensitivity and uncertainty analyses of the most relevant criticality safety parameters of detailed complex reactor designs from the neutronic point of view, i.e., k_{eff}, β_{eff}, Λ_{eff} and reactivity coefficients, using state-of-the-art nuclear data libraries and covariances.

A detailed description of SUMMON, including the methodologies used to calculate the sensitivity coefficients and perform the uncertainty quantification analyses, can be found in Ref. [7, 13].

Recently, the Data Assimilation With summoN (DAWN) module has been developed to perform data assimilation using integral experiments from public databases, with the aim of providing adjusted nuclear data, not only capable of predicting reactor properties within the target design accuracy, but also statistically consistent with the various differential measurements.

DAWN is based on the generalised least squares technique [14], in which the evaluation of the posterior expectation of the cross sections, σ', and covariances, M'_σ, is done by finding the minimum of the following function:

$$\chi^2 = \left(\frac{\sigma' - \sigma}{\sigma}\right)^T M_\sigma^{-1} \left(\frac{\sigma' - \sigma}{\sigma}\right) + \left(\frac{E - C}{C}\right)^T M_{EC}^{-1} \left(\frac{E - C}{C}\right) \tag{1}$$

where σ and M_σ are the a priori expectation and covariance matrix of the cross sections, E is the experimental value of the measured integral parameter, C is the a priori calculated value of the integral parameter and M_{EC} is the relative integral parameter covariance matrix.

The cross sections modifications that minimize the χ^2 thanks to the adjustment are:

$$\left(\frac{\sigma' - \sigma}{\sigma}\right) = M_\sigma S^T G^{-1} \left(\frac{E - C}{C}\right) \tag{2}$$

and the associated a posteriori covariance matrix is:

$$M'_\sigma = M_\sigma - M_\sigma S^T G^{-1} S M_\sigma = M_\sigma (I - S^T G^{-1} S M_\sigma) \tag{3}$$

being S the sensitivities of nuclear data to the integral parameter and G the relative total integral covariance matrix.

https://doi.org/10.1051/epjconf/201921105002

3 Sensitivity and uncertainty analyses

MYRRHA's updated core design is described in detail in Ref. [4]. For this work a simplified model [15], homogenised on fuel assembly level, of the critical core configuration in nominal conditions at Beginning of Life (BoL) has been used. The layout of the core is shown in Fig. 1.

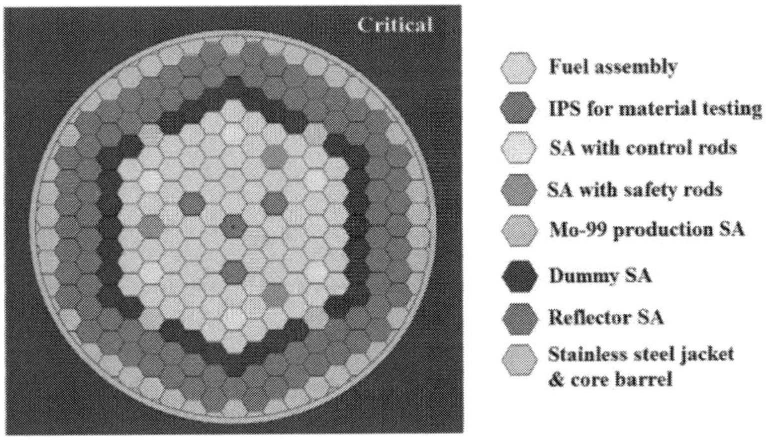

Fig. 1. Critical MYRRHA core layout.

Sensitivity calculations were performed in order to obtain the sensitivity coefficients for uncertainty propagation and to identify the most important nuclear data for neutron induced reactions in MYRRHA. A comparison between the Integrated Sensitivity Coefficients (ISC) obtained using JEFF-3.3 and ENDF/B-VIII.0 nuclear data libraries of the most relevant reactions for k_{eff} is shown in Table 1. It can be seen that the sensitivity coefficients obtained with both libraries are practically the same, therefore for the uncertainty quantification, the uncertainties in nuclear data have been propagated using only the sensitivities derived with JEFF-3.3 library.

Table 1. ISC of the top 5 reactions for MYRRHA using JEFF-3.3 and ENDF/B-VIII.0 libraries.

MYRRHA – ISC (%/%)

Quantity		**JEFF-3.3**	**ENDF/B-VIII.0**
^{239}Pu	ν	0.696	0.698
^{239}Pu	(n,f)	0.482	0.486
^{238}U	(n,γ)	-0.112	-0.115
^{240}Pu	ν	0.081	0.080
^{239}Pu	(n,γ)	-0.053	-0.056

Using the "Sandwich Rule" of the "Propagation of Moments" method [7], the uncertainties were propagated with both JEFF-3.3 and ENDF/B-VIII.0 covariance data. The top 10 contributors, as well as the total uncertainty in k_{eff} due to these contributors, are presented in Table 2. A good agreement can be observed in the total k_{eff} uncertainty derived by JEFF-3.3 and ENDF/B-VIII.0, even though the contributors differ. This can be attributed to differences in the evaluations, such as the case of ^{240}Pu(n,f), to missing correlations, such as ^{240}Pu(n.f)-^{240}Pu(n,γ) in ENDF/B-VIII.0, and to missing covariance evaluations, such as ^{209}Bi(n,el.) in JEFF-3.3.

Table 2. UQ for MYRRHA using different covariance evaluations.

MYRRHA									
JEFF-3.3					**ENDF/B-VIII.0**				
Quantity				$\Delta k_{eff}/k_{eff}$ **(%)**	**Quantity**				$\Delta k_{eff}/k_{eff}$ **(%)**
^{240}Pu	(n,f)	^{240}Pu	(n,f)	0.543	^{239}Pu	(n,f)	^{239}Pu	(n,f)	0.550
^{240}Pu	(n,f)	^{240}Pu	(n,γ)	-0.420	^{239}Pu	(n,γ)	^{239}Pu	(n,γ)	0.227
^{239}Pu	ν	^{239}Pu	ν	0.321	^{209}Bi	(n,el)	^{209}Bi	(n,el)	0.222
^{239}Pu	(n,f)	^{239}Pu	(n,f)	0.295	^{239}Pu	ν_p	^{239}Pu	ν_p	0.187
^{239}Pu	χ	^{239}Pu	χ	0.261	^{238}U	(n,γ)	^{238}U	(n,γ)	0.139
^{240}Pu	(n,γ)	^{240}Pu	(n,γ)	0.197	^{239}Pu	χ	^{239}Pu	χ	0.128
^{239}Pu	(n,f)	^{239}Pu	(n,γ)	0.174	^{238}U	(n,in)	^{238}U	(n,f)	-0.095
^{238}U	(n,γ)	^{238}U	(n,γ)	0.167	^{240}Pu	(n,γ)	^{240}Pu	(n,γ)	0.091
^{239}Pu	(n,γ)	^{239}Pu	(n,γ)	0.151	^{238}U	ν_p	^{238}U	ν_p	0.087
^{238}U	(n,in)	^{238}U	(n,f)	-0.138	^{238}U	(n,in)	^{238}U	(n,γ)	0.081
Uncertainty in k_{eff}				**0.691**	**Uncertainty in k_{eff}**				**0.698**

Design target accuracies for a wide range of innovative systems were provided in 2008 by the SG-26 of the Working Party on International Nuclear Data Evaluation Co-operation of OECD/NEA [16]. For fast reactors at BoL, a target accuracy of 300 pcm in the effective neutron multiplication factor was specified. Nearly 700 pcm of uncertainty in k_{eff} have been obtained in the UQ using the state-of-the-art JEFF-3.3 and ENDF/B-VIII.0 libraries; therefore, the design target accuracy is still exceeded by more than a factor of two for the considered modern nuclear data evaluations.

4 Data assimilation

In order to constrain the uncertainties, an assimilation on the main contributors to the uncertainty in k_{eff}, i.e., ^{240}Pu, ^{239}Pu and ^{238}U nuclear data, was performed using only JEFF-3.3 cross sections and covariances as a prior.

However, before nuclear data can be adjusted, the first step is to select a comprehensive set of complementary experiments, representative of the target application. In practice, this means calculating correlations among the reactor concept and experiments to determine how representative is the latter of the former. This was done by means of the representativity factor, f_{RE}:

$$f_{RE} = \frac{S_R^T M_\sigma S_E}{\sqrt{(S_R^T M_\sigma S_R)(S_E^T M_\sigma S_E)}} \tag{4}$$

where S_R is the sensitivity vector for the targeted reactor and S_E is the sensitivity vector for the experiment. The closer f_{RE} is to one, the more similarity exists between the reactor and the experiment.

Publicly available critical mass experiments from the ICSBEP database were assessed and a set of three experiments that provide information on separated individual physics effects related to the isotopes with highest sensitivity were chosen to perform data assimilation. However, correlation coefficient data of uncertainties in criticality cases only exist for a limited number of integral experiments. In particular, only for the ones derived from IPPE and from ANL for the ZPR/ZPPR values are publicly available in the ICSBEP database [17]. Therefore, if all three experiments were used in the data adjustment without taking into account correlations, possible compensation effects may occur. Consequently, three separate data adjustments have been carried out. The experiments selected were:

- JEZEBEL (PMF001) [18]: bare sphere of ^{239}Pu metal, sensitive to ^{239}Pu;
- ^{240}Pu JEZEBEL (PMF002) [19]: bare sphere of ^{239}Pu metal with 20.1 at% ^{240}Pu, sensitive to ^{240}Pu;
- PU-MET-FAST-006 (PFM006) [20]: plutonium sphere reflected by uranium, sensitive to ^{238}U.

In Table 3, the representativity factors between MYRRHA and the three experiments and among the experiments themselves are shown. While the most representative experiment is PMF002, a close similarity between the experiments is observed, due to all of them employing plutonium as fissile material and having similar energy spectrums.

Table 3. Representativity factor of the experiments used in the assimilation.

f_{RE}	MYRRHA	PMF001	PMF002	PMF006
MYRRHA	1.00	0.71	0.88	0.52
PMF001	0.71	1.00	0.78	0.98
PMF002	0.88	0.78	1.00	0.75
PMF006	0.52	0.98	0.75	1.00

Once the assimilation was performed, the posterior covariance matrices and original sensitivity coefficients were used to propagate the uncertainties. The consistency of the assimilation was assessed against experimental data. An example of the consistency checks is given in Fig. 2, where the prior and posterior ^{240}Pu(n,f) cross sections after the assimilation with PMF002 integral experiment are represented. It can be seen that the central value of the cross section has been modified in the fast energy range and is compatible with the various experimental differential data.

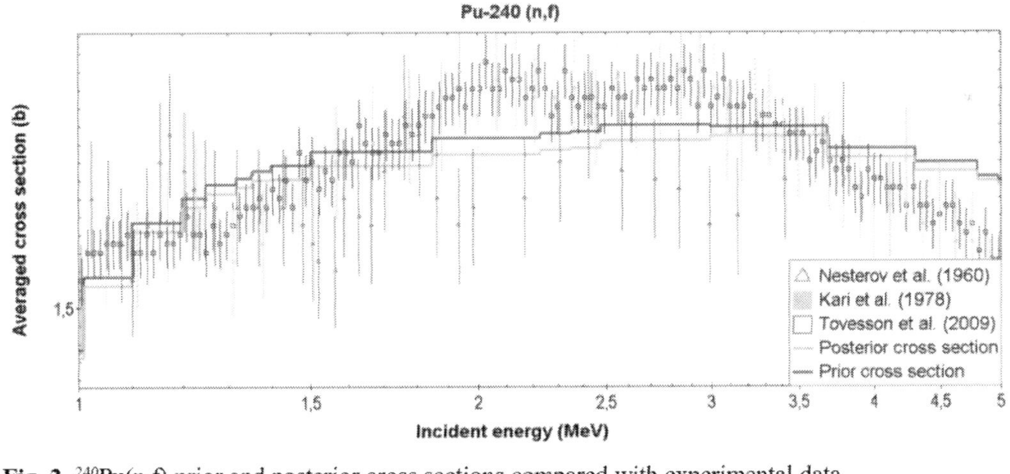

Fig. 2. ^{240}Pu(n,f) prior and posterior cross sections compared with experimental data.

The results from the UQ using the posterior covariance matrices are presented in Table 4. The highest reduction in MYRRHA k_{eff} uncertainty is given by the assimilation of PMF002 experiment (the most sensible to ^{240}Pu and most representative of MYRRHA) due to the substantial reduction in the uncertainty of the two major contributors, i.e., ^{240}Pu(n,f) and ^{240}Pu(n,f)- ^{240}Pu (n,γ). Furthermore, new strong negative cross-correlations between isotopes and reactions appear after the adjustment, such as ^{239}Pu ν_p-^{240}Pu(n,f), which are the main responsible for the significant reductions in the total uncertainty using PMF001 and PMF002 integral experiments. On the other hand, the use of the PMF006 for the assimilation produces negligible adjustments that have insignificant impact when the uncertainties are propagated. This can be attributed to the low similarity of PMF006 with MYRRHA, in spite of the high sensitivity of this experiment to ^{238}U in the fast energy range.

Table 4. UQ for MYRRHA using adjusted covariance matrices.

MYRRHA								
Adjusted JEFF-3.3 with PMF001			**Adjusted JEFF-3.3 with PMF002**			**Adjusted JEFF-3.3 with PMF006**		
Quantity		**$\Delta k_{eff}/k_{eff}$ (%)**	**Quantity**		**$\Delta k_{eff}/k_{eff}$ (%)**	**Quantity**		**$\Delta k_{eff}/k_{eff}$ (%)**
^{240}Pu (n,f)	^{240}Pu (n,f)	0.525	^{240}Pu (n,f)	^{240}Pu (n,f)	0.333	^{240}Pu (n,f)	^{240}Pu (n,f)	0.539
^{240}Pu (n,f)	^{240}Pu (n,γ)	-0.405	^{239}Pu χ	^{240}Pu (n,f)	-0.314	^{240}Pu (n,f)	^{240}Pu (n,γ)	-0.416
^{239}Pu (n,f)	^{239}Pu (n,f)	0.292	^{239}Pu ν_p	^{239}Pu ν_p	0.303	^{239}Pu ν_p	^{239}Pu ν_p	0.300
^{239}Pu ν_p	^{239}Pu ν_p	0.259	^{239}Pu ν_p	^{240}Pu (n,f)	-0.302	^{239}Pu (n,f)	^{239}Pu (n,f)	0.291
^{239}Pu ν_p	^{239}Pu χ	-0.259	^{239}Pu (n,f)	^{239}Pu (n,f)	0.294	^{239}Pu χ	^{239}Pu χ	0.229
^{239}Pu ν_p	^{240}Pu (n,f)	-0.231	^{240}Pu (n,f)	^{240}Pu (n,γ)	-0.254	^{240}Pu (n,γ)	^{240}Pu (n,γ)	0.196
^{239}Pu χ	^{240}Pu (n,f)	-0.223	^{239}Pu χ	^{239}Pu χ	0.234	^{239}Pu ν_p	^{239}Pu χ	-0.169
^{239}Pu χ	^{239}Pu χ	0.193	^{239}Pu (n,f)	^{239}Pu (n,γ)	0.173	^{239}Pu (n,f)	^{239}Pu (n,γ)	0.169
^{240}Pu (n,γ)	^{240}Pu (n,γ)	0.192	^{239}Pu χ	^{240}Pu (n,γ)	0.173	^{238}U (n,γ)	^{238}U (n,γ)	0.167
^{239}Pu (n,f)	^{239}Pu (n,γ)	0.173	^{238}U (n,γ)	^{238}U (n,γ)	0.167	^{239}Pu (n,γ)	^{239}Pu (n,γ)	0.150
Uncertainty in k_{eff}		**0.444**	**Uncertainty in k_{eff}**		**0.422**	**Uncertainty in k_{eff}**		**0.658**

5 Conclusions

A target accuracy assessment of the effective neutron multiplication factor for MYRRHA innovative lead-bismuth cooled fast reactor has been carried out using the JEFF-3.3 and ENDF/B-VIII.0 state-of-the-art nuclear data libraries and the SUMMON system. Similar sensitivities and total k_{eff} uncertainty have been derived using both libraries, however, the contributors to the uncertainty significantly differ. This have been attributed to differences in the evaluations, to missing correlations and to missing covariance evaluations. Moreover, the uncertainties in k_{eff} due to uncertainties in nuclear data have been assessed against the target accuracies provided by SG-26 of the WPEC of OECD/NEA in 2008 for LFR. Results show that k_{eff} target accuracy is still exceeded by more than a factor of two using the latest nuclear data evaluations released in 2018.

In order to constrain the uncertainties in k_{eff}, an assimilation on the main contributors to the uncertainty has been performed using only JEFF-3.3 nuclear data as a prior and publicly available critical mass experiments from the International Criticality Safety Benchmark Evaluation Project. The consistency of the nuclear data adjustment has been checked against differential experimental data and good agreement has been found. A significant reduction in k_{eff} uncertainty has been obtained using the experiments most representative of MYRRHA, due to the reduction in the uncertainty of the major contributors and to the presence a posteriori of strong cross-correlations between isotopes and reactions that did not exist a priori. Results show that a reduction of nearly 300 pcm can be achieved performing an assimilation with the most sensitive experiment to the major contributor to the uncertainty, proving that the combination of experimental covariance data and integral experiments together with Generalised Least Squares technique, can provide adjusted nuclear data capable of predicting reactor properties with lower uncertainty and consistent with differential data.

References

1. Sustainable Nuclear Energy Technology Platform. http://www.snetp.eu
2. I. Kodeli, Report OECD/NEA (2007)
3. H. Harada, A. Plompen, NEA/NSC/WPEC/DOC(2014)446 (2014)
4. G. Van den Eynde et al., J. Nucl. Sci. Technol. **52**, 1053 (2015)
5. A. Plompen, OECD/NEA JEF/DOC-1864 (2017)

6. D.A. Brown et al., Nucl. Data Sheets **148**, 1-142 (2018)
7. P. Romojaro et al., *M&C 2017* (2017)
8. C.J. Díez et al., Nucl. Data Sheets **118**, 526-518 (2014)
9. A. Stankovskiy et al., Nucl. Data Sheets **118**, 513-515 (2014)
10. T. Sugawara et al., Ann. Nucl. Energy **38**, 1098-1108 (2011)
11. P. Romojaro, Ann. Nucl. Energy **101**, 330-338 (2017)
12. OECD/NEA, NEA/NSC/DOC(95)03/I (2016)
13. P. Romojaro et al., *PHYSOR 2018* (2018)
14. C. De Saint Jean et al., OECD/NEA NEA/NSC/WPEC/DOC(2010)429 (2011)
15. A. Stankovskiy et al., SCK•CEN/4463803 (2014)
16. M. Salvatores et al., OECD/NEA NEA/WPEC-26 (2008)
17. OECD/NEA, NEA/NSC/DOC(95)03/II (2016)
18. OECD/NEA, NEA/NSC/DOC(95)03/I Volume I PU-MET-FAST-001 (2016)
19. OECD/NEA, NEA/NSC/DOC(95)03/I Volume I PU-MET-FAST-002 (2016)
20. OECD/NEA, NEA/NSC/DOC(95)03/I Volume I PU-MET-FAST-006 (2016)

EPJ Web of Conferences **211**, 06001 (2019) https://doi.org/10.1051/epjconf/201921106001
WONDER-2018

Validation of Thermal Neutron Scattering Cross Sections for Heavy Water based on Molecular Dynamics Simulation

Haelee Hyun[1,*], *Do Heon* Kim[1] , and *Young-Ouk* Lee[1]

[1]Korea Atomic Energy Research Institute, 111, Daedeok-daero 989beon-gil, Yuseong-gu, Daejeon, 34057, Republic of Korea

Abstract. Recently, the thermal scattering libraries of ENDF/B-VIII.0 and JEFF-3.3 for light and heavy water were released with a new water model (CAB model) proposed by Damian. For the CAB model, the molecular dynamics code GROMACS was used to more accurately describe the realistic motions of water molecules. In this paper, to consider the coherent component we also generated the thermal scattering cross section of the deuterium and oxygen bound in the heavy water molecules using the GROMACS code and EPSR code. In addition, the frequency spectrum was also calculated using the GROMACS code. Thermal scattering cross sections based on the newly calculated Sköld correction factor and the frequency spectrum were generated by NJOY2016 code. Finally, the performance of the generated thermal scattering cross sections were validated by performing an ICSBEP benchmark simulation using MCNPX code.

1 Introduction

Light and heavy water are two of the most important materials used as a moderator in thermal-neutron reactors. In the thermal energy region (~5 eV), to perform neutronics simulations based on nuclear data, it is important to accurately describe changes in the reaction cross section and the energy-angular distribution of secondary neutrons, using a thermal scattering library. Recently, the newly evaluated nuclear data files ENDF/B-VIII.0 and JEFF-3.3 have been released, which includes measurable improvements with respect to the thermal scattering libraries of light and heavy water. A new water model of the CAB model, which was developed at the Neutron Physics Department at Centro Atómico Bariloche (CAB model) [1] has been applied in the thermal scattering libraries of ENDF/B-VIII.0 and JEFF-3.3. The significant difference between the CAB model and existing models is that the CAB model was produced using a molecular dynamics (MD) simulation with GROMACS code [2] and NJOY code [3] which are comprehensive computer code packages for producing nuclear cross sections in the ENDF format. In particular, in the CAB model for heavy water, the Sköld correction factor [4] and frequency spectrum of the deuterium and oxygen bound in the heavy water molecules were calculated in the MD

* Corresponding author: hyunhl@kaeri.re.kr

simulation using GROMACS code, taking the coherency and vibration of the molecules of heavy water into consideration.

In this paper, we also generated the Sköld correction factors and frequency spectra of deuterium and oxygen to generate a thermal scattering library for heavy water using GROMACS v.5.1.4. and EPSR (Empirical Potential Structure Refinement) code [5] and compared the results against an ICSBEP [6] benchmark simulation using the MCNPX 2.7.0 code [7].

2 Thermal scattering law for heavy water

In the ENDF/B notation, the thermal incoherent scattering cross section is given by Eq. (1).

$$\sigma^{inc}(E,E',\mu) = \frac{\sigma_b}{2\pi kT}\sqrt{\frac{E'}{E}}e^{-\beta/2}S(\alpha,\beta) \tag{1}$$

where E and E' are the incident and secondary neutron energies, σ_b is the characteristic bound cross section, k is the Boltzmann constant and T is the Kelvin temperature of the material. Also, the thermal scattering law depends on parameters of α and β. Here α is the momentum transfer parameter defined as

$$\alpha = \frac{E'+E-2\sqrt{EE'}\cos\theta}{AkT} \tag{2}$$

and β is the energy transfer parameter defined as

$$\beta = \frac{E'-E}{kT} \tag{3}$$

where A is the mass ratio of the scattering nuclide to the neutron and $\cos\theta$ is the scattering angle in the laboratory system.

2.1 Sköld approximation for heavy water

In comparison with existing libraries, the thermal scattering libraries of ENDF/B-VIII.0 and JEFF-3.3 for heavy water for the CAB model have two significant improvements. Firstly, it has considered the coherent scattering cross section of heavy water by introducing the Sköld approximation with Sköld correction factors. Secondly, the thermal scattering cross section dealing with the oxygen atom in the heavy water molecule (i.e. the bound O in the D_2O library) is newly generated to take into account the interference effects of oxygen atoms.

To generate the scattering cross section which is considered the coherent component of the deuterium and oxygen, the thermal scattering cross section for heavy water uses the Sköld approximation:

$$S(\alpha,\beta) = S^{inc}(\alpha,\beta) + S^{coh}(\alpha,\beta) \tag{4}$$

$$S^{D}_{coh}(\alpha,\beta) = S^{D}_{inc}\left(\frac{\alpha}{\tilde{S}^D},\beta\right)\tilde{S}^D(Q) \tag{5}$$

$$S^{O}_{coh}(\alpha,\beta) = S^{O}_{inc}\left(\frac{\alpha}{\tilde{S}^O},\beta\right)\tilde{S}^O(Q) \tag{6}$$

where $\tilde{S}^D$ and $\tilde{S}^O$ are the Sköld correction factors for deuterium and oxygen, respectively.

$$\tilde{S}^D(Q) = 1 + \frac{2}{3}[S_{DD}(Q) - 1] + \frac{1}{3}\frac{\sigma^O_{coh}}{\sigma^D_{coh}}[S_{DO}(Q) - 1] \tag{7}$$

$$\tilde{S}^O(Q) = 1 + \frac{1}{3}[S_{OO}(Q) - 1] + \frac{2}{3}\frac{\sigma^D_{coh}}{\sigma^O_{coh}}[S_{DO}(Q) - 1] \tag{8}$$

The $S_{DD}(Q)$, $S_{DO}(Q)$ and $S_{OO}(Q)$ are the static structure factors for each atom in the heavy water molecules.

2.1.1 GROMACS and EPSP code simulations

To apply the Sköld approximation, the CAB model introduced the Sköld correction factors by calculating the static structure factors using the MD simulation with the GROMACS code. In this paper, we also calculated the Sköld correction factors using not only GROMACS but also EPSR code. The GROMACS code is a versatile package for performing the MD simulation, which simulates the motions of a system composed of hundreds to millions of particles using Newtonian equations. In sequence, EPSR uses a Monte Carlo method which evolved from the Reverse Monte Carlo, which was developed to build a structural model of disordered materials such as a glass or liquid.

As a result, Fig.1 shows the Sköld correction factors. When the Sköld correction factor from the GROMACS and EPSR codes were compared with the ENDF/B-VIII.0 data, both Sköld correction factors for deuterium showed high consistency. In case of the oxygen, however, the Sköld correction factors from the GROMACS and EPSR code showed lower peaks than the ENDF/B-VIII.0 data.

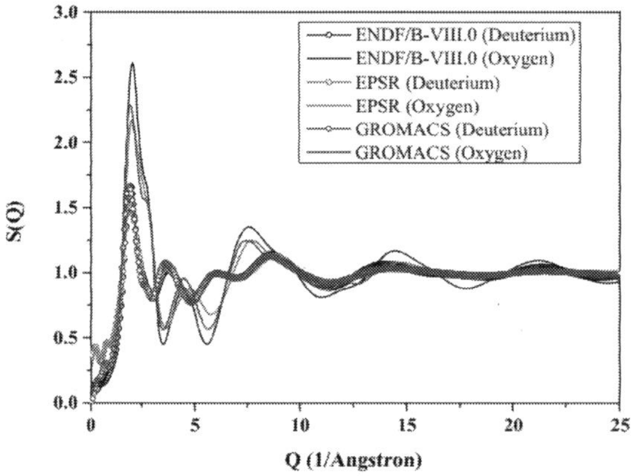

Fig. 1. Comparison of Sköld correction factors obtained from the ENDF/B-VIII.0 data and calculated by the EPSR and GROMACS code (293.6K).

2.2 Frequency spectrum

The thermal scattering law which contains dynamic and structural information about the target system and determines the energy and angular distribution of the secondary neutrons can be written by Eq. (9).

https://doi.org/10.1051/epjconf/201921106001

$$s(\alpha,\beta) = \frac{1}{2\pi}\int_{-\infty}^{\infty} e^{\beta\,\hat{t}} e^{-\gamma(\hat{t})}\, d\hat{t}\,, \tag{9}$$

where $\hat{t}$ is time measured in the unit of $\hbar/kT$ seconds. The intermediate scattering function $\gamma(\hat{t})$ is given by Eq. (10).

$$\gamma(\hat{t}) = \alpha\int_{-\infty}^{\infty} P(\beta)[1 - e^{-\beta\,\hat{t}}]e^{-\beta/2} d\beta\,, \tag{10}$$

where

$$P(\beta) = \frac{\rho(\beta)}{2\beta \sinh\,(\beta/2)}\,, \tag{11}$$

where $\rho(\beta)$ is the frequency spectrum. In the case of heavy water, it consists of a diffusion spectrum, a continuous spectrum and discrete oscillators, which is used as an input of the LEAPR module in the NJOY code for generating the thermal scattering cross section.

Figure 2 shows the difference in the continuous spectra of the deuterium and oxygen in heavy water. The solid line represents the continuous spectrum used in the thermal scattering cross section in ENDF/B-VIII.0, and the dotted line is the spectrum calculated by GROMACS in this paper. The frequency spectrum was calculated by using just the GROMACS code because the EPSR code doesn't have a function for calculating the frequency spectrum.

As shown in Fig. 2, although both spectra are based on the GROMACS code, they show some differences. The differences might be from differences in the simulation conditions, such as simulation time, applied ensemble, conditions of equilibration and so on. Nevertheless, the frequency spectrums in this work have a shape analogous to frequency spectrum used in the thermal scattering library in ENDF/B-VIII.0. In particular, the peaks of both spectra show substantial accordance. The peaks at around 0.6 meV and 2~3 meV represent the motions of intermolecular bending and stretching of the deuterium and oxygen, respectively. In contrast with the oxygen, the deuterium has a third peak at around 4~6 meV, which indicates the motion of the librations.

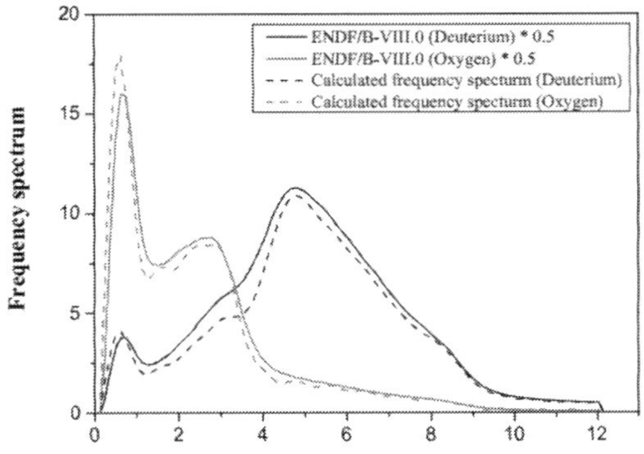

Fig. 2. Comparison of continuous spectra for deuterium and oxygen between the data used in ENDF/B-VIII.0 and calculated by GROMACS code (293.6 K).

3 Calculation results

3.1 Comparison of thermal scattering cross sections

Finally, we generated the thermal scattering cross sections for heavy water using the NJOY2016 code with the new Sköld correction factors and the frequency spectrum calculated by MD simulations. As shown in Fig.3, three kinds of thermal scattering cross sections are compared with the ENDF/B-VIII.0 data, which is used as a reference. The red lined scattering cross sections (FS+ENDF/B-VIII.0) are generated by using the same Sköld correction factor and the newly calculated frequency spectrum (FS) in this work. The green lined thermal scattering cross sections (FS+EPSR) were generated using the Sköld correction factor newly calculated with the EPSR code and the calculated frequency spectrum. Lastly, the blue lined scattering cross sections (FS+GROMACS) were generated using the Sköld correction factor newly calculated using the GROMACS code and the calculated frequency spectrum.

As a result, when comparing the reference and FS+ENDF/B-VIII.0, both scattering cross sections agreed well above the 1 meV range. At energy ranges lower than 1 meV, the FS+ENDF/B-VIII.0 scattering cross sections of the deuterium and oxygen had a slightly higher value than the reference. However, the differences were not outstanding.

In case of the scattering cross sections of the FS+EPSR and FS+GROMACS, the differences from the reference are more remarkable, which means the scattering cross sections were more significantly affected by the Sköld correction factors. Also, Fig. 3 shows the tendency that the scattering cross sections of the deuterium generated using the GROMACS and EPSR Sköld correction factors have larger values than the reference below 3meV.

However, for the scattering cross section for oxygen from 2 meV to 9 meV, values lower than the reference were observed. Also, all scattering cross sections had a common feature, in that two dips were observed around 1 meV and from 10 to 30 meV. The first dip comes from the coherence of each atom and the second dip is caused by the effect of oxygen interference in the heavy water molecules. In contrast with the second dip, each scattering cross section showed considerable differences in the first dip range. However, because the neutron flux in thermal systems is much higher in the energy range of the second dip, the description of the second dip is more important.

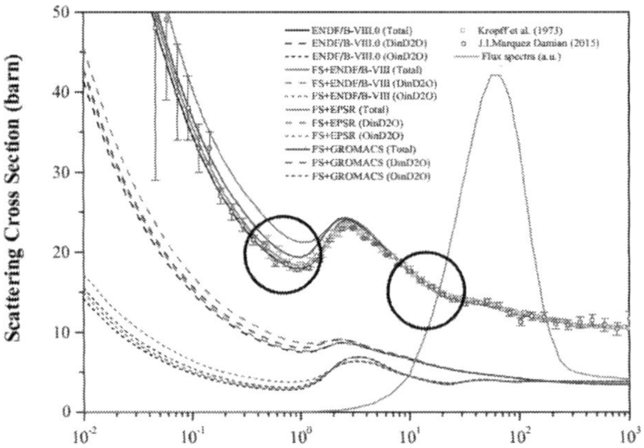

Fig. 3. Scattering cross sections for heavy water and the flux spectrum in the thermal system (293.6K).

3.2 Criticality benchmark problems

To estimate the effects of the thermal scattering libraries, 59 heavy water moderated/reflected experiments were taken from the International Handbook of Evaluated

Criticality Safety Benchmark Problems (ICSBEP). All of benchmark calculations were performed using Monte Carlo Transport code MCNPX 2.7.0. The ENDF/B-VII.1-based KNE71 library [8] was used for all nuclides except for the thermal scattering cross sections of D in D_2O and/or O in D_2O. All the MCNP benchmark simulations were carried out at 293.6K. Also, the MCNP runs were terminated after statistical uncertainty was reduced to below 20 pcm.

Fig. 4 shows the benchmark uncertainties and differences between the k_{eff} calculated from the benchmark k_{eff} with the thermal scattering libraries of ENDF/B-VIII.0, and generated thermal scattering libraries in the categories of low enriched uranium (LEU) and highly enriched uranium (HEU). The performance of the generated thermal scattering cross sections are similar to the thermal scattering cross section of ENDF/B-VIII.0.

As mentioned in Section 3.1, because the thermal neutron flux spectrum is biased to the energy range of the second dip, the benchmark results were comparable although the generated scattering cross sections showed relatively large differences in the low energy range.

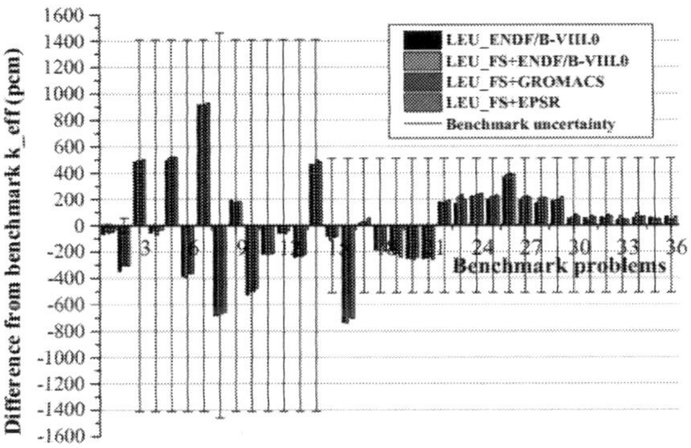

(a) LEU category

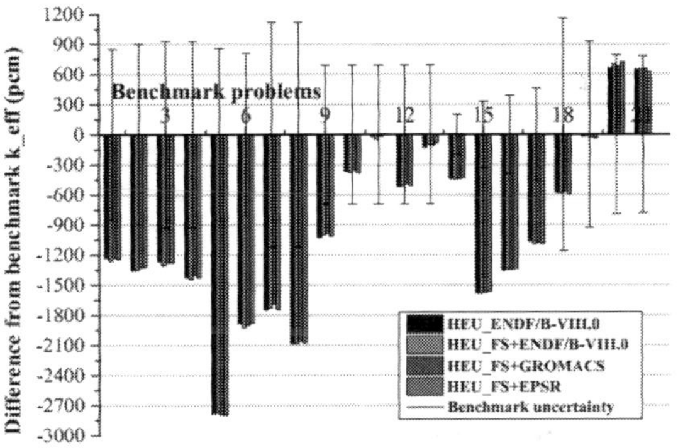

(b) HEU category

Fig. 4. Differences in the k_{eff} calculated from the benchmark k_{eff} and the thermal scattering libraries of the LEU and HEU categories.

Also, the results of the root mean square (RMS) errors for heavy water problems are described in Table 1. The total RMS errors of ENDF/B-VIII.0 and the generated libraries were 0.801~0.808%, which also shows the generated libraries have performance similar to the ENDF/B-VIII.0 library in aggregate.

Table 1. Comparison of RMS errors relative to benchmark k_{eff} values among different libraries (Unit: %)

Category (# of cases)	ENDF/B-VIII.0	FS+ENDF/B-VIII.0	FS+GROMACS	FS+EPSR
HEU (21)	1.273	1.279	1.265	1.273
IEU (1)	0.139	0.167	0.172	0.178
LEU (36)	0.327	0.328	0.328	0.328
U233 (1)	0.570	0.549	0.558	0.528
TOTAL (59)	0.805	0.808	0.801	0.805

4 Conclusions

In this work, we generated thermal scattering libraries for heavy water using the MD simulation and NJOY code. To take into account the coherent scattering cross section which is included in the scattering cross section of heavy water, we calculated the Sköld correction factors of the deuterium and oxygen bound in heavy water molecules, using the MD simulation with the GROMACS and EPSR code. In sequence, the frequency spectra of each atom were also calculated using GROMACS code to consider the intermolecular vibration of heavy water molecules. To generate the thermal scattering cross section for heavy water, finally, the Sköld correction factor and frequency spectrum were used as inputs in the NJOY code. In comparison with the ENDF/B-VIII.0 data, the generated thermal cross sections showed some discrepancies at energy ranges lower than 3 meV. Nevertheless, we confirmed that the performance of the generated libraries was similar to the thermal scattering libraries of ENDF/B-VIII.0 for heavy water by comparison with the results of the criticality benchmark simulation of ICSBEP.

References

1. J. I. Marquez Damian, J. R. Granada, D. C. Malaspina, CAB models for water: a new evaluation of the thermal neutron scattering laws for light and heavy water in ENDF-6 format, Ann. Nucl. Energy, **65**, 280 (2014)
2. D. Van Der Spoel, E. Lindahl, B. Hess, G. Groenhof, A. E. Mark, H. J. C. Berendsen, GROMACS: Fast, Flexible, and free, Journal of Computational Chemistry, **26(16)**, 1701-1718 (2005)
3. A. C. Kahler et al., The NJOY Nuclear Data Processing System, Version 2016, Los Alamos National Laboratory, LA-UR-17-20093 (2016)
4. K. Sköld, Small energy transfer scattering of cold neutrons from liquid argon, Phys. Rev. Lett, **19(18)**, 1023-1025 (1967)
5. A. K. Soper, EPSRshell User Manual (2009)
6. OECD Nuclear Energy Agency, International handbook of evaluated criticality safety benchmark experiments, OECD Nuclear Energy Agency NEA/NSC/ DOC(95)03 (2009)
7. D. B. Pelowitz, ed. MCNPX USER'S MANUAL, Version 2.7.0, Los Alamos National Laboratory, LA-CP-11-00438 (2011)
8. D. H. Kim, C. S. Gil, Y. O. Lee, Current Status of ACE Format Libraries for MCNP at Nuclear Data Center of KAERI, Journal of Radiation Protection and Research, **41(3)**, 191-195 (2016)

Influence of nuclear data parameters on integral experiment assimilation using Cook's distance

D. Kumar[1,2], *S. B.* Alam[1], *H.* Sjöstrand[2], *J. M.* Palau[1], and *C.* De Saint Jean[1,*]

[1]CEA, DEN, Cadarache, F-13108 Saint Paul les Durance, France
[2]Department of Physics and Astronomy, Uppsala University, Sweden

Abstract. Nuclear data used in designing of various nuclear applications (*e.g.*, core design of reactors) is improved by using integral experiments. To utilize the past critical experimental data to the reactor design work, a typical procedure for the nuclear data adjustment is based on the Bayesian theory (least-square technique or Monte-Carlo). In this method, the nuclear data parameters are optimized by the inclusion of the experimental information using a Bayesian inference. The selection of integral experiments is based on the availability of well-documented specifications and experimental data. Data points with large uncertainties or large residuals (outliers) may affect the accuracy of the adjustment. Hence, in the adjustment process, it is very important to study the influence of experiments as well as of the prior nuclear data on the adjusted results. In this work, the influence of each individual reaction (related to nuclear data) is analyzed using the concept of Cook's distance. First, JEZEBEL (Pu239, Pu240 and Pu241) integral experiment is considered for data assimilation and then the transposition of results on ASTRID fast reactor concept is discussed.

1 Introduction

Nuclear data contains all the nuclear physics needed for the mathematical modeling of a nuclear application (*e.g.*, for reactor design). Nuclear data is evaluated using nuclear reaction models in combination with differential data and is improved by using integral experiments [1, 2]. The improvement using integral experiments is called nuclear data adjustment which is typically based on the Bayesian theory, where the nuclear data parameters are optimized by the inclusion of experimental information using least square analysis or Monte-Carlo approach [2]. Parameters in the context of nuclear data can refer to both, the underlying nuclear model parameters [3], or to the nuclear data itself. In the context of this paper, nuclear data parameters only refer to the multi-group values of the nuclear data defined by isotope, energy and nuclear observable (cross-section of a certain reaction, $\bar{\nu}$, etc).

The so-called "adjusted data" has been used in the design of fast reactors such as Phenix and Super-Phenix in France [4, 5]. However, the adjustment process has raised several questions about the physical meaning of the individual adjustment (*i.e.* by isotopes, reaction types and energy ranges). There is no clear definition of the application domain of the adjusted multi-group data. For a new reactor concept, it is difficult to say if a previously adjusted data can be used or a new adjustment will be needed with a different set of integral experiments.

*Corresponding author: e-mail: cyrille.de-saint-jean@cea.fr

How to choose a set of integral experiments for a particular reactor concept is also not obvious. A detailed analysis is needed to estimate the influence of input data on the adjustment. These data points could be a given observation (integral experiment) or a given nuclear data parameter. The Cook's distance is used to estimate the influence of a data point in performing the least square fit [6, 7]. In this work, the influence of input parameters on the Bayesian fit is analysed using the concept of Cook's distance.

2 Nuclear data adjustment using Bayesian inference

The Bayesian inference is a mathematical framework that allows one to assimilate information from measurements (microscopic and integral) and from a prior knowledge of the parameters, in order to reduce the uncertainties of these parameters and modify their values if necessary.

2.1 General principle

The Bayes' theorem gives the relationship between the probability densities of the posterior, prior knowledge and measurements. If $\vec{y} = \vec{y_i}(i = 1, 2, ...N_y)$ are some experimentally measured variables and $\vec{x}$ are the parameters defining the model M, which is used to simulate these variables, the probability of the adjusted parameters can be estimated using the Bayesian inference. Using the Bayes' theorem, one can write the following relationship between the posterior, prior and the likelihood:

$$p(\vec{x}|M, \vec{y}, U) = \frac{p(\vec{x}|M, U).p(\vec{y}|\vec{x}, M, U)}{\int d\vec{x}.p(\vec{x}|M, U).p(\vec{y}|\vec{x}, M, U)} \tag{1}$$

where the probability density of these parameters $p(\vec{x}|M, U)$ is called a prior of the background (the prior knowledge U). $p(\vec{y}|\vec{x}, M, U)$ is the likelihood of the measurements knowing $\vec{x}$ and $p(\vec{x}|M, \vec{y}, U)$ is the posterior probability density of the parameters. The denominator term in the above equation is a normalization constant.

This relation can therefore be understood as the advancing of the a prior knowledge by a likelihood function of new measurements, *i.e.*,

$$posterior\left[p(\vec{x}|\vec{y}, U)\right] \propto prior\left[p(\vec{x}|U)\right].likelihood\left[p(\vec{y}|\vec{x}, U)\right] \tag{2}$$

A fitting procedure can be used (*e.g.*, generalized least square fit) to estimate the first two moments of the posterior or to obtain the posterior distribution of these parameters. In nuclear data field, this is a standard procedure for nuclear data adjustment [2]. In the context of this paper, x is the multi-group nuclear data, M is a reactor model and y is the result of an integral experiment.

2.2 Analytical resolution

The above problem can be solved analytically by considering some necessary assumptions. In particular, the prior probability density and the likelihood are considered as Gaussian multi-parameterized distributions. Thus, if the prior data and its covariance matrix are respectively $\vec{x_m}$ and M_x, and $\vec{y}$ is a set of integral experiments with M_y the covariance matrix, the posterior density of $\vec{x}$ can be written as:

$$posterior\left[p(\vec{x}|\vec{y},U)\right] \propto e^{-(1/2)((\vec{x}-\vec{x_m})^T M_x^{-1}(\vec{x}-\vec{x_m})+(\vec{y}-\vec{t})^T M_y^{-1}(\vec{y}-\vec{t}))} \tag{3}$$

where $\vec{t}$ is the theoretical model prediction (or result from computer simulation), which is functions of $\vec{x}$. For posterior distribution, the Laplace approximation is used for the deterministic resolution of the least-square equation which consists in considering that this density is also of the Gaussian type is used.

Now, if the mean and the covariance matrix of the prior parameters, and the experimental covariance matrix are provided, the parameters after adjustment (and the posterior covariance matrix) can be estimated by looking for the minimum the following cost function:

$$\chi^2_{GLS} = (\vec{x} - \vec{x_m})^T M_x^{-1}(\vec{x} - \vec{x_m}) + (\vec{y} - \vec{t})^T M_y^{-1}(\vec{y} - \vec{t}) \tag{4}$$

Using the least square fitting technique, the solution of this minimization gives the new adjusted values :

$$\vec{x} - \vec{x_m} = M_x.G^{(n)T}(M_y + G^{(n)}.M_x.G^{(n)T})^{-1}.(\vec{y} - t(\vec{x_m})) \tag{5}$$

and the a posterior covariance matrix:

$$\tilde{M}_x = \left[M_x^{-1} + G^{(n)T} M_y^{-1} G^{(n)}\right]^{-1} \tag{6}$$

where G is the sensitivity matrix, *i.e.*, a matrix of derivatives. n is associated with the nth iteration of the algorithm corresponding to a given numerical convergence. The entries of the sensitivity matrix are calculated as:

$$(G)_{ij} = \left(\frac{\partial t_i}{\partial x_j}\right) \tag{7}$$

with t_i the theoretical calculation corresponding to the experimental value. The coefficients G_{ij} are computed using sensitivity analysis (perturbation theory). For the adjustment of multi-group nuclear data, the model parameters x and the covariance matrix M_x in the above equations can be replaced by the nuclear data (σ_p) and its covariance matrix (M_{σ_p}) respectively.

3 Effect of influential data using Cook's distance

It is commonly accepted that the effect of input data on regression models should be analysed. A strongly influential case dominates the regression model in such a way, that the estimated regression line lies close to this case. The analysis of residuals cannot be used to detect influential cases as the cases with high residuals are indicated as outliers. The influence of an individual data on the overall regression fit can be thought of in terms of how much the fit would differ if the individual data is not included in the fitting process. If the predictions are the same with or without the individual data, the latter has no influence on the regression model. If the predictions differ significantly when the observation is not included in the analysis, the data is highly relevant.

Cook's distance provides an overall measurement of the change in all parameter estimates. The concept of Cook's distance is a good measure of the influence of observation and is proportional to the sum of the squared differences between predictions made with all observations in the analysis and predictions made by removing the observation in question. A formula for the Cook's Distance is provided in [6, 7]. In this work, the effect of the inputs is analysed using the concept of Cooks distance.

If σ_p is the original posterior nuclear data and σ_p^i is the adjusted posterior nuclear data when the ith parameter is discarded in the adjustment process, the Cook's distance can be calculated as (see [7]):

$$D_i = (\sigma_p^i - \sigma_p)M_{\sigma_p}^{-1}(\sigma_p^i - \sigma_p) \tag{8}$$

where M_{σ_p} is the covariance matrix of the posterior parameters. The M_{σ_p} matrix is not re-estimated when a subset of data is removed in calculating the Cook's distance. The covariance matrix is calculated using uncertainties (the standard deviation) of the nuclear data parameters and their correlation matrix. However, due to the high dimensional nuclear data used in the adjustment process, currently, several issues (such as singularity, negative definite) are encountered when inverting the covariance matrix. For those reasons, currently, a modified covariance matrix based on the nuclear data uncertainties is considered in calculating the Cook's distance for this work.

4 Test case

For this study, the prior nuclear data uncertainties and their correlations are taken from CEA files COMAC-V1 in 33 energy groups [2]. The JEZEBEL experiment is considered as an integral experiment for the adjustment [8]. The JEZEBEL experiment was a small spherical assembly of plutonium alloyed with gallium (3.41% Ga, 91.95% Pu239, 4.35% Pu240 and 0.29% Pu241). Pu239 was the major component of the core in this experiment. It was used to determine the critical mass of spherical and homogeneous Pu-alloy. Using Cook's distance, the influence of isotopes and nuclear reactions on the adjustment is investigated in this section.

4.1 Influence of Pu isotopes on the adjustment

To analyse the effect of Pu isotopes on the adjustment, Pu239, Pu240 and Pu241 isotopes are considered in the adjustment process. It means that the JEZEBEL (integral experiment) sensitivities of k_{eff} with respect to all Pu239, Pu240 and Pu241 nuclear reactions are provided in the adjustment process. The Cook's distance is plotted in Figure 1. The Cook's distance for Pu239 is very large in comparison to Pu240 and Pu241 isotopes. It indicates that the major contribution on the adjustment is due to the k_{eff} sensitivities of the Pu239 isotope, which is justified from the core composition of the JEZEBEL experiment. It is also important to investigate the effect of individual nuclear reaction cross-section of Pu239 isotope on the adjustment, which is analysed in the next section.

4.2 Influence of Pu239 nuclear reactions on the adjustment

To analyse the effect of Pu239 nuclear reactions on the adjustment, the JEZEBEL sensitivities of k_{eff} with respect to all Pu239 nuclear reactions are provided in the adjustment process.

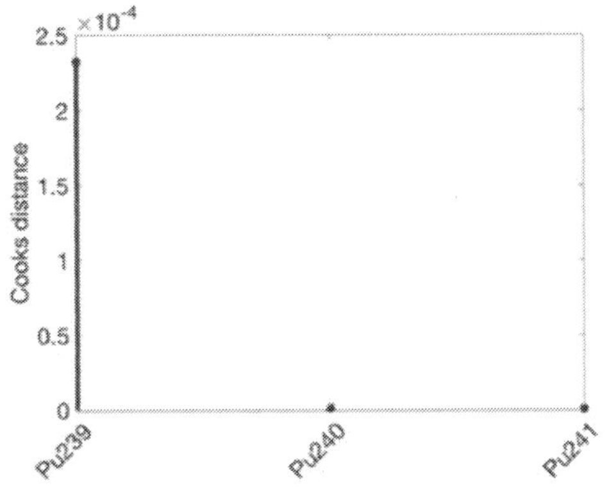

Figure 1. Cook's distance for Pu isotopes

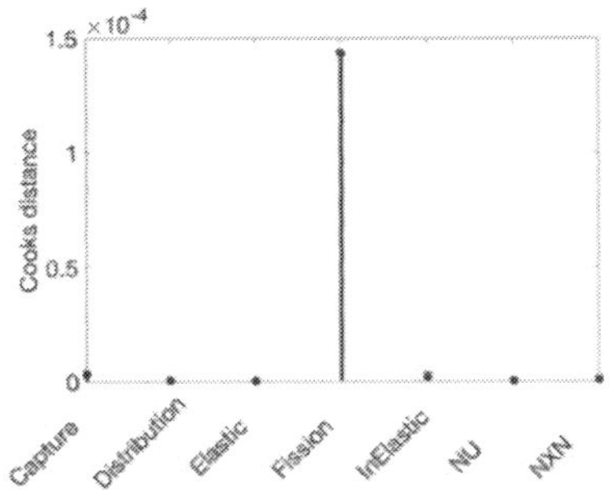

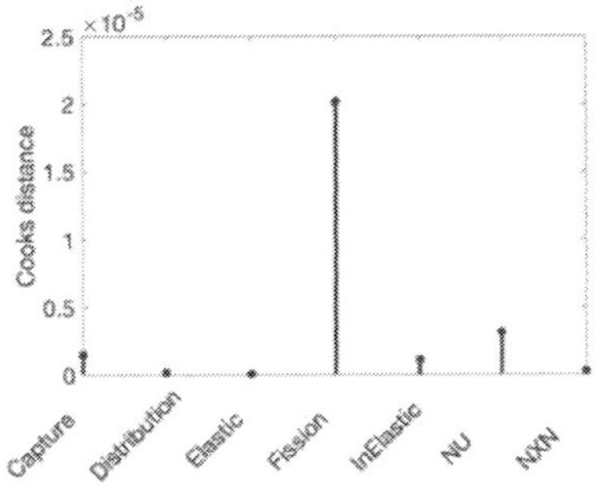

Figure 2. Cook's distance for Pu239 nuclear data (left) and Cook's distance with increased uncertainty in the prior $\bar{\nu}$ (right)

It means that only the nuclear data related to Pu239 are adjusted here using the integral experiment.

The Cook's distance for this case is plotted in Figure 2 (left). The Cook's distance for the fission reaction is very large in comparison to the other nuclear reactions. It indicates that the major contribution on the adjustment is due to the k_{eff} sensitivities related to the Pu239 fission cross-section. However, from theoretical aspect and from expert opinions, it is known that for JEZEBEL experiment, both fission and $\bar{\nu}$ are very influential for k_{eff}. This contradiction may suggest that there is a possibility that the prior uncertainties are not evaluated properly (too low). Thus, when the prior uncertainty on Pu239 $\bar{\nu}$ data is increased by 1% and the adjustment process is repeated with JEZEBEL, the influence of Pu239 $\bar{\nu}$ can now be seen on Cook's distance (see Figure 2 (right)). The Cook's distance is plotted for the case where the prior uncertainty of $\bar{\nu}$ is increased by 1%. Now, it can be seen that the Cook's distance is larger for both (the fission and $\bar{\nu}$) comparing to other nuclear reactions.

Further, in the Cook's distance calculation, if Case i is the case where the ith parameter is discarded in the adjustment process (see Equation 8), *i.e.* Case 1 (all data-Capture), Case 2 (all data-Distribution), Case 3 (all data-Elastic), Case 4 (all data-Fission), Case 5 (all data-Inelastic), Case 6 (all data-$\bar{\nu}$) and Case 7 (all data-nXn). In Figure 3, the trends and their uncertainties are plotted for posterior Pu239 capture cross-section for all cases. For Case 1 (where all JEZEBEL Pu239 reaction cross-sections excluding the capture are considered for the fit), no trend is visible for the adjusted values of Capture data as JEZEBEL Pu239

Capture cross-section is excluded from the adjustment. For Case 4 and Case 6, trends are very much different from the others. It indicates that the major difference is coming from fission and $\bar{\nu}$ data. Similarly, for Case 4 and Case 6, trends and uncertainties are plotted for Pu239 fission (in Figure 4) and for Pu239 $\bar{\nu}$ (in Figure 5) simultaneously. For these 2 cases also, the major contribution in the adjustment can be seen from the fission and $\bar{\nu}$. Major differences in uncertainties for some cases can be seen but it is not discussed here since the uncertainties do not make a significant contribution to the Cook's distance.

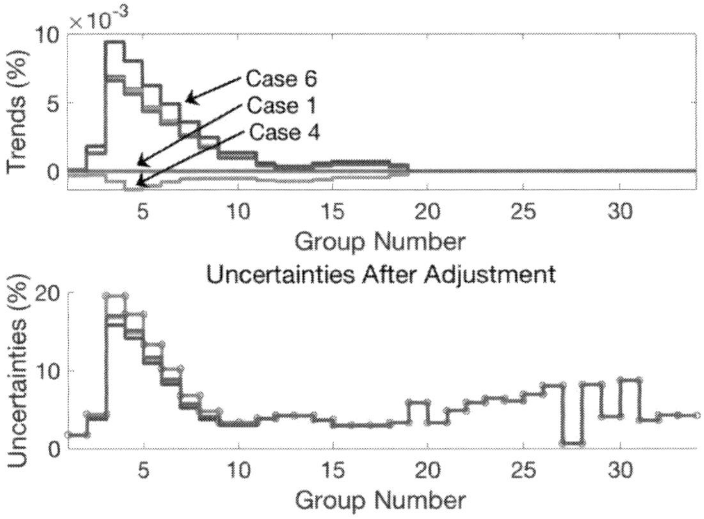

Figure 3. Adjusted Pu239 Capture data (trends and uncertainties) for all 7 sub-cases

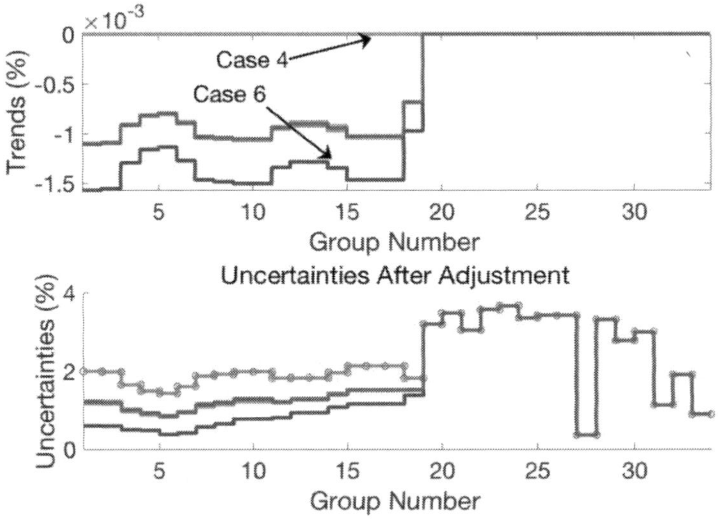

Figure 4. Adjusted Pu239 Fission data (trends and uncertainties) for all 7 sub-cases

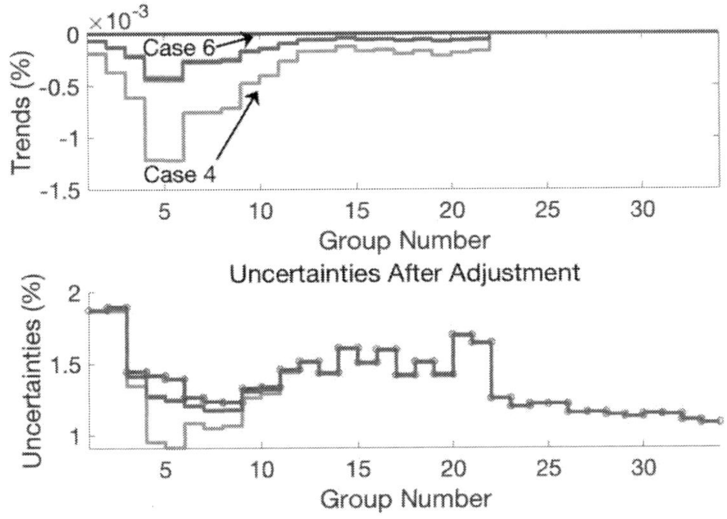

Figure 5. Adjusted Pu239 $\bar{\nu}$ data (trends and uncertainties) for all 7 sub-cases

Table 1. k_{eff} uncertainties for ASTRID

	using prior data	using prior data +1% in $\bar{\nu}$ unc.
Initial unc.	1.3227	1.3639
Final unc (fit: Pu239)	1.1365	1.1728
Individual fit effect:		
Pu239 Capture	1.32239	1.36369
Pu239 Distribution	1.31796	1.35939
Pu239 Elastic	1.32266	1.36395
Pu239 Fission	1.14785	1.19518
Pu239 InElastic	1.32245	1.36374
Pu239 Nu	1.32213	1.34499
Pu239 NXN	1.32265	1.36392

5 Transposition to ASTRID

Adjusted nuclear data using JEZEBEL integral experiment is used to compute the k_{eff} uncertainties of ASTRID GEN-IV concept reactor. A reduction in uncertainties can be seen when the posterior data is used (see Table 1). The initial uncertainty using the prior data is 1.3227% and the final uncertainty with the adjusted posterior data (when Pu239 all reaction parameters from JEZEBEL are used) is 1.1365%. It can also be seen from the Table 1 that the major reduction in the ASTRID k_{eff} uncertainty is due to the Pu239 fission reaction parameters. The final uncertainty due to Pu239 fission uncertainty is 1.14785% which is approximately equal to the final uncertainty due to Pu239 all reactions. Other reaction parameters of Pu239 isotopes are not making any significant contribution to the uncertainty reduction. However, when the prior uncertainty in $\bar{\nu}$ is increased by 1%, an additional contribution of Pu239 $\bar{\nu}$ data is also seen in the uncertainty reduction of ASTRID k_{eff}.

6 Conclusions

In this work, the influence of isotopes and nuclear reactions on adjustment is investigated using Cook's distance. For this study, the prior nuclear data uncertainties and their correlations are taken from CEA files. The JEZEBEL experiment is used as an integral experiment for nuclear data adjustment. It is concluded that the major contribution on the adjustment is due to the k_{eff} sensitivities related to the Pu239 isotope, more precisely due to Pu239 fission and Pu239 $\bar{\nu}$ reaction. The transposition of the results on ASTRID fast reactor concept is also discussed.

Acknowledgements

This study was performed in the framework of an on-going collaboration between Uppsala University and CEA. The first author gratefully acknowledges financial support from the Swedish Research Council.

References

[1] P. Leconte et al., Feedback on Pu239 and Pu240 nuclear data and associated covariances matrix through the CERES integral experiments, Journal of Nuclear Science and Technology, 52:7-8, 1044-1052, 2015.

[2] C. De Saint Jean et al., Evaluation of Neutron-induced Cross Sections and their Related Covariances with Physical Constraints, Nuclear Data Sheets, Volume 148, Pages 383-419, 2018.

[3] A. Koning, S. Hilaire, S. Goriely, User Manual, Talys-1.8, A Nuclear Reaction Program, 2015.

[4] G. Palmiotti et al., Integral Experiments Analysis for Validation and Improvement of Minor Actinide Data for Transmutation Needs, AIP Conference Proceedings 769, 1436, 2005.

[5] J. Tommasi et al., Analysis of Superphenix and Phenix neutron physics experiments with the ERANOS code and data system, Operational and decommissioning experience with fast reactors. Proceedings of a technical meeting, Cadarache, France, 2002.

[6] T. A. B. Snijders, and R. J. Bosker, Multilevel Analysis: An Introduction to Basic and Advanced Multilevel Modeling, second edition, Sage Publishers, ISBN 9781849202008, 2012.

[7] Methods and Issues for the Combined Use of Integral Experiments and Covariance data, A report by the working party of International Nuclear Data Evaluation Co-operation of NEA Nuclear Science Committee.

[8] International Handbook of Evaluated Criticality Safety Benchmark Experiments, 2018.

EPJ Web of Conferences **211**, 07002 (2019) https://doi.org/10.1051/epjconf/201921107002
WONDER-2018

Covariance Matrices for Differential and Angle-Integrated Neutron-Induced Elastic and Inelastic Scattering Cross Sections of ^{56}Fe

Shengli Chen[1,2,a], *David* Bernard[1], *Pascal* Archier[1], *Cyrille* De Saint Jean[1], *Gilles* Noguere[1], and *Pierre* Tamagno[1]

[1] CEA, Cadarache, DEN/DER/SPRC/LEPh, 13108 Saint Paul Les Durance, France
[2] Université Grenoble Alpes, I-MEP2, 38402 Saint Martin d'Hères, France

Abstract. Correlations between neutron inelastic scatterings angular distributions are not included in the Joint Evaluated Fission and Fusion (JEFF) nuclear data library, while they are key quantities for uncertainty propagation of nuclear data. By reproducing the angle-integrated cross sections and uncertainties of JEFF-3.1.1, the present work obtains covariance matrix between high energy model parameters using the least square method implemented in the CONRAD code. With this matrix, it is possible to generate correlations between angle-integrated cross sections and angular distributions, which are usually presented by Legendre coefficients. As expected, strong correlations are found, for example, between the Legendre coefficients of elastic and first-level-inelastic scatterings and the angle-integrated total, elastic, total inelastic cross sections.

1. Introduction

In the nuclear industry, both Reactor Pressure Vessels (RPVs) for Pressurized Water Reactors (PWRs) and fuel cladding in Sodium-cooled Fast Reactors (SRFs) are Stainless Steel (SS), which is also a candidate cladding material for PWR Accident Tolerant Fuel (ATF) [1], [2]. SS is mainly constituted with iron, in which the abundance of ^{56}Fe is 91.8%. Therefore, the analyses of covariance matrices of ^{56}Fe are necessary for both neutronic investigations and studies of neutron irradiation in materials.

The angular distribution of scattering reactions, called also as the differential scattering cross section, is one of the most important quantities in nuclear data evaluation. The angular distribution of neutrons determines directly the neutron spectrum. On the other hand, the recoil energy of a Primary Knock-on Atom (PKA) is deduced from the angular distributions, while the former is the basis for investigating the irradiation damage of materials.

However, the complete covariance matrices of differential scattering cross sections in the Joint Evaluated Fission and Fusion (JEFF) nuclear data libraries are not given. The present work focuses on the determination of the covariance matrix between high energy model parameters for ^{56}Fe. This matrix can be used to calculate complete covariance matrix between angle-integrated cross sections and Legendre coefficients.

[a] Corresponding author: shengli.chen@cea.fr

2. Methods

The neutron inelastic scattering threshold is 862 keV for ^{56}Fe in both JEFF-3 libraries and ENDF/B-VII libraries. The second level of inelastic scattering of ^{56}Fe starts at 2.1 MeV. Due to few experimental data for each level of inelastic scattering, the R-matrix theory [3] is no longer the best choice for neutron energy above 2 MeV. Therefore, the Optical Model (OM) and the Statistical Model (SM) are used in the present work. The Bayes' theorem is used to determine the optimized model parameters and the corresponding covariance matrix.

2.1 Optical Model

In high energy region, the OM is usually used to calculate cross sections. Morillon and Romain proposed parameters for a Dispersive Optical Model Potential (DOMP) for neutrons with incident energies from 1 keV to 200 MeV [4]:

$$U(r,E) = [V_V(E) + iW_V(E)]f(r,R,a) + [V_S(E) + iW_S(E)]g(r,R,a)$$
$$+[V_{SO}(E) + iW_{SO}(E)]\frac{1}{r}\left(\frac{h}{m_\pi c}\right)^2 g(r,R,a)\mathbf{1}\cdot\boldsymbol{\sigma}, \quad (1)$$

where V and W respectively represent the real and imaginary terms of potential, the subscript V, S, SO respectively refer to the Volume, Surface, and Spin-Orbit parts in the DOMP. The volume shape function f is a Wood-Saxon form and g is its partial deviation to position r:

$$f(r,R,a) = \frac{1}{1+\exp[(r-R)/a]}. \quad (2)$$

$$g(r,R,a) = -4a\frac{d}{dr}f(r,R,a). \quad (3)$$

The same radius in Woods-Saxon form is used for different parts in the DOMP. For ^{56}Fe, the radius and the diffuseness proposed by Morillon and Romain are [4]:

$$R = 1.268A^{1/3}\ \text{(fm)}, \quad (4)$$

$$a = 0.566 + 5\times 10^{-9}A^3\ \text{(fm)}, \quad (5)$$

where A is the nuclear mass number.

All prior parameters in the DOMP can be found in the Reference Input Parameter Library (RIPL) [5]. The OM and SM calculations are performed with the nuclear reaction code TALYS [6], in which the OM code ECIS [7] is included.

2.2 Optimization and Uncertainties of Parameters

The COde for Nuclear Reaction Analysis and Data assimilation (CONRAD) [8] is used in the present work for the data assimilation. The Bayes' theorem implicates that the posterior probability density is:

$$p(\vec{x}|\vec{E},U) = \frac{p(\vec{E}|\vec{x},U)p(\vec{x},U)}{\int p(\vec{E}|\vec{x},U)p(\vec{x},U)d\vec{x}}, \quad (6)$$

where vector $\vec{x}$ represents the parameters in physical model, $\vec{E}$ denotes the experimental data, and U refers to the prior information. Under the hypothesis of Gaussian distribution for the probability density of $\vec{x}$, maximization of the posterior probability density is equivalent to minimization of the Generalized Least Square (GLS) cost function:

$$\chi^2_{GLS} = (\vec{x}-\vec{x}_0)^T M_x^{-1}(\vec{x}-\vec{x}_0) + (\vec{C}-\vec{E})^T M_E^{-1}(\vec{C}-\vec{E}), \quad (7)$$

where $\vec{x}_0$ represents the vector containing prior values, $\vec{C}$ and $\vec{E}$ denote the calculated and experimental data, respectively. M_x (M_E resp.) stands for the covariance matrix of $\vec{x}$ ($\vec{E}$ resp.). In the present work, the cross sections in JEFF-3.1.1 [9] are used as the experimental data in order to obtain the results corresponding to other information in JEFF-3.1.1. The Gauss-Newton scheme (known as Newton method for one-dimension solution) is used to find the minimum of the GLS cost function by iteration. The criterion of the convergence judgment is the relative variation of χ^2_{GLS}. It is set to 10^{-5} in the present work.

The parameters contained in the vector $\vec{x}$ are observable parameters. Posterior $\vec{x}$ and M_x are determined in the fitting procedure by iteration. On the other hand, nuisance parameters have also contributions to the posterior covariance matrix between observable parameters $\vec{x}$. The posterior covariance matrix between observable parameters can be expressed by [10]:

$$\Sigma_{11} = M_x + (G_x^T G_x)^{-1} G_x^T G_\theta M_\theta G_\theta^T G_x (G_x^T G_x)^{-1}. \quad (8)$$

where the matrices G contain partial derivation of calculated values. For $k = x, \theta$,

$$G_{k,(i,j)} = \partial C_i / \partial k_j. \quad (9)$$

More information about the data assimilation applied in the generation of covariance matrices for nuclear data can be found in Ref. [10].

An important remark is that the elastic cross sections from 850 keV up to 3.905 MeV in JEFF-3.1.1 are direct interpolations of experimental data of natural iron (Kinney's measurements [11] between 850 keV and 2.5 MeV and Smith's results [12] from 2.5 MeV to 3.905 MeV) [13]. Therefore, the OM and the SM cannot reproduce the structure in the angle-integrated cross sections of JEFF-3.1.1. The least square method implemented in CONRAD code permits to determine high energy model parameters that "mimic" the JEFF-3.1.1 evaluation.

2.3 Angular Distribution

In ENDF-6 format, the angular distribution is given by:

$$\sigma(\mu, E) = \frac{\sigma(E)}{2\pi} f(\mu, E), \quad (10)$$

where $\sigma(E)$ is the angle-integrated cross section at incident neutron energy E, and the normalized probability distribution is expressed by:

$$f(\mu, E) = \sum_{l=0}^{NL} \frac{2l+1}{2} a_l(E) P_l(\mu), \quad (11)$$

where μ is the cosine of neutron emission angle, P_l is the l-th order Legendre polynomial, a_l is the coefficient of Legendre polynomial stored in the ENDF, NL ($\leq$64) is the highest order of Legendre polynomial. NL increases with incident energy due to more forward-oriented angular distribution for higher energy (e.g. 4 at low incident energy and up to 20 at high energy in JEFF-3.1.1). Because the integral of P_l ($l > 0$) over [-1,1] is null and that of P_0 is 2, one can conclude that $a_0(E) = 1$. It is not necessary to store a_0 or to analyze its uncertainty.

3. Results and Discussion

3.1 Reproduction of cross sections

Figure 1 shows the ^{56}Fe total and first-level-inelastic scattering cross sections of JEFF-3.1.1 (dashed line) and the results with corresponding uncertainties calculated by the OM and the SM with optimized parameters. The fluctuations from experimental data of JEFF-3.1.1 cannot be reproduced by the OM and the SM, but the global correspondence between our results and JEFF-3.1.1 is ensured, especially for incident energy above 6 MeV. It is noticeable that the OM parameters should be improved to compute the cross sections below 4 MeV for ^{56}Fe. However, such an approach is sufficient for producing suitable covariance information for nuclear applications on atomic displacement and fast neutron flux in RPVs.

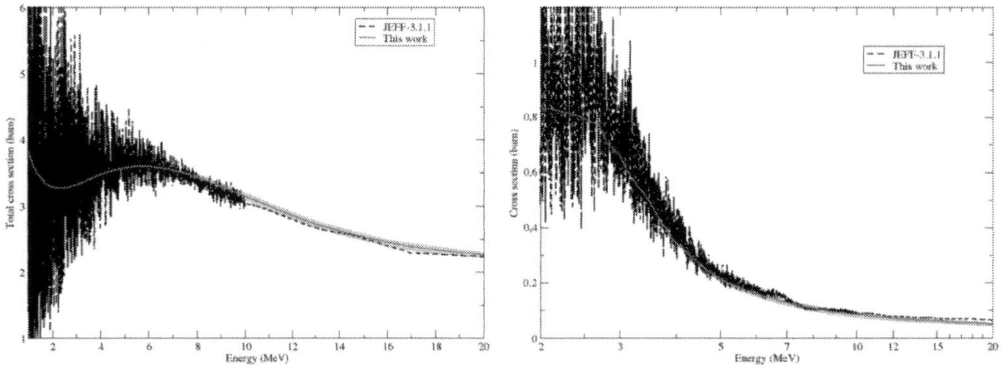

Figure 1. Total (left) and first-level-inelastic scattering (right) cross sections of JEFF-3.1.1 (dashed line) and the cross sections with corresponding uncertainties obtained in the present work

3.2 Correlation matrices

The correlation matrices presented in the present work are from neutron energy of 860 keV up to 20 MeV with logarithm scale. Figure 2 shows the correlation matrices between angle-integrated cross sections, including the total cross section (MT1), elastic scattering cross section (MT2), total inelastic scattering cross sections (MT4), first- to third-level-inelastic scattering cross sections (MT51 to MT53), and the continuum inelastic scattering (MT91). Figure 3 (Figure 4 resp.) illustrates the correlations between angle-integrated cross sections and the Legendre coefficients of elastic (first-level-inelastic resp.) scattering. Figure 5 presents the correlations between Legendre coefficients from the 1st order to the 5th order for MT51. Figure 6 reveals the correlations of Legendre coefficients between MT2 and MT51.

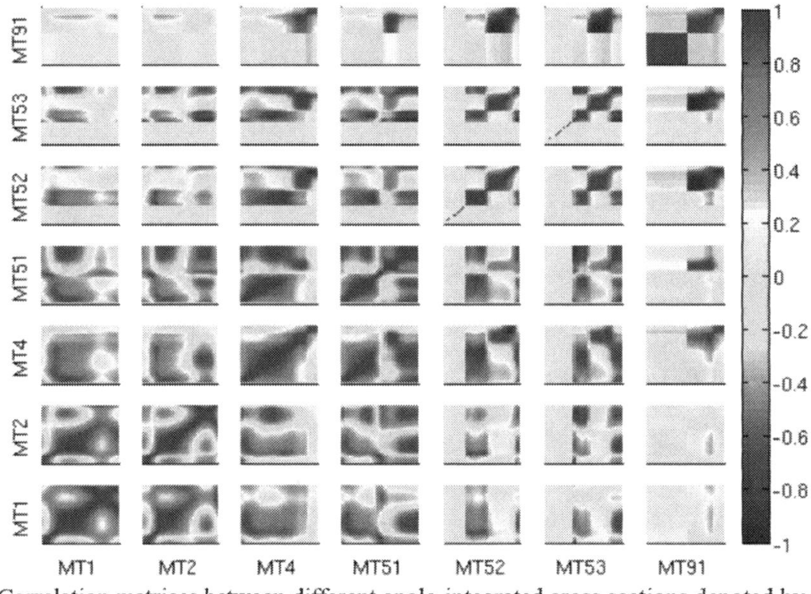

Figure 2. Correlation matrices between different angle-integrated cross sections denoted by the corresponding MT numbers

In Figure 2, the existence of correlations with MT91 at low energy is due to the low incident threshold energy of the MT91 reaction in TALYS. It has no influence on the application because of the negligible MT91cross sections between the threshold in TALYS and that in evaluated libraries. The results are in agreement with the expectation because: (a) MT1 is the sum of all cross sections and MT2 is predominant at low energy; (b) MT4 is the sum of MT51 to MT80 plus MT91, while MT51 has the most important contribution at low energy and MT91 dominates at high energy.

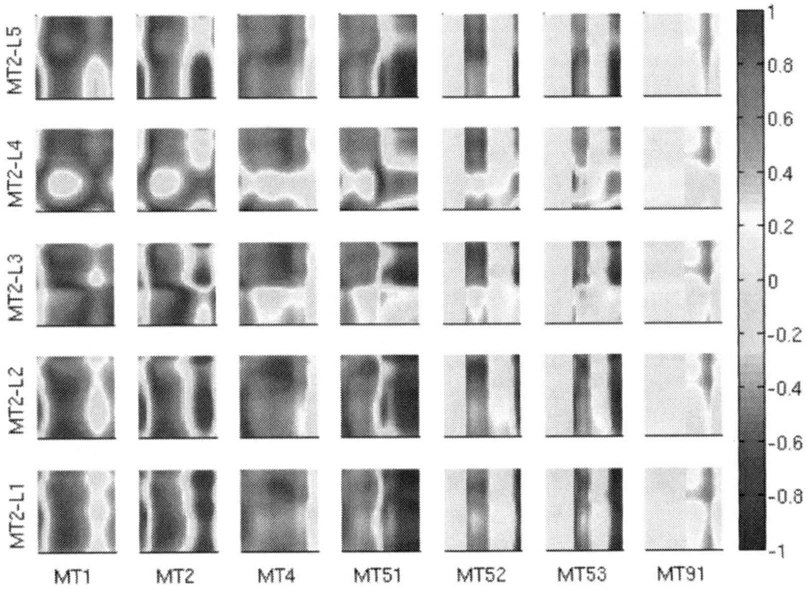

Figure 3. Correlation matrices between cross sections and Legendre coefficients of elastic scattering

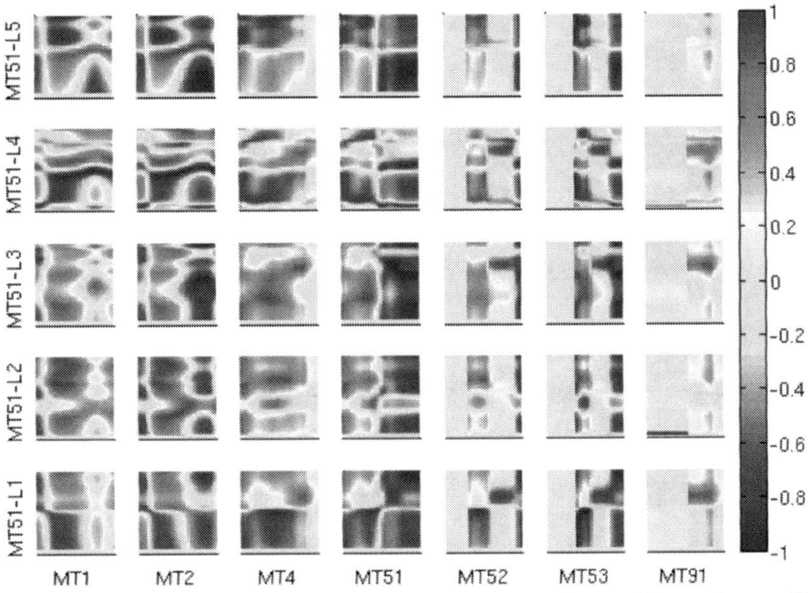

Figure 4. Correlation matrices between angle-integrated cross sections and Legendre coefficients of first-level-inelastic scattering

Strong correlations are found between angle-integrated cross sections and low-order Legendre coefficients for MT2 (Figure 3). The correlations of Legendre coefficients of MT2 and angle-integrated cross sections of MT4 are important within the OM and the SM. However, weak correlations exist between MT2 and MT91 for both differential and angle-integrated cross sections (Figure 2 and Figure 3). Comparing with MT2 (Figure 3), the

Legendre coefficients of MT51 have less strong correlations with angle-integrated cross sections (shown in Figure 4).

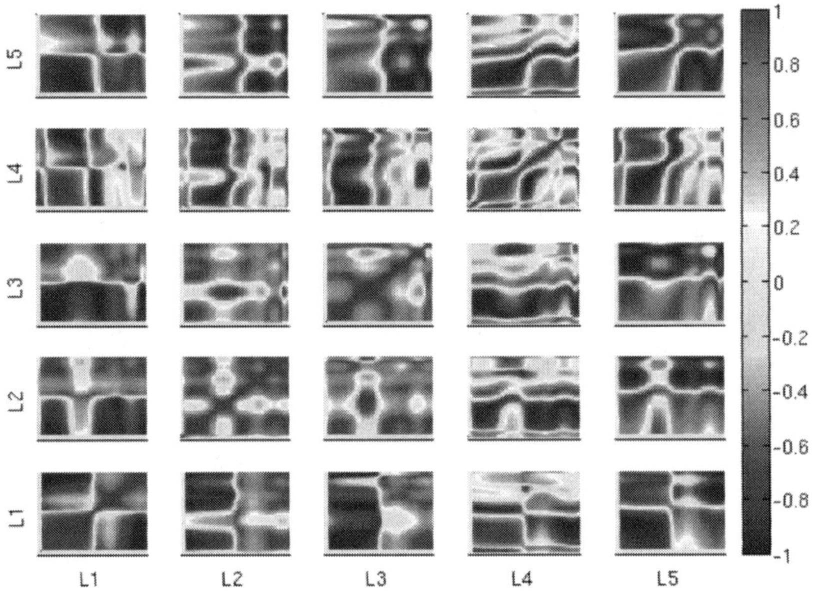

Figure 5. Correlation matrices between Legendre coefficients from the 1st to the 5th order for the first-level-inelastic scattering

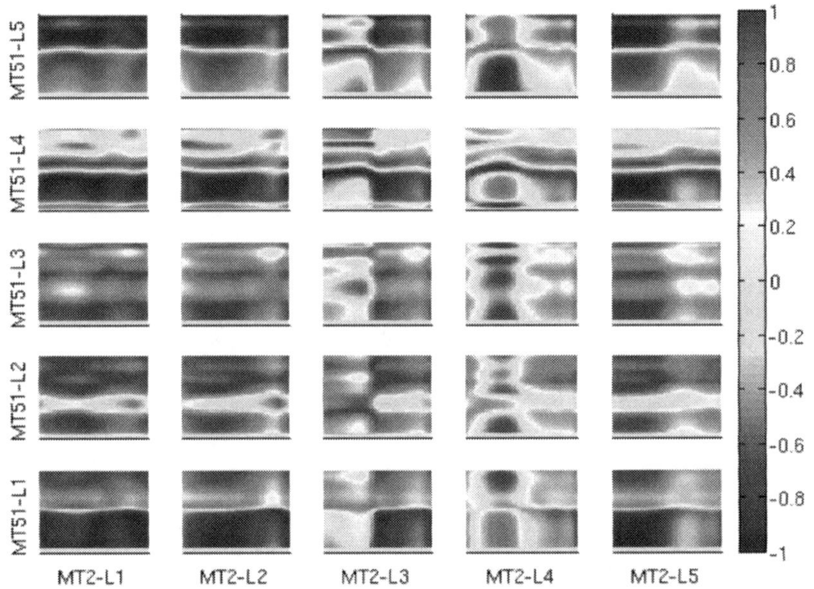

Figure 6. Correlation matrices of Legendre coefficients between the elastic scattering and the first-level-inelastic scattering

From Figure 5, one can conclude that the correlations between Legendre coefficients are of great importance for the first-level-inelastic scattering. It is noticeable that the correlations between Legendre coefficients for elastic scattering is even stronger. Figure 6 shows the strong correlations between low-order Legendre coefficients of MT2 and those of MT51. Therefore, correlations between differential and angle-integrated cross sections should be considered in nuclear applications, such as the uncertainty quantification for fast neutron fluence investigated in Ref. [14].

4. Conclusions

Thanks to the least square method implemented in the CONRAD code, we succeed to reproduce the n+^{56}Fe angle-integrated cross sections and uncertainties of JEFF-3.1.1 with the optical model (ECIS) and the statistical model (TALYS). Correlations between all angle-integrated cross sections and differential scattering cross sections are obtained. Strong correlations are found between angle-integrated cross sections and differential cross sections. The obtained covariance matrices between differential and angle-integrated cross sections are useful for further applications on uncertainty quantification of neutron fluence in reactor pressure vessels and neutron-induced irradiation damage of materials containing ^{56}Fe.

5. References

[1] S. Chen and C. Yuan, "Neutronic Analysis on Potential Accident Tolerant Fuel-Cladding Combination U_3Si_2-FeCrAl," *Sci. Technol. Nucl. Install.*, vol. 2017, p. 3146985, 2017.

[2] S. Chen, C. Yuan, and D. Guo, "Radial distributions of power and isotopic concentrations in candidate accident tolerant fuel U_3Si_2 and UO_2/U_3Si_2 fuel pins with FeCrAl cladding," *Ann. Nucl. Energy*, vol. 124, pp. 460–471, Feb. 2019.

[3] A. M. Lane and R. G. Thomas, "R-Matrix Theory of Nuclear Reactions," *Rev. Mod. Phys.*, vol. 30, no. 2, pp. 257–353, Apr. 1958.

[4] B. Morillon and P. Romain, "Dispersive and global spherical optical model with a local energy approximation for the scattering of neutrons by nuclei from 1 keV to 200 MeV," *Phys. Rev. C*, vol. 70, no. 1, p. 014601, Jul. 2004.

[5] R. Capote *et al.*, "RIPL – Reference Input Parameter Library for Calculation of Nuclear Reactions and Nuclear Data Evaluations," *Nucl. Data Sheets*, vol. 110, no. 12, pp. 3107–3214, Dec. 2009.

[6] A. J. Koning, S. Hilaire, and M. C. Duijvestijn, "TALYS-1.8," in *International Conference on Nuclear Data for Science and Technology*, 2007, pp. 211–214.

[7] J. Raynal, "ECIS code, distributed by the NEA DATA Bank," Paris, France, 2003.

[8] P. Archier *et al.*, "CONRAD Evaluation Code: Development Status and Perspectives," *Nucl. Data Sheets*, vol. 118, no. Supplement C, pp. 488–490, 2014.

[9] A. Santamarina *et al.*, "The JEFF-3.1.1 Nuclear Data Library," OECD/NEA, JEFF Report 22, NEA No. 6807, 2009.

[10] G. Noguere, P. Archier, C. D. S. Jean, and B. Habert, "Zero Variance Penalty Model for the Generation of Covariance Matrices in Integral Data Assimilation Problems," *Nucl. Sci. Eng.*, vol. 172, no. 2, pp. 164–179, 2012.

[11] W. E. Kinney and J. W. McConnell, "High Resolution Neutron Scattering Experiments at ORELA," in *Int.Conf.on Interact.of Neutr.with Nuclei*, Lowell, USA, 1976, p. 1319.

[12] A. Smith and P. Guenther, "Scattering of MeV Neutrons from Elemental Iron," *Nucl. Sci. Eng.*, vol. 73, p. 186, 1980.

[13] C. Vaglio-Gaudard, "Validation of iron nuclear data for the neutron calculation of nuclear reactors," PhD Thesis, Grenoble INP, 2010.

[14] L. Clouvel, P. Mosca, and J. M. Martinez, "Uncertainty propagation of double-differential scattering cross section in fast fluence calculation for the PWR surveillance capsules," presented at the PHYSOR 2018, Cancun, Mexico, 2018.

JEFF-3.3 covariance application to ICSBEP using SANDY and NDAST

Luca Fiorito[1,*], *James* Dyrda[1], and *Michael* Fleming[1]

[1]OECD / Nuclear Energy Agency, 45 Quai Alphonse le Gallo, Boulogne-Billancourt, 92100, France

Abstract. Providing reliable estimates of the nuclear data contribution to the uncertainty of well-known integral benchmarks is fundamental to the validation and verification process for a nuclear data library. The Nuclear Energy Agency has produced and maintains the NDaST sensitivity tool, which integrates the DICE sensitivities and nuclear data covariances. This system has been used to rigorously and efficiently provide direct feedback to evaluators and streamline validation. For its future evolution and to identify high-priority development areas, NDaST is continuously compared against state-of-the-art codes that use different uncertainty propagation methodologies.

In this work, NDaST was compared to the nuclear data sampling code SANDY for several ICSBEP criticality benchmarks using the JEFF-3.3 evaluated data. Despite excellent overall agreement for cross sections and fission neutron multiplcities, discrepancies due to processed covariance descriptions for angular distributions and prompt fission neutron spectra have identified areas where coordinated development of nuclear data covariance descriptions should be prioritised.

1 Introduction

The International Criticality Safety Benchmark Evaluation Project (ICSBEP) [1] provides a compilation of critical and subcritical benchmark experiment data that are maintained and continuously updated by the OECD/Nuclear Energy Agency (NEA) Working Party on Nuclear Criticality Safety (WPNCS). Nuclear data evaluation projects such as JEFF and ENDF/B regularly use these benchmarks to validate their nuclear data libraries. The validation process includes assessing the impact of nuclear data on the uncertainty of the benchmark integral responses, often with a focus on k_{eff}.Nuclear data evaluators have made significant efforts to include covariances for several data types into the official general-purpose nuclear data libraries. For example, the latest released library of the Joint Evaluated Fission and Fusion (JEFF) project, JEFF-3.3, contains 562 evaluations including:

- 50 covariance files for fission neutron multiplicities;
- 352 covariance files for resonance parameters;
- 442 covariance files for cross sections;
- 359 covariance files for angular distributions;

*e-mail: luca.fiorito@oecd-nea.org

- 36 covariance files for energy distributions; and
- 286 covariance files for activation cross sections.

In parallel to the development of more complete nuclear data evaluations and covariances, computer codes have been developed to process and propagate these covariances to be able to quantify the uncertainty on various neutronics parameters, including k_{eff}. Two uncertainty propagation approaches have been widely implemented in these codes: linear perturbation theory and Monte Carlo uncertainty propagation. The first method requires calculating the sensitivity of simulated quantities to the nuclear data, a feature that has been added to several neutron transport codes such as MCNP [2], SCALE (TSUNAMI) [3], ERANOS [4] and SERPENT [5]. The second refers to computational approaches that rely on the repeated random sampling of chosen parameters to obtain statistical outcomes of selected responses. Several codes have been developed to sample the nuclear data evaluations, including SANDY [6], SAMPLER[7], XSUSA [8], NUDUNA [9] and NUSS [10]. Semi-empirical model codes such as TALYS [11] are also used to generate sampled nuclear data evaluations based on the direct sampling of parameters used within physics models. This has the added benefit of allowing complete freedom in the distributions of quantities in nuclear data that are often assumed to have Gaussian distributions.

Today, the NEA Nuclear Data Sensitivity Tool (NDaST) is one of the most efficient codes to propagate nuclear data uncertainties into ICSBEP benchmarks. In this work, the performances of NDaST and SANDY were compared on a suite of ICSBEP cases to cross-validate their methods and identify any potential shortcomings. The study focused on newly implemented features using sensitivities to angular distributions and the propagation of these uncertainties.

2 Uncertainty propagation tools

For this work the JEFF-3.3 data library was processed into BOXER format with NJOY. The BOXER multigroup covariance files were produced on the SCALE 238-energy-group structure and were included in the JANIS database.

All available cross section covariances, P1 elastic scattering[1], total fission multiplicities and prompt fission neutron spectra (PFNS) covariances were utilised. Although the evaluations for PFNS contain covariance data for different incident-neutron energies, only those for 1 MeV were considered due to limitations in the current NJOY-2016.

Perturbed files were generated with SANDY for all evaluations in the JEFF-3.3 library. Each of these was processed with NJOY into ACE format to be used in MCNP calculations.

2.1 NDaST

NDaST is the nuclear data sensitivity tool developed and maintained by the OECD/Nuclear Energy Agency [12]. Amongst is well-known features, NDaST can propagate nuclear data uncertainties and covariances to model responses, such as the k_{eff}, for most of benchmark models recommended by the ICSBEP. To carry out this task, NDaST interfaces with the JANIS [13] and DICE [14] databases to retrieve nuclear data covariance matrices (C_{ND}) and k_{eff} sensitivity profiles (S), respectively. Uncertainties are propagated using the so-called *sandwich* formula,

$$C_{keff} = S^t C_{ND} S \ . \quad (1)$$

[1] P1 is the first Legendre polynomial coefficient for the angular distribution of elastic scattering. In the rest of the paper we will use the nomenclature Pn to refer to the coefficient of the Legendre polynomial of order n.

The sensitivity profiles stored in DICE were calculated on a 238-multigroup structure with MCNP6 and/or TSUNAMI and cover most of the reaction cross section channels. Sensitivities are also often available for total fission neutron multiplicities, PFNS and P1 elastic scattering.

2.2 SANDY

SANDY is a nuclear data sampling tool that, for a given evaluated nuclear data and covariance file formatted according to the standard ENDF-6 rules [15], can produce *perturbed* copies of the file that statistically reflect the covariance information. The perturbed files are generated by applying random perturbation coefficients sampled from the multivariate Gaussian distribution defined by the evaluated data. SANDY can perturb any type of nuclear data for which covariances are provided, including cross sections, and fission neutron multiplicities, as well as emitted energy and/or angular distributions. SANDY enforces consistency with the conservation rules by renormalizing the probability distribution functions and recalculating derived cross sections[2].

The perturbed files are suitable for Monte Carlo uncertainty propagation calculations, where a given model is solved repetitively, each time adopting a different file. From the statistical analysis of the resulting output predictions one can calculate the distributions of any calculated quantity, including their mean, variance and any other moments.

3 Uncertainty propagation with ICSBEP benchmarks

The two codes and methodologies were cross-validated and compared against a selection of ICSBEP cases. The Jezebel benchmark was selected as a simple but realistic model for comparison, as well as the full set of 19 PMF[3] cases within the so-called *Mosteller* suite. The Mosteller suite is a well-known suite of MCNP input files for 119 ICSBEP benchmark models that cover several applications including different fuel types, neutron spectra and reflective materials.

3.1 Jezebel

For this test case, only the uncertainties on ^{239}Pu data were propagated, which include covariances for cross sections, fission neutron multiplicities (nubar) and PFNS, but angular distribution covariances were excluded. Table 1 reports the k_{eff} uncertainty values calculated with SANDY and NDaST. SANDY was used to generate 1000 perturbed files, which produced a distribution of k_{eff} values from which the standard deviation was calculated. A variance of variance estimator was calculated to determine the confidence of this uncertainty value.

An excellent agreement was observed between NDaST and SANDY for cross sections and nubar. The difference found for PFNS generates from the way the PFNS covariances were processed by NJOY into BOXER-formatted files. SANDY perturbed files can be correctly used to propagate the PFNS covariances using the fission spectrum of the system as weighing function without the intermediate data format restrictions imposed by the BOXER format. SANDY reproduces NDaST results by sampling from the PFNS covariance matrix for incident-neutron energy of 1 MeV and extending its validity range to the entire energy domain.

[2]Derived cross sections are those that can be reconstituted from other existing reactions using standard summation rules

[3]PU-MET-FAST follows the DICE nomenclature to identify the ICSBEP cases by fissile material (PU=plutonium), fuel physical form (MET=metallic) and neutron spectrum (FAST=fast). PU-MET-FAST is also often shortened as PMF.

 https://doi.org/10.1051/epjconf/201921107003

Table 1: k_{eff} uncertainty prediction for Jezebel using SANDY and NDaST. Uncertainties calculated with SANDY are given together with an estimate of the variance of variance.

data type	NDaST uncertainty (pcm)	SANDY uncertainty (pcm)
All ^{239}Pu nuclear data	604	562 ± 13
^{239}Pu cross sections	221	221 ± 5
^{239}Pu fission neutron multiplicities	414	418 ± 9
^{239}Pu prompt fission neutron spectra	381	269 ± 6

3.2 PU-MET-FAST

Similarly to what was done for Jezebel, NDaST and SANDY were used to quantify the nuclear data contributions to the k_{eff} uncertainty for 19 PMF benchmarks. For all the isotopes reported in the selected benchmarks we generated 300 perturbed files with SANDY, which were then used in parallel MCNP calculations. All available covariance matrices and sensitivity profiles for the 19 benchmarks were employed for the NDaST calculation.

Figure 1 displays an overall agreement between the two codes for most cases, although some results were still affected by the BOXER format of the PFNS covariances used by NDaST. The k_{eff} uncertainties for 16 of the 19 benchmarks range from 500 pcm up to 1000 pcm, while three benchmarks (PMF6, PMF10 and PMF20) exhibited relatively large uncertainties that exceeded 1500 pcm in both SANDY and NDaST calculations.

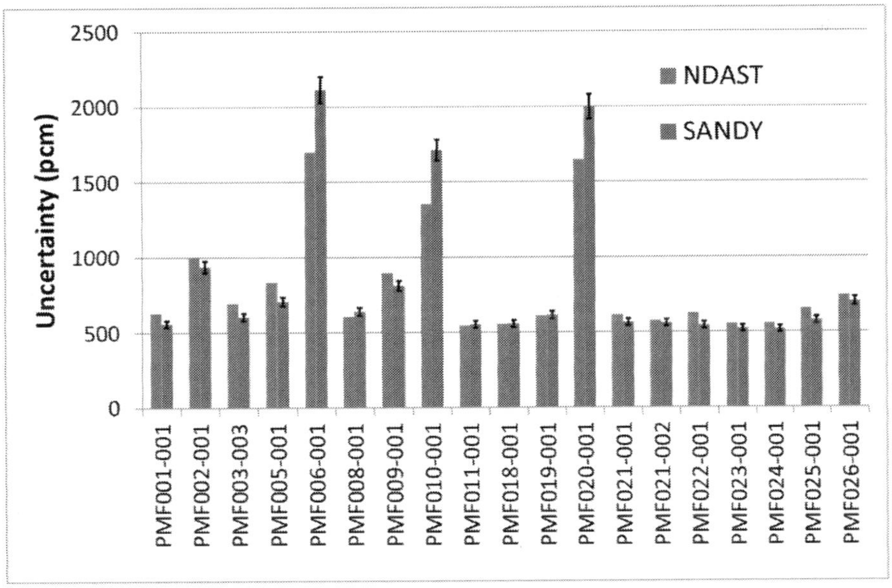

Figure 1: k_{eff} uncertainties for ICSBEP PU-MET-FAST benchmarks calculated with NDaST and SANDY.

The inter-case k_{eff} correlations — also known in the nuclear industry as C_k values — originate from the nuclear data covariances and were computed using the SANDY simulations, as shown in Figure 2. A strong correlation > 99 % was identified between the same three benchmarks (PMF6, PMF10 and PMF20) with significantly higher nuclear data uncertainty. All three of these benchmarks inlclude a heavy ^{238}U reflector. It is common for systems with important reflectors to have a large k_{eff} sensitivity to the neutron leakage term, and consequently to the angular distribution of the neutrons scattered out of the reflected material. In the following section we analyzed this contribution and its impact to the total k_{eff} uncertainty for the PMF6 case.

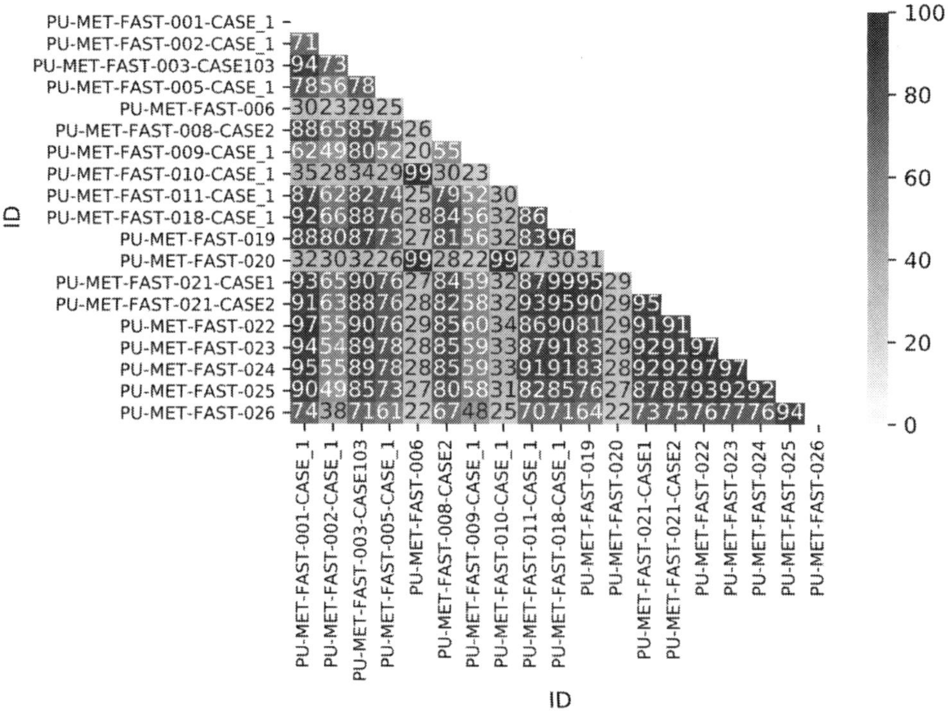

Figure 2: k_{eff} correlation matrix for ICSBEP PU-MET-FAST benchmarks generated by nuclear data.

3.3 PU-MET-FAST-006

The PMF6 benchmark model is a delta-phase plutonium-alloy sphere with a mass of 6.06 kg reflected by about 19 cm of natural uranium. The contribution to the PMF6 k_{eff} uncertainty of the angular distributions of neutrons scattered in ^{238}U was assessed with NDaST and SANDY. For the latter, we relied upon 1000 perturbed files that were processed into ACE format and used in MCNP simulations. The covariance matrix taken from JEFF-3.3 includes covariances for the Legendre polynomial coefficients (including cross-coefficient correlations) up to the 6th order. DICE currently contains sensitivity profiles for the first Legendre polynomial coefficient , i.e. P1. As a consequence, NDaST does not propagate the available covariances from P2 to P6.

The uncertainties calculated with SANDY and NDaST were 2056±46 and 1382 pcm, respectively. Not only do the SANDY and NDaST results disagree on the uncertainty prediction, but the resulting k_{eff} distribution is significantly non-Gaussian.

3.3.1 Propagation of covariances for Legendre polynomial coefficients

To further investigate the issue, SANDY was used to sample perturbations from:

1. a covariance matrix that includes all six Legendre polynomial coefficients (the *default* SANDY method); and

2. only the covariance matrix for P1 (to mimic NDaST)[4].

These perturbed files were used to derive uncertainties and energy-correlations for the tabulated angular distribution of scattered neutrons for an incident-neutron energy of 1.8 MeV, where the fission neutron flux is the highest. This data is shown in Figure 3.

At this energy, the distribution is strongly peaked both forward and backward, a shape that requires a large number of polynomials to be accurately reproduced. As a consequence, the k_{eff} sensitivity to P1 represents only a fraction of the total sensitivity to angular distributions, as it is described by a series of polynomial coefficients. Figure 3 shows that the contribution of P1 to the distribution uncertainty is only a small fraction of the collective uncertainty coming from all the terms, including their cross correlations.

The use of covariances over Legendre polynomials to represent uncertainties on angular distributions implicitly imposes strong correlations between the scattering angles. These reflect precisely the shape of the polynomials. Figure 3 shows that the use of only the P1 covariance matrix results in a full anti-correlation between all the forward and backward scattering angles. We confirmed in Figure 3 that the uncertainty and correlation matrix for the angular distribution are much different, both in uncertainties and correlations, when the full set of polynomial covariances are considered.

3.3.2 k_{eff} sensitivity to the Legendre polynomial coefficients

The non-linearity of the PMF6 k_{eff} with respect to variation of Legendre coefficients was investigated by sampling 1000 files with SANDY where only one Legendre polynomial coefficient was perturbed. To isolate the effects in each pertubation, a 10% standard deviation with full energy correlation was used for one coefficient in each study. The scatterplots of the results for P1 and P2 variation are shown in Figure 4, together with the k_{eff} distributions. From the quantile-quantile plot in Figure 5 one can see that the null-hypothesis of a Gaussian distribution is rejected, as confimed by the Shapiro-Wilks test.

The regression analysis of the scatterplots in Figure 4 shows that linearity is not an accurate approximation for these coefficients. The k_{eff} results for P2 variation are even more notable in that an increase *or decrease* in the coefficient causes approximately the same increase in k_{eff}. The use of a linear surrogate approach for P2 in this problem can give the false impression that k_{eff} is not sensitive to the second Legendre polynomial coefficient, when the distribution is highly non-normal and the result of the unperturbed calculation is at one extreme of the distribution! Similar results were produced for all higher order coefficients that are present in the JEFF-3.3 library.

[4]The results obtained from this sampling are coherent with what NDaST predicted.

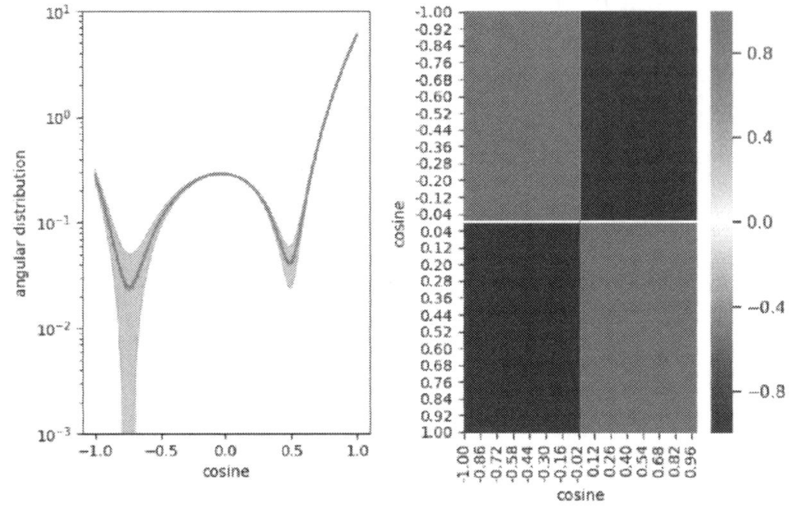

(a) Covariances were propagated for P1 only.

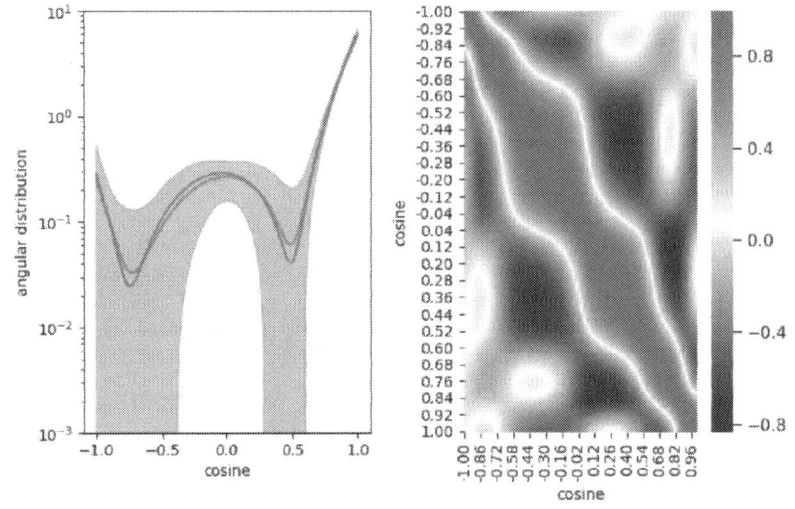

(b) Covariances were propagated for all Legendre polynomial coefficients.

Figure 3: ^{238}U elastic scattering angular distibution uncertainties and correlation matrices for incident-neutron energy E_{in}= 1.8 MeV. On the left, the green line represents the best estimate distribution, as derived from the evaluated file. The orange line is the average of the sampling, while the grey band is the associated standard deviation.

4 Conclusions

The JEFF-3.3 nuclear data library was processed with NJOY and SANDY to produce a suite of derived files for nuclear data uncertainty propagation. Uncertainties and other moments were quantified for a selection of ICSBEP cases using the NEA sensitivity tool NDaST and the Monte Carlo sampling code SANDY in combination with MCNP. The results show an excellent agreement between the two methodologies for the propagation of cross sections

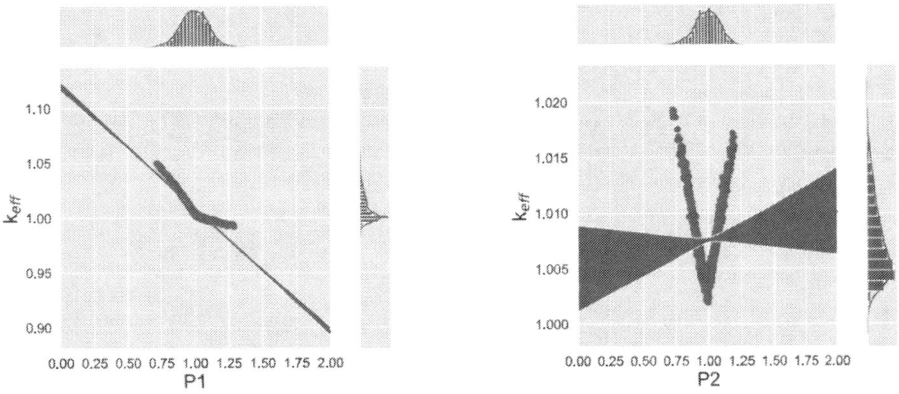

Figure 4: Estimates of the PMF6 k_{eff} sensitivity to the P1 and P2 coefficients for elastic scattering in ^{238}U.

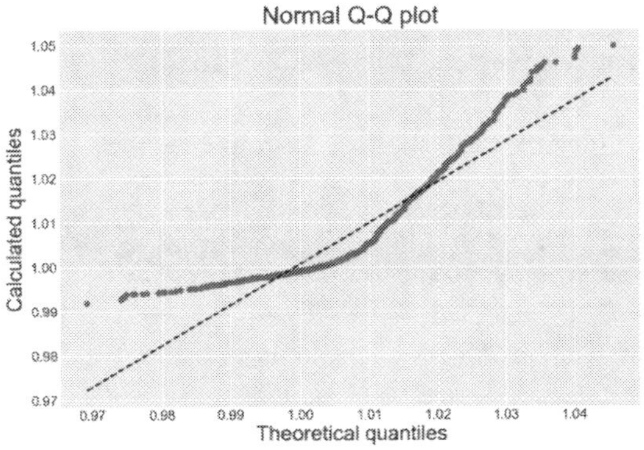

Figure 5: Quantile-quantile plot (Q-Q plot) of k_{eff} produced with 300 perturbed P1 coefficients. The dashed line represents the Q-Q plot for a Normal PDF.

and fission neutron multiplicities covariances. Due to the limitations of the NJOY processing of prompt fission neutron spectra covariances that NDaST relies upon, differences were found with some benchmarks due to the propagation of this uncertainty.

Also, it was proved that the k_{eff} of strongly reflected systems is highly sensitive to the neutron angular distribution in the reflective material, and in particular to Legendre polynomial coefficients above the first order. The propagation of only P1 covariances largely underestimates the total effect of the elastic scattering angular distribution and introduces strong correlations between forward and backward scattering angles that are an artifact of the format and not based on experiment. In addition, the linearity assumption cannot be accurately employed for angular distribution in these and many other systems.

References

[1] J. Briggs, J. Bess, J. Gulliford, Nuclear Data Sheets **118**, 396 (2014)
[2] C. Werner, Tech. Rep. LA-UR-17-29981 (2017)

[3] B.T. Rearden, D.A. Reed, R.A. Lefebvre, D. Mueller, W.B.J. Marshall (2011)
[4] J.M. Ruggieri et al., *ERANOS 2.1: International Code System for GEN IV Fast Reactor Analysis*, in *Proceedings of ICAPP 06* (2006), pp. 2432–2439
[5] J. Leppänen, M. Pusa, T. Viitanen, V. Valtavirta, T. Kaltiaisenaho, Annals of Nuclear Energy **82**, 142 (2015), joint International Conference on Supercomputing in Nuclear Applications and Monte Carlo 2013, SNA + MC 2013. Pluri- and Trans-disciplinarity, Towards New Modeling and Numerical Simulation Paradigms
[6] L. Fiorito, G. Žerovnik, A. Stankovskiy, G.V. den Eynde, P. Labeau, Annals of Nuclear Energy **101**, 359 (2017)
[7] M. Williams, G. Ilas, M. A. Jessee, B. Rearden, D. Wiarda, W. Zwermann, L. Gallner, M. Klein, B. Krzykacz-Hausmann, A. Pautz, Nuclear technology **183**, 515 (2013)
[8] W. ZWERMANN, A. AURES, L. GALLNER, V. HANNSTEIN, B. KRZYKACZ-HAUSMANN, K. VELKOV, J. MARTINEZ, Nuclear Engineering and Technology **46**, 343 (2014)
[9] O. Buss, A. Hoefer, J. Christian Neuber (2011), international Conference on Nuclear Criticality Safety (ICNC 2011)
[10] T. Zhu, A. Vasiliev, H. Ferroukhi, A. Pautz, S. Tarantola, Journal of Nuclear Science and Technology **52**, 1000 (2015)
[11] A. Koning, D. Rochman, Nuclear Data Sheets **113**, 2841 (2012), special Issue on Nuclear Reaction Data
[12] J. Dyrda, I. Hill, L. Fiorito, O. Cabellos, N. Soppera, EPJ Nuclear Sci. Technol. **4**, 14 (2018)
[13] N. Soppera, M. Bossant, E. Dupont, Nuclear Data Sheets **120**, 294 (2014)
[14] I. Hill, J. Gulliford, N. Soppera, M. Bossant, *Dice 2013 new capabilities and data* (2015)
[15] M. Herman, Members of the Cross Sections Evaluation Working Group (2009)

Integral data assimilation of the MERCI-1 experiment for the nuclear data associated with the PWR decay heat computation

J. Huyghe[1,*], C. De Saint-Jean[1], D. Lecarpentier[2], C. Reynard-Carette[3], C. Vaglio-Gaudard[1], V. Vallet[1].

[1]CEA, DEN, DER Cadarache, F-13108 Saint Paul les Durance, France
[2]EDF Research and Development, 7 Boulevard Gaspard Monge, Palaiseau, France
[3]Aix Marseille Univ, Université de Toulon, CNRS, IM2NP, Marseille, France

Abstract. Nuclear decay heat is a crucial issue for PWR in-core safety after reactor shutdown and back-end cycle. It is a dimensioning parameter for safety injection systems (SIS) to avoid a dewatering of the reactor core. The decay heat uncertainty needs to be controlled over the largest range of applications. The assimilation of the MERCI-1 experiment was studied to provide feedbacks on nuclear data. This experiment consisted in the measurement of the decay heat of a PWR UOX fuel sample irradiated in the OSIRIS reactor, for cooling times between 45 minutes and 42 days. More specifically, the consideration of several experimental values of MERCI-1 at different cooling times was tested. This raised issues about correlations to consider between different measurements. Besides, the impact of considering correlations between independent fission yields in covariance matrices on the decay heat uncertainty calculation and on the feedbacks on nuclear data is discussed.

1 Introduction

Decay heat is released by both radioactive decay of unstable fuel and material structure isotopes after reactor shutdown. It is a key challenge for in-core safety and back-end cycle. The decay heat depends on about 40,000 nuclear data, and mainly from fission yields. With the currently used JEFF-3.1.1 nuclear data evaluation, independent fission yield uncertainties are strong (mainly between 10 and 35% at 1 standard deviation (std)) and no correlation matrices are associated with the independent fission yields. New evaluations are proposed in the JEFF-3.3 evaluation, recently released. In the framework of nuclear data adjustment, the assimilation of the MERCI-1 integral experiment [1] for a feedback on the nuclear data involved in the nuclear decay heat calculation is discussed in this paper. In particular, the impact of considering correlations between independent fission yields [2] on the integral data assimilation process is studied. The second aspect discussed in the paper is the use of several measurements from the MERCI-1 experiment at the same time in the assimilation process. To do this, an experimental correlation matrix needs to be determined.

* Corresponding author: jordan.huyghe@cea.fr

2 MERCI-1 experiment overview

The MERCI-1 integral experiment [1] represents a significant contribution for the experimental validation of short cooling time decay heat calculation tools. Carried out in 2008, it consisted in irradiating a PWR UOX fuel rod sample (3.7w.% ^{235}U enrichment) in the French OSIRIS experimental reactor (CEA Saclay, France), and then in measuring the decay heat released by the sample with the MOSAIC calorimeter [3]. The OSIRIS reactor core was loaded with U_3Si_2Al plates (e_{235U} = 19.75%). The burnup of the sample reached at the end of irradiation was 3.6GWd/t. The MOSAIC calorimeter is based upon the heat pipe principle. The decay heat is assessed by a heat balance measurement on the secondary system using an instrumentation designed to have a decay heat uncertainty of 0.5% at 1 std and its measurement goes from 45 minutes to 42 days of cooling time.

3 Control of the uncertainties associated with the MERCI-1 experiment

To be used within the framework of the integral assimilation (cf. section 4), both the experimental uncertainty associated with MERCI-1 and the calculation uncertainty associated with the interpretation of the experiment need to be rigorously estimated and controlled.

3.1 MERCI-1 experimental uncertainty

Different sources of experimental uncertainties can be identified, related to the use of the MOSAIC calorimeter [3] for the measurement of the irradiated fuel's decay heat. The sources are the temperature difference between the inlet and outlet of the condenser of the calorimeter (ΔT), the heat capacity of water (C_p), the water mass flow measured between the inlet and outlet of the condenser (Q_m). All these sources of uncertainties are propagated to the decay heat uncertainty. The decay heat is obtained through the formula (1) as the image of the enthalpy balance between the inlet and the outlet of the condenser:

$$DH = Q_m \, C_p \, \Delta T \qquad (1)$$

The different sources of experimental uncertainty are assumed independent: thus the global uncertainty is obtained by combining them in a quadratic summation. The different terms in the quadratic summation are the products of the sensitivity of the decay heat to each term in equation (1) with the relative uncertainty of each parameter (Q_m, C_p, ΔT), given the fact that there are no correlations between Q_m and C_p and between Q_m and ΔT, and that the correlation between C_p and ΔT was determined and is negligible. The resulting uncertainty on the decay heat is 0.5% at 1 std, for cooling times between 45 minutes and 42 days except between 16 and 21 days and between 23 and 25 days where it is 1.0% at 1 std, due to a valve failure leading to a discontinuity in the heat measurement.

3.2 MERCI-1 calculation uncertainty

The calculation uncertainty associated with the MERCI-1 experiment comes from different sources:

- the gamma leakage and activation of the cladding and overcladding: the gamma leakage was determined with a stochastic code and the activation of the cladding was taken into account in the decay heat calculation. The main contributors to the gamma decay heat and activation power for the cooling times considered are ^{140,142}La, ^{132,134}I and ^{138}Cs for the former (responsible for the gamma leakage given their gamma spectra) and ^{56}Mn, ^{60}Co and ^{51}Cr for the latter. Judging from these isotopes' associated nuclear data uncertainties (gamma decay energies and periods), the resulting uncertainties on the gamma leakage and activation power of the cladding and overcladding is negligible with respect to the other sources of uncertainty.

- the adjustment of the burnup reached at the end of irradiation (the adjustment was done by minimizing the calculation-over-experiment discrepancies of the neodymium concentrations (^{145,146,148,150}Nd) ; this adjustment depends on the cumulated fission yield uncertainties of ^{235}U to the neodymium isotopes. The resulting uncertainty is 1.1% at 1 std on the decay heat.

- the irradiation conditions (fuel temperature, coolant temperature and ^{235}U initial enrichment whose uncertainties are evaluated by direct perturbation in the transport calculation and results in in a global uncertainty of 0.1% at 45 minutes of cooling and 1.1% at 42 days of cooling at 1 std).

The sources of calculation uncertainties together with the experimental uncertainties for the MERCI-1 experiment were propagated to the decay heat by quadratic summation, assuming that they are all independent, and result in a global decay heat uncertainty going from 1.2 to 1.6% at 1 std according to the cooling time considered.

4 Integral data assimilation of the MERCI-1 experiment

4.1 CONRAD code

The evaluation code used for the data assimilation of the MERCI-1 experiment is CONRAD (**CO**de for **N**uclear **R**eaction **A**nalysis and **D**ata assimilation), developed at CEA, Cadarache [4]. CONRAD enables, *inter alia*, the assimilation of integral experiments to provide feedback on nuclear data. It is based on a prior set of nuclear data that will be updated through the assimilation of the integral data. The adjustment is done within the framework of the Bayesian inference, by minimizing the Generalized Least Squared function (cf. equation (2)), giving the posterior value of each parameter x fitted (bias adjustment), and their posterior covariances M_x^P [4], that will be referred to as trends in the paper.

$$S(\mathbf{x}) = (\mathbf{x}-\mathbf{x_0})^T M_x^{0^{-1}} (\mathbf{x}-\mathbf{x_0}) + (\mathbf{C}(\mathbf{x})-\mathbf{E})^T M_{C/E}^{-1} (\mathbf{C}(\mathbf{x})-\mathbf{E}) \quad (2)$$

where $\mathbf{x}$ are the parameters (nuclear data), $\mathbf{x_0}$ are the prior values of the parameters, M_x^0 is the prior covariance matrix of the parameters, $\mathbf{C}(\mathbf{x})$ are the calculation values of the decay heat (function of the nuclear data, which explains why sensitivities of the decay heat to nuclear data need to be computed), $\mathbf{E}$ are the experimental values of the decay heat considered for the adjustment, and $M_{C/E}$ is the experimental covariance matrix, containing the propagated calculation and experimental (C/E) decay heat uncertainties.

4.2 Assimilation of one particular MERCI-1 experimental value

Let us consider the case of a standard PWR UOX fuel irradiated up to 15 GWd/t (about one cycle in reactor) at around one year of cooling (for fuel transport issues for instance). Within the framework of the adjustment of nuclear data of importance for the decay heat calculation, the first study consisted in assimilating the last measurement of the MERCI-1 experiment, at 42 days of cooling, whose associated calculation/experiment discrepancy (C/E -1) is -0.83% ± 1.6% (calculation performed with the DARWIN2.3 package [5] and the JEFF-3.1.1 nuclear data evaluation [6]). Sensitivity studies performed on the MERCI-1 fuel rod meshing and the number of depletion steps showed that the biases due to the calculation scheme are negligible [1]. Therefore, it is possible to use the integral data assimilation theory with the MERCI-1 C/E discrepancies. The list of parameters to fit with CONRAD to perform the adjustment was established from the major parameters the decay heat was sensitive to, at 42 days of cooling for the MERCI-1 measurement, and at around 1 year of cooling for the PWR UOX fuel. The corresponding parameters are the ^{235}U thermal fission cross section and the independent thermal fission yields of ^{235}U to ^{140}Xe, ^{140}Cs, ^{95}Sr, ^{95}Y, ^{144}La, and ^{144}Ba. The results of the adjustment for these parameters are shown in Fig. 1. The JEFF-3.1.1 uncertainty values of each parameter are drawn in green on Fig. 1. The trends for each parameter obtained with CONRAD (respectively planned with JEFF-3.3 [7]) are plotted in blue (respectively red). The results show that no actual significant changes are observed with the use of one experiment, except for the uncertainties of the fission yields of ^{235}U to ^{140}Cs and ^{95}Y which are reduced. However, the trends obtained are consistent with JEFF-3.3 at 1 std.

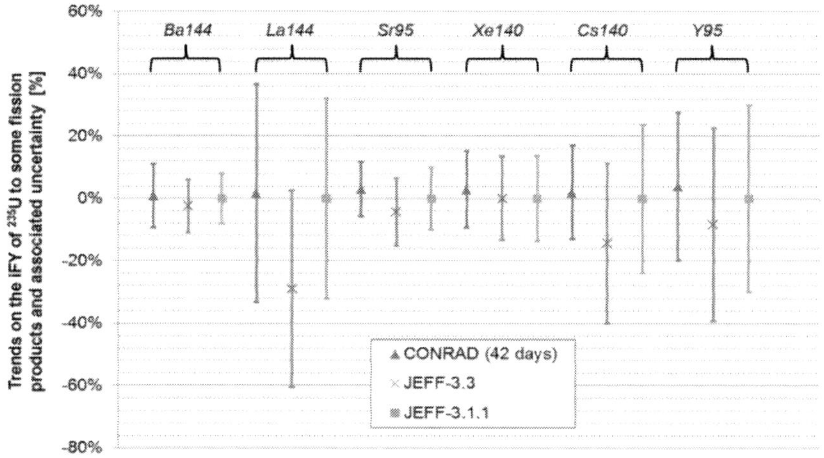

Fig. 1. Feedbacks on JEFF-3.1.1 independent fission yield (iFY) data after assimilation of the MERCI-1 experimental at 42 days of cooling with CONRAD and comparison to JEFF-3.3 trends

4.3 Impact of the use of correlations between independent fission yield data

The main nuclear data for the decay heat computation are fission yield data. JEFF-3.1.1 does not provide covariance matrices for independent fission yields, therefore covariance matrices for both fissile systems of ^{235}U and ^{239}Pu were produced at the CEA, associated with JEFF-3.1.1 fission yield data [2]. The covariance matrices produced are stored in the COMAC covariance matrix database [8]. Without these correlations, the decay heat uncertainty at around one year of cooling for the standard PWR UOX fuel is around 4%. When using these correlations, the decay heat uncertainty is reduced to 1%. The results of using these correlations in the integral data assimilation of the MERCI-1 experimental value mentioned in subsection 4.2 are shown in Fig. 2, compared to JEFF-3.3 and the assimilation previously performed without correlations between independent fission yield data (cf. subsection 4.2). Results show that using correlations between independent fission yields enables more consistent trends with JEFF-3.3 of the fission yield data of ^{235}U to the major contributors of the decay heat at the cooling times considered. In particular, the trends obtained for the independent fission yields of ^{235}U to ^{144}La and ^{140}Cs are more in accordance with JEFF-3.3 than what was obtained in Fig. 1. Moreover, the associated uncertainties of these fission yields are reduced.

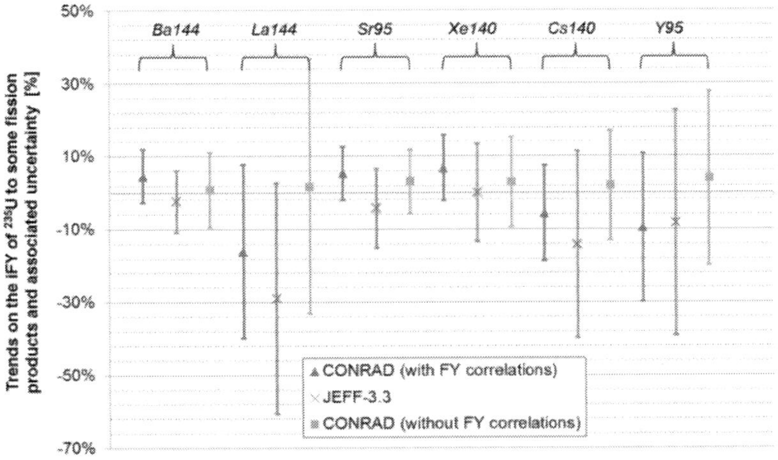

Fig. 2. Feedbacks on JEFF-3.1.1 correlated iFY data after assimilation of one MERCI-1 experimental value (42 days) with CONRAD

4.4 Experimental correlations and assimilation of several MERCI-1 experimental values

In order to improve the data assimilation process, several experimental data could be used at the same time, using correlated fission yield data, for this study, given the positive results obtained in subsection 4.3. In this paragraph, measurements at different cooling times coming from the MERCI-1 experiment are taken into account simultaneously. In order to select which experimental data to use, the relative contribution of the fission products of interest are shown in Fig. 3. as a function of the cooling time.

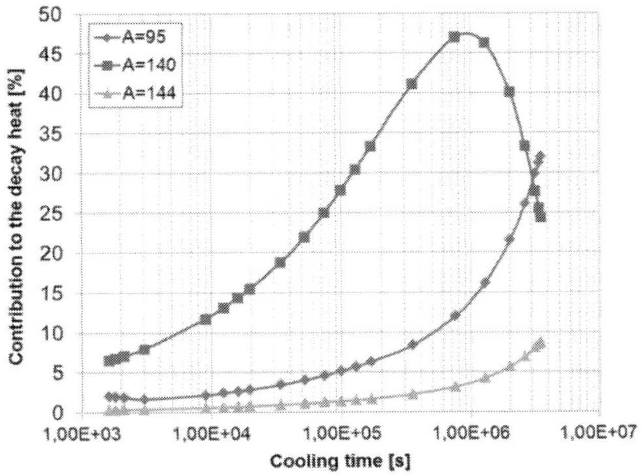

Fig. 3. Contributions of the main contributing masses to the decay heat [%] over the cooling time [s] for the MERCI-1 experiment

Judging from Fig. 3, it seems judicious to select cooling times rather close to the right side of the plot, *i.e.* cooling times above 10^5s for example, since the contributions of isotopes of masses 140 and 95 (a little less for mass 144) vary quite much on this range. Nine experimental values are then considered, corresponding to cooling times between 2 ($1.7.10^5$s) and 42 days (whose associated (C/E – 1) discrepancies are shown on Fig. 4).

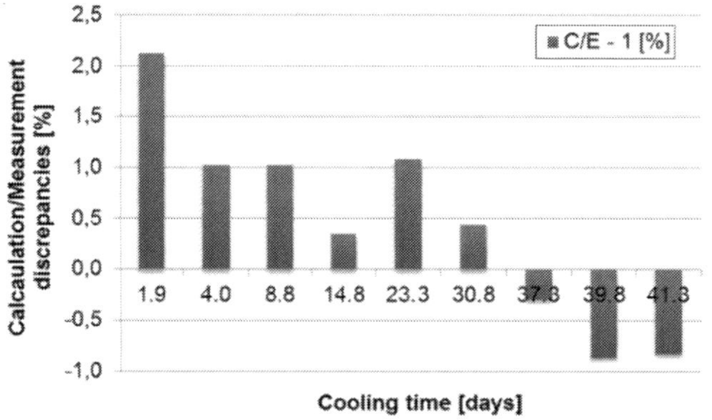

Fig. 4. MERCI-1 calculation/experiment discrepancies (C/E -1) for the nine cooling times considered

The simultaneous use of several experimental values coming from the same experiment raises indeed issues of experimental correlations to consider for the assimilation. Experimental correlations for each couple of experimental values (a,b) were determined by calculating the covariance between each couple of measured decay heat values (DH_a,DH_b) (at different cooling times) using equation (1) (cf. subsection 3.1). The resulting covariance is the following (3):

$$cov(DH_a,DH_b) = \alpha * cov(Q_{m_a},Q_{m_b}) + \beta * cov(T_a,T_b) + \gamma * cov(C_{p_a},C_{p_b}) \quad (3)$$

$cov(T_a,T_b)$ is the covariance between temperatures, $cov(Q_{m_a},Q_{m_b})$ is the covariance between two water mass flows and $cov(C_{p_a},C_{p_b})$ is the covariance between water heat capacities. The two first covariance terms could not be assessed, due to a lack of information thereon in the experimental process. The last covariance term is determined by performing a polynomial regression (3rd order) of the heat capacity against the temperature. The different coefficients of each power of the temperature in this polynomial are determined with associated uncertainties and covariances. It is thus possible to determine $cov(C_{p_a},C_{p_b})$. $\gamma * cov(C_{p_a},C_{p_b})$ is numerically negligible in front of the two other terms because of the values of α, β and γ, involving C_p, T and Q_m values.

The assimilation results with the use of the nine experimental values abovementioned are presented on Fig. 5. Different tests were performed with different experimental correlation values to try and measure the sensitivity of the assimilation results due to the experimental correlations considered between each couple of decay heat values (0.1, 0.5, 0.8 and 0.9, cf. Fig.5). The trends obtained with CONRAD for each independent fission yield are consistent at 1 std with what is planned for JEFF-3.3, regardless of the experimental correlations considered. The sensitivity of the assimilation results due to the experimental correlations considered is lower than the sensitivity due to the inclusion of the correlations between iFY. However, depending on the iFY studied, the trends obtained with the simultaneous use of the nine experimental values get closer to JEFF-3.3 with lower experimental correlations for ^{144}Ba, ^{140}Xe, ^{95}Sr and ^{95}Y, but with higher correlations for ^{144}La and ^{140}Cs, hence the importance of ensuring the experimental correlations to consider, to be able to propose real feedbacks on nuclear data when using more than a single experiment.

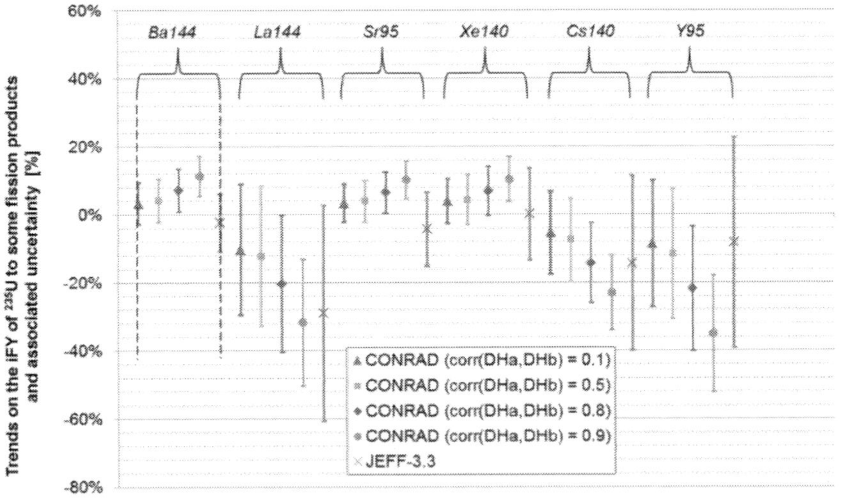

Fig. 5. Assimilation of nine correlated MERCI-1 experimental values (with different correlations)

5 Conclusions

The assimilation studies presented in this paper show that assimilating several correlated experimental values (instead of a single one) and adding covariance information on independent fission yield data of ^{235}U result in trends (on nuclear data of interest for the decay heat computation) in better accordance with JEFF-3.3. Even if it was not possible yet to determine the actual experimental correlations between the different MERCI-1 experimental values at different cooling times, it was possible to carry out a sensitivity study of the assimilation results obtained for several experimental correlation test values. It can be noticed that the results are in better accordance with JEFF-3.3 regardless of the experimental correlations considered between each couple of decay heat measurements. Other studies over the simultaneous assimilation of different integral experimental data and/or fission burst experiments could be performed, to verify the consistency between the different feedbacks thus obtained on nuclear data of interest for the decay heat computation.

References

1. J.C. Jaboulay, S. Bourganel, Nuclear Technology, **177**, 73-82 (2012)
2. N. Terranova, *Covariance evaluation for nuclear data of interest to the reactivity loss estimation of the Jules Horowitz Material Testing Reactor* (PhD thesis, 2016)
3. Ch. Blandin et al., *International Group Operating Research Reactors IGORR12* (2009)
4. C. De Saint Jean et al., *Int. Conf. On Nuclear Data for Science and Technology* (2007)
5. L. San Felice et al., Nuclear Technology, **184**, 217-232 (2013)
6. A. Santamarina et al., *JEFF Report 22 NEA No. 6807 OECD* (2009)
7. OECD, *JEFF-3.3, https://www.oecd-nea.org/dbdata/jeff/jeff33/*
8. P. Archier et al., *Proc. Int. Conf. PHYSOR* (2014)

EPJ Web of Conferences **211**, 07005 (2019) https://doi.org/10.1051/epjconf/201921107005
WONDER-2018

A first sketch: Construction of model defect priors inspired by dynamic time warping

Georg Schnabel[1,*] and *Henrik* Sjöstrand[1]

[1]Division of Applied Nuclear Physics, Uppsala University

Abstract. Model defects are known to cause biased nuclear data evaluations if they are not taken into account in the evaluation procedure. We suggest a method to construct prior distributions for model defects for reaction models using neighboring isotopes of ^{56}Fe as an example. A model defect is usually a function of energy and describes the difference between the model prediction and the truth. Of course, neither the truth nor the model defect are accessible. A Gaussian process (GP) enables to define a probability distribution on possible shapes of a model defect by referring to intuitively understandable concepts such as smoothness and the expected magnitude of the defect. Standard specifications of GPs impose a typical length-scale and amplitude valid for the whole energy range, which is often not justified, e.g., when the model covers both the resonance and statistical range. In this contribution, we show how a GP with energy-dependent length-scales and amplitudes can be constructed from available experimental data. The proposed construction is inspired by a technique called dynamic time warping used, e.g., for speech recognition. We demonstrate the feasibility of the data-driven determination of model defects by inferring a model defect of the nuclear models code TALYS for (n,p) reactions of isotopes with charge number between 20 and 30. The newly introduced GP parametrization besides its potential to improve evaluations for reactor relevant isotopes, such as ^{56}Fe, may also help to better understand the performance of nuclear models in the future.

1 Introduction

Evaluated nuclear data are important input for all kinds of nuclear physics applications. It has been shown in the past, e.g., [1], that evaluation techniques without proper account of potential model defects tend to underestimate uncertainties and bias results. Acknowledging the problem and being in need of a solution, Bayesian procedures have been often pragmatically tuned to deal with the problem, e.g., by ad-hoc adjustments of the likelihood function or by rescaling the obtained posterior uncertainties to make them plausible. A methodological elegant way is to introduce the information about model defects already as prior knowledge. In the field of nuclear data evaluation, this approach was pioneered by Pigni and Leeb [2]. Since then, various improvements and suggestions have been made, e.g., [3] and it was recognized [4] that these developments essentially deal with the design of a covariance function for a Gaussian process (GP), e.g., [5]. GPs are flexible tools in the domain of non-parametric

*e-mail: georg.schnabel@physics.uu.se

Bayesian statistics but commonly used specifications, such as the squared exponential covariance function, may not always be optimal for scenarios we encounter in nuclear data evaluation, e.g., quickly rising cross sections near thresholds or various degrees of predictive power of a nuclear model depending on the energy region where it is employed. So far this problem has been addressed by tailoring the covariance function to the specific situation, e.g., [6–8], whose shape can be adjusted by a few so-called hyperparameters.

Our suggestion in this paper is to replace the typically very structured covariance function by a very flexible one and to infer its shape by taking into account reaction systems that can be considered similar, i.e., the same reaction channels of neighboring isotopes. Formally, the amplitude and length-scale parameter—two numbers—of a standard squared exponential covariance function are replaced by an amplitude function and a metric function. The introduction of the latter function was inspired by a technique called dynamic time warping used, e.g., in speech recognition [9] where the time axis is locally stretched and shrank to align different voice samples as good as possible. In our case, it is exactly the same idea: To modify the effective distance between energies in order to align the Gaussian process as good as possible to the experimental data of all reaction systems at the same time. We remark that a certain construction of a model defect based on the information of neighboring isotopes has already been suggested in [3]. A distinctive feature of our approach is the determination of the model defect by the maximization of a score function, i.e., the marginal likelihood.

This paper is structured as follows. In section 2 we introduce the new parametrization of a GP which we call *dynamic time warping GP* or DTW GP for short. In section 3 we derive the criterion to determine the amplitude function and metric function. In section 4 we demonstrate the feasibility of the data-driven approach to infer the model defect of TALYS for (n,p) reactions using data for ^{56}Fe and neighboring isotopes. Finally, in section 5 we summarize and conclude.

2 DTW GP

The relation between experimental data and the model prediction can be written as

$$\vec{\sigma}_{\text{exp}} = \mathcal{M}(\vec{p}) + \vec{\varepsilon}_{\text{def}} + \vec{\varepsilon}_{\text{exp}} \tag{1}$$

where $\vec{\sigma}_{\text{exp}}$ contains the experimental measurements, $\mathcal{M}(\vec{p})$ is the corresponding model prediction based on the set of model parameters $\vec{p}$, the vector $\vec{\varepsilon}_{\text{def}}$ contains the deficiency of the model, and $\vec{\varepsilon}_{\text{exp}}$ contains the errors of the measurements. The unobservable truth $\vec{\sigma}_{\text{true}}$ is therefore given by both $\mathcal{M}(\vec{p}) + \vec{\varepsilon}_{\text{def}}$ and $\vec{\sigma}_{\text{exp}} - \vec{\varepsilon}_{\text{exp}}$. The only accessible quantity in this model is $\vec{\sigma}_{\text{exp}}$. The values in all other quantities are uncertain and we have to assign probability distributions to express our belief about the likelihood of possible realizations. Once all prior probability distributions are defined, we can use Bayesian statistics to obtain estimates of all quantities involved. In this section, we discuss the specification of a probability distribution for $\vec{\varepsilon}_{\text{def}}$. The specifications of probability distributions for the other quantities follows in section 3.

For the following discussion, we assume that $\vec{\sigma}_{\text{exp}}$ in eq. (1) contains angle-integrated cross sections $\sigma_{\text{exp},i}$ of a single reaction channel measured at incident energies E_i. The corresponding model defect term $\vec{\varepsilon}_{\text{def}}$ can be specified as a Gaussian process, e.g., [5]. A Gaussian process is the generalization of a multivariate normal distribution to functions and is characterized by the marginal distributions over finite sets of function values: Any finite set of function values at arbitrary locations is governed by a multivariate normal distribution. A Gaussian process is therefore uniquely defined by a mean function $\mu(E)$ and a covariance function $k(E, E')$. As the names imply, the former provides a mean value for any energy E

and the latter covariances between pairs of function values at arbitrary energies. Expressed in terms of our setting: For any choice of incident energies $E_1, E_2, \ldots$ the probability distribution of associated cross sections follows a multivariate normal distribution. We assume that the model prediction is a priori the most probable option, i.e., $\mu(E) = 0$. A common choice for the covariance function is the so-called squared exponential form, e.g., [5], introduced in [4] for nuclear data evaluation,

$$k(E, E') = \delta^2 \exp\left(-\frac{1}{2}\frac{(E - E')^2}{\lambda^2}\right). \tag{2}$$

This form is a reasonable default but depending on the energy range and observable other forms may be more suitable. Some alternatives have been employed and studied in [6–8]. All of these alternatives, however, are still rather rigid in terms of their structure and therefore foremost good solutions for the observable and energy range for which they have been designed. In this paper, we aim to construct a very flexible covariance function that is capable to adapt to any setting. Therefore we suggest the covariance function

$$k(E, E') = \delta(E)\delta(E') \exp\left(-\frac{1}{2}\Big(m(E) - m(E')\Big)^2\right) \tag{3}$$

where both the amplitude δ and length-scale λ of the squared exponential form are replaced by an energy-dependent amplitude and metric function, respectively. We call $m(.)$ a metric function because a distance (i.e. difference) between two energies gets transformed to another one. We demand the triangle inequality to hold, which is a defining criterion for a metric. Therefore, the function $m(.)$ must be monotonically increasing. The idea we pursue in this contribution is the determination of the functions $\delta(.)$ and $m(.)$ in a data-driven way by taking into account the same reaction channel of neighboring isotopes. The result for the (n,p) channel is displayed in fig. 2. The remainder of this and the next section deals with the parametrization of $\delta(.)$ and $m(.)$ and the method to infer its shape from the data.

We want a maximum of flexibility for the functions $\delta(.)$ and $m(.)$ and therefore suggest to define them as continuous piecewise linear functions. A continuous piecewise linear function can be written as

$$f(E \,|\, \{z_i\}_{i=1..N}) = \sum_{i=1}^{N-1} \left(\frac{E_{i+1} - E}{E_{i+1} - E_i}\mathbf{z_i} + \frac{E - E_i}{E_{i+1} - E_i}\mathbf{z_{i+1}}\right) \mathcal{I}_i(E). \tag{4}$$

where the factor $\mathcal{I}_i(E) = 1$ if $E_i \leq E < E_{i+1}$ and zero otherwise. The defining parameters of this function are the function values $\{z_i\}_{i=1..N}$ at energies $\{E_i\}_{i=1..N}$, i.e., $f(E_i) = z_i$. With a dense enough energy grid, arbitrary functions can be well approximated. The flexibility of functions parametrized as in eq. (4) requires reasonable constraints on features of these functions, which can be regarded as prior knowledge. Plausible constraints can be formulated in terms of minimal and maximal allowed changes per energy unit. In order to properly formulate such constraints, we can make the variable transformation

$$z_i(z_0, \Delta z_1, \cdots, \Delta z_N) = z_0 + \sum_{k=1}^{i} \Delta z_k. \tag{5}$$

Using this variable transformation, we obtain

$$g(E \,|\, z_0, \Delta z_1, \ldots, \Delta z_N) = f(E \,|\, \{z_i(z_0, \Delta z_1, \cdots, \Delta z_N)\}_{i=1..N}). \tag{6}$$

With this definition, the piecewise linear functions parametrized in terms of differences of function values for $\delta(.)$ and $m(.)$ can be written as

$$\delta(E) = g(E \mid u_0, \cdots, u_N) \text{ and } m(E) = g(E \mid v_0, \ldots, v_N) \,. \tag{7}$$

In the next section, we discuss how to determine the parameters u_i and v_i in a data-driven way by optimization.

3 Marginal likelihood maximization

The determination of the parameters $\{u_i\}_{i=1..N}$ and $\{v_i\}_{i=1..N}$ in eq. (7) requires a score function to assess the performance of possible solutions. In this section we derive one within the framework of Bayesian statistics. First, we have to complete the statistical model in eq. (1) by specifying the missing probability distributions for the model parameters $\vec{p}$, and the experimental errors $\vec{\varepsilon}_{\text{exp}}$. We assume multivariate normal distributions for both of them, i.e.,

$$\vec{\varepsilon}_{\text{exp}} \sim \mathcal{N}(\vec{0}, K_{\text{exp}}) \text{ and } \vec{p} \sim \mathcal{N}(\vec{p}_0, K_{\text{par}}) \,. \tag{8}$$

For the model defect we have $\vec{\varepsilon}_{\text{def}} \sim \mathcal{N}(\vec{0}, K_{\text{def}})$ with the elements of the covariance matrix K_{def} determined by eq. (3). The assumption of a priori independence of $\vec{p}$, $\vec{\varepsilon}_{\text{def}}$, and $\vec{\varepsilon}_{\text{exp}}$ completes the specification of the statistical model. In order to obtain analytic solutions, we further substitute the nuclear model by a first-order Taylor approximation

$$\mathcal{M}_{\text{lin}}(\vec{p}) = \vec{\sigma}_{\text{mod}} + S\,(\vec{p} - \vec{p}_0) \tag{9}$$

with

$$\vec{\sigma}_{\text{mod}} = \mathcal{M}(\vec{p}_0) \text{ and } S = \left.\frac{\partial \mathcal{M}(\vec{p})}{\partial \vec{p}}\right|_{\vec{p}=\vec{p}_0} . \tag{10}$$

Now we can compute the evidence, i.e., the probability distribution associated with $\vec{\sigma}_{\text{exp}}$. This variable is multivariate normal because it is given as a sum of multivariate normal random variables. Therefore it suffices to calculate the associated mean vector and covariance matrix,

$$\begin{aligned} \mathbb{E}[\vec{\sigma}_{\text{exp}}] &= \vec{\sigma}_{\text{mod}} \,, \\ \text{Var}[\vec{\sigma}_{\text{exp}}] &= M = S\,K_{\text{par}} S^T + K_{\text{def}} + K_{\text{exp}} \,, \end{aligned} \tag{11}$$

to obtain the result that $\vec{\sigma}_{\text{exp}} \sim \mathcal{N}(\vec{\sigma}_{\text{mod}}, M)$ which corresponds to the probability density function $\pi(\vec{\sigma}_{\text{exp}})$, given for a specific realization $\vec{\sigma}'_{\text{exp}}$ in eq. (12). Noteworthy, this distribution provides prior probabilities of possible outcomes of the experiment. As an aside, we remark that the underlying statistical computation—called marginalization—has been used for nuclear data evaluation before [10]. Marginalization serves the purpose to get rid of variables not of interest by themselves but nevertheless to properly account for the extra uncertainty they introduce. Here we marginalized over nuclear model parameters.

As soon as we have made the experimental measurements, we know the values in $\vec{\sigma}_{\text{exp}}$. This vector of experimental measurements must be compatible with the theoretical distribution parameters in eq. (11). Without the covariance matrix K_{def} it is usually not because too many things can go wrong. For instance, some error sources may have been overlooked for the construction of the experimental covariance matrix or the model is not able to adapt perfectly to the experimental data. Acknowledging the fact that we need statistical compatibility between the realization of $\vec{\sigma}'_{\text{exp}}$ coming from real experiments and the a priori theoretical distribution, we remove the inconsistency by adjusting the structure of K_{def}. We achieve this by adjusting the parameters $\{u_i\}_{i=0..N}$ and $\{v_i\}_{i=1..N}$ of eq. (7) which define the form of eq. (3) and consequently also K_{def} to maximize $\pi(\vec{\sigma}'_{\mathrm{exp}})$,

$$\log \pi(\vec{\sigma}'_{\mathrm{exp}}) = -\frac{N}{2}\log(2\pi) - \frac{1}{2}\log|M| - \frac{1}{2}\chi^2, \text{ with} \tag{12}$$

$$\chi^2 = \left(\vec{\sigma}'_{\mathrm{exp}} - S\,\vec{\sigma}_{\mathrm{mod}}\right)^T M^{-1}\left(\vec{\sigma}'_{\mathrm{exp}} - S\,\vec{\sigma}_{\mathrm{mod}}\right). \tag{13}$$

In other words, we seek to adjust $\delta(.)$ and $m(.)$ of the covariance function to maximize the probability of the experimental measurements $\vec{\sigma}'_{\mathrm{exp}}$. As a reminder, eq. (12) is the logarithmized multivariate normal pdf characterized by the mean vector and covariance matrix defined in eq. (11). Noteworthy, a linear transformation of the occurring vectors and corresponding transformations of S and M leave the value of the third term (i.e. the χ^2 term) invariant but alter the value of the second term containing the determinant. It seems more reasonable to us to penalize complexity on a relative scale and hence we used the transformation $\tilde{\sigma}_i = (\sigma_i - \sigma_{\mathrm{mod},i})/\max(0.1, \sigma_{\mathrm{mod},i})$ to transform $\vec{\sigma}_{\mathrm{exp}}$ and $\vec{\sigma}_{\mathrm{mod}}$ and also performed the corrsponding transformation on the sensitivity matrix S and the covariance matrix M. We expect the model to be less predictive for very low cross sections. To avoid unreasonable large uncertainty bands for such cross sections, we revert back to an absolute scale for cross sections below 0.1 millibarn, which explains the maximum in the denominator. The model defect is applied additively on this transformed scale. The amplitude function $\delta(E)$ therefore roughly amounts to the relative deviation of the truth from the model if the underlying prior model cross section is greater than 0.1 mbarn.

We are interested to learn about the *global* predictive power of the nuclear physics model. Thus we take reaction data of several neighboring isotopes into account for the maximization of eq. (12). We aim to find a set of $\{u_i\}_{i=0..N}$ and $\{v_i\}_{i=0..N}$ that is suited for a specific reaction channel, e.g., (n,p), for all the isotopes. Assuming experimental data and model parameters associated with different isotopes to be independent, we obtain the following form of the marginal likelihood $\pi(\vec{\sigma}'_{\mathrm{exp}})$:

$$\log \pi(\vec{\sigma}'_{\mathrm{exp}}) = -\frac{N}{2}\log(2\pi) - \frac{1}{2}\sum_{i=1}^{N_{\mathrm{iso}}}\log|M_i| + \log\left(\sum_{i=1}^{N_{\mathrm{iso}}}\exp\left(-\frac{1}{2}\chi_i^2\right)\right) \tag{14}$$

The sums run over the number of isotopes N_{iso} and contain the covariance matrices M_i and χ^2 values with respect to the individual isotopes.

4 Exemplary application

We determined the model defect covariance function by the outlined approach for (n,p) reactions in combination with the nuclear models code TALYS [11]. Because we want to use the model defects for an evaluation of ^{56}Fe, we included angle-integrated cross section data for isotopes with charge numbers between 20 and 30 and incident energies between 0.1 and 30 MeV available in the EXFOR database [12]. We assigned an uncorrelated uncertainty of 10% to all the data points, which is sufficient for the purpose of demonstrating the outlined approach. The included experimental data for (n,p) reactions is displayed in fig. 1. As can be seen in the figure, we did not attempt to remove experimental outliers. Unfortunately, given space restrictions, we are not able to give proper credit to the authors of these measurements due the fact that the data is spread over 207 entries in the database. References to the data can be provided upon request.

We employed an equidistantly spaced energy grid with 100 points covering the range from 0.1 to 30 MeV for the construction of $\delta(E)$ and $m(E)$ given in eq. (7). This point density

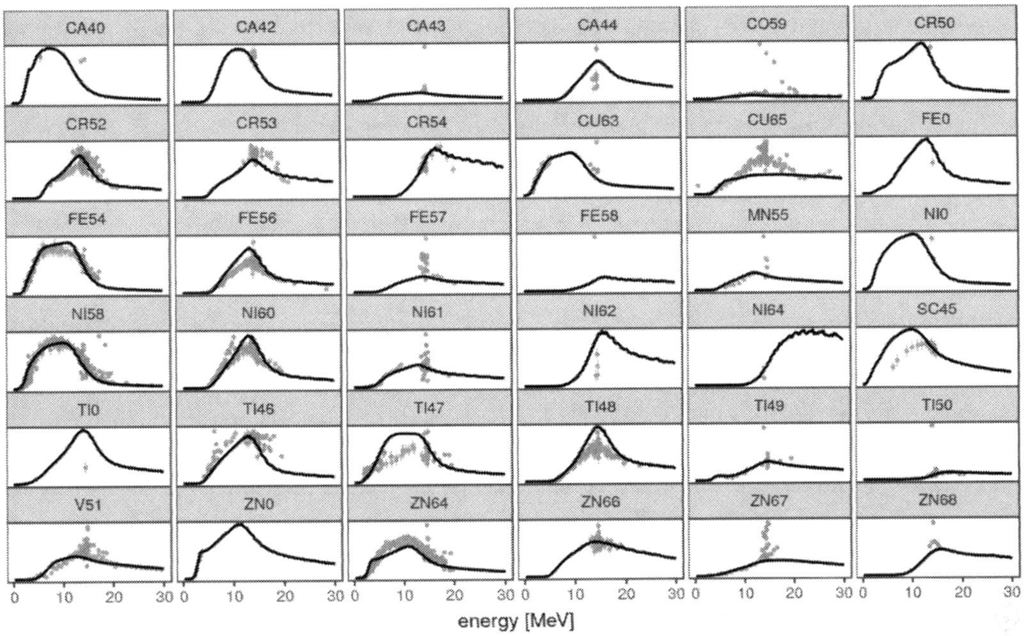

Figure 1. Experimental data for angle-integrated (n,p) cross sections of isotopes with charge numbers between 20 and 30 used for the marginal likelihood maximization. Uncertainties of experiments are assumed to be 10% and independent across data points. An energy of zero corresponds to the threshold energy of the reaction. The mass number zero indicates natural composition. No efforts have been made to identify outliers or assess the quality of the data sets. The data were retrieved from 207 entries of the EXFOR database.

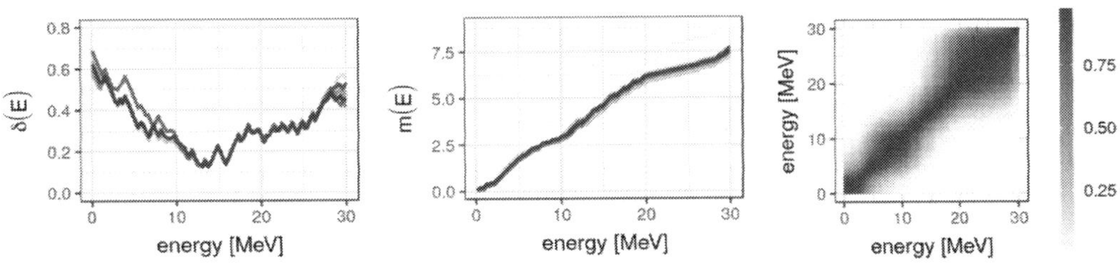

Figure 2. The dynamic time warping GP specification resulting from the marginal likelihood maximization: (left) the amplitude function $\delta(.)$; (middle) the metric function $m(.)$; and (right) the corresponding model defect correlation matrix. Different curves in the left and middle figure correspond to different local maxima of the marginal likelihood. The majority of the ten optimization results coincide and indicate the global maximum. Due to the transformation of the cross sections described in the paragraph below eq. (12), the function $\delta(E)$ can be interpreted as relative model defect expected a priori.

corresponds roughly to an energy resolution of 0.3 MeV and gives rise to 200 hyperparameters (i.e. the u_i's and v_i's in eq. (7)) that have to be optimized. Importantly, we shifted the origin of the energy grid to the threshold of the reactions. This measure ensures that cross sections near thresholds of different isotopes are mapped to the same grid points of the Gaussian process and also peaks are potentially better aligned.

We used the L-BFGS-B algorithm [13] for the maximization of eq. (14), which also permits the specification of box constraints. Applying box constraints on the transformed variables in eq. (7) enabled us to regularize the shape of the solution. We enforced that the amplitude function may not change more than about three percent between consecutive grid points. In other words, we constrained the rate of change of the amplitude function to be below roughly 10% per MeV. As the metric function has to be monotonically increasing, we applied a lower limit of 0.03 and an upper limit of 0.15 for its change between consecutive grid points. This specification translates to the prior knowledge that the effective length-scale should be at least roughly 1 MeV and not larger than 10 MeV at any energy. The lower limit on the effective length-scale protects against discrepant data, which would elsewise drive the effective length-scale to unreasonable low values. We emphasize here again that the dynamic amplitude and length-scale determined in a data-driven way is the real novelty of our approach. We derived analytic expressions for the gradient of the marginal likelihood, see, e.g., [14], which can be exploited by the L-BFGS-B algorithm. We performed the optimization ten times with different starting values and achieved in all cases convergence. Averaged over the individual optimization runs, it took the algorithm about 630 iterations to converge—tens of minutes on a decent personal computer with eight cores. Even though there were two local maxima, the majority of calculations converged to the global one. The learned amplitude and metric function as well as the correlation matrix of the model defect are shown in fig. 2.

5 Summary and conclusion

We introduced a new parametrization of the covariance function based on the concept of an amplitude and metric function and showed how these functions can be inferred from experimental data for neighboring isotopes by maximizing the marginal likelihood. We demonstrated the feasibility of the approach with the nuclear models code TALYS [11] and (n,p) reaction data for isotopes with charge numbers between 20 and 30. The obtained amplitude function and metric functions are interesting by themselves because they potentially allow us to gain not only qualitative but also quantitative insight into model performance on a per-energy basis. In the future, these model defects constructed by taking a global view on the predictive performance of nuclear models can be included in evaluation procedures to obtain more robust and reliable results.

References

[1] P. Helgesson, D. Neudecker, H. Sjostrand, M. Grosskopf, D.L. Smith, R. Capote, *Assessment of Novel Techniques for Nuclear Data Evaluation*, in *ASTM 16th International Symposium on Reactor Dosimetry* (Santa Fe, New Mexico, 2017)

[2] M.T. Pigni, H. Leeb, *Uncertainty Estimates of Evaluated ^{56}Fe Cross Sections Based on Extensive Modelling at Energies Beyond 20 MeV*, in *Proc. Int. Workshop on Nuclear Data for the Transmutation of Nuclear Waste. GSI-Darmstadt, Germany* (2003)

[3] H. Leeb, D. Neudecker, T. Srdinko, Nuclear Data Sheets **109**, 2762 (2008)

[4] G. Schnabel, Ph.D. thesis, Technische Universität Wien, Vienna (2015)

[5] C.E. Rasmussen, C.K.I. Williams, *Gaussian Processes for Machine Learning* (MIT Press, Cambridge, Mass., 2006), ISBN 0-262-18253-X 978-0-262-18253-9

[6] G. Schnabel, H. Leeb, EPJ Web of Conferences **111**, 09001 (2016)

[7] G. Schnabel, arXiv eprint arXiv:1803.00928 (2018)

[8] P. Helgesson, H. Sjöstrand, Review of Scientific Instruments **88**, 115114 (2017)

[9] C. Myers, L. Rabiner, A. Rosenberg, IEEE Transactions on Acoustics, Speech, and Signal Processing **28**, 623 (1980)

[10] Cyrille De Saint Jean, G. Noguere, B. Habert, B. Iooss, Nuclear Science & Engineering **161**, 363 (2009)

[11] A.J. Koning, S. Hilaire, S. Goriely, *TALYS-1.6 - A Nuclear Reaction Program*, http://www.talys.eu (2013)

[12] N. Otuka, E. Dupont, V. Semkova, B. Pritychenko, A. Blokhin, M. Aikawa, S. Babykina, M. Bossant, G. Chen, S. Dunaeva et al., Nuclear Data Sheets **120**, 272 (2014)

[13] R.H. Byrd, P. Lu, J. Nocedal, C. Zhu, SIAM Journal on Scientific Computing **16**, 1190 (1995)

[14] G. Schnabel, arXiv eprint arXiv:1803.00960 (2018)

EPJ Web of Conferences **211**, 07006 (2019) https://doi.org/10.1051/epjconf/201921107006
WONDER-2018

Formulation of Model Defects Suitable for the Resonance Regime

Benedikt Raab[1,*], *Thomas* Srdinko[1,**], and *Helmut* Leeb[1,***]

[1]Atominstitut, TU Wien, Wiedner Hauptstrasse 8-10, 1040 Vienna, Austria

Abstract. A method to account for model deficiencies in nuclear data evaluations in the resonance regime is proposed. The method follows the ideas of Schnabel and coworkers and relies on Gaussian processes with a novel problem-adapted ansatz for the covariance matrix of model uncertainties extending the formalism to the energy region of resonances. The method was used to evaluate a set of schematic but realistic neutron reaction data generated by an R-matrix code and a well defined model defect. Using the extended ansatz for model defects the Bayesian evaluation successfully recovered the built-in model defect in size and structure thus demonstrating the applicability of the method.

1 Introduction

The availability of reliable evaluated nuclear data is an important prerequisite for developments in several fields of science and technology. At present, Bayesian evaluation techniques are considered as the proper tool to consistently combine experimental data with model information. Because of the complexity of nuclear systems, nuclear models can only describe certain properties of the system, but not the complete system. Consequently there exist always model defects and they have an important impact on evaluations [1, 2]. Several attempts for their inclusion were proposed in the past [3, 4]. Recently Schnabel [1] presented a statistically consistent formulation in terms of Gaussian processes, which was applicable for smooth energy dependence of reaction cross sections and more recently for angle differential cross sections [5].

In this contribution we propose an extension of the formalism of model defects for the resonance regime, especially suited for light nuclear systems. The essential step, presented in Sect. 2, is the formulation of a new ansatz for the covariance matrix of the model defects and its inclusion in the evaluation process. In Sect. 3 the schematic example, the details of the evaluation procedure and the results are presented and discussed. The work is summarized in Sect. 4 and an outlook is given.

*e-mail: benedikt.raab@tuwien.ac.at
**e-mail: thomas.srdinko@tuwien.ac.at
***e-mail: helmut.leeb@tuwien.ac.at

2 Evaluation including Model Defects

2.1 Concept of Formalism

The mathematical basis of up-to-date evaluation techniques is provided by Bayes theorem

$$\pi\left(\vec{p} \mid \vec{\sigma}_{\mathrm{exp}}\right) = \frac{1}{\int \mathrm{d}^d p \; l\left(\vec{\sigma}_{\mathrm{exp}} \mid \vec{p}\right)\pi\left(\vec{p}\right)} \; l\left(\vec{\sigma}_{\mathrm{exp}} \mid \vec{p}\right)\pi\left(\vec{p}\right) . \tag{1}$$

Thereby the a-priori distribution $\pi(\vec{p})$ of model parameters $\vec{p}$ is updated with experimental information represented by the likelihood function $l\left(\vec{\sigma}_{\mathrm{exp}} \mid \vec{p}\right)$, where $\vec{\sigma}_{\mathrm{exp}}$ is the vector of experimental data points. The fraction in Eq. (1) acts as normalization. The result is an a-posteriori distribution $\pi\left(\vec{p} \mid \vec{\sigma}_{\mathrm{exp}}\right)$ of model parameters taking into account the experiments.

In standard Bayesian evaluations it is assumed that the model is perfect, i.e. $\vec{\sigma}_{\mathrm{exp}}=\vec{\sigma}_{\mathrm{mod}} + \vec{\epsilon}_{\mathrm{exp}}$, where $\vec{\sigma}_{\mathrm{mod}}$ and $\vec{\epsilon}_{\mathrm{exp}}$ are the vectors of model values and experimental errors, respectively. In general model defects always occur and their neglect affects the evaluation [1, 2]. At present it is well established that model defects must be taken into account in nuclear data evaluation. There have been several attempts to account for model defects. Recently Schnabel [1, 5] developed a statistically well defined approach based on Gaussian processes. The evaluation starts from the ansatz

$$\vec{\sigma}_{\mathrm{exp}} = \vec{\sigma}_{\mathrm{mod}} + \vec{\epsilon}_{\mathrm{mod}} + \vec{\epsilon}_{\mathrm{exp}} , \tag{2}$$

where $\vec{\epsilon}_{\mathrm{mod}}$ is the model defect. All quantities are assumed to be random variables. For simplicity we assume multivariate normal distributions, i.e.

$$\vec{\sigma}_{\mathrm{mod}} \sim \mathcal{N}(\vec{\sigma}_0, \mathbf{A}_0) , \quad \vec{\epsilon}_{\mathrm{mod}} \sim \mathcal{N}(\vec{0}, \mathbf{K}_0) , \quad \vec{\epsilon}_{\mathrm{exp}} \sim \mathcal{N}(\vec{0}, \mathbf{B}) , \tag{3}$$

where the first argument refers to the mean value and the second one to the associated covariance matrix. In addition we assume that $\vec{\sigma}_{\mathrm{mod}}$, $\vec{\epsilon}_{\mathrm{mod}}$ and $\vec{\epsilon}_{\mathrm{exp}}$ are mutually independent.

2.2 Model defects for the Resonance Regime

The inclusion of model defects via a Gaussian process assumes a vanishing a-priori mean value of $\vec{\epsilon}_{\mathrm{mod}}$ and an a-priori covariance matrix $\mathbf{K}_0$ which must be chosen according to the expectation. The covariance function for angle-integrated and angle-differential cross sections introduced in [1] and [5], respectively, is not suited for the resonance region. Frequently phenomenological R-matrix descriptions of the resonant part are considered to be perfect. However, the achievable description depends on several quantities which are chosen prior to a fit, i.e. the chosen matching radii, the number of poles, the choice of background poles etc. Of course one can consider these quantities as parameters of the fit, but high correlations and the strong non-linear relation to the observables will hamper reliable estimates of their uncertainties. Therefore a treatment as model defects of a specifically chosen model is preferable.

In order to account for model defects of the cross section $\sigma(E)$ in the resonance regime we propose for model defects the following covariance matrix,

$$\begin{aligned}\mathbf{K}_0(E_i, E_j) =& \sigma(E_i)\sigma(E_j)\,\delta_1^2 \cdot \exp\left[-\frac{1}{2\lambda_1^2}\left(E_i - E_j\right)^2\right] \\ &+ \delta_2^2 \cdot \alpha\frac{\mathrm{d}\sigma}{\mathrm{d}E}(E_i) \cdot \alpha\frac{\mathrm{d}\sigma}{\mathrm{d}E}(E_j) \cdot \exp\left[-\frac{1}{2\lambda_2^2}\left(E_i - E_j\right)^2\right] .\end{aligned} \tag{4}$$

The first term is equivalent to the expression introduced by Schnabel [1] for smooth cross sections, while the second one is sensitive to energy regions with strongly varying cross sections. The quantity α can be considered as mean widths of the resonances, λ_2 as correlation length between the resonances and δ_2 as a relative uncertainty with respect to the resonance height. Usually relative high values are chosen for δ_2 because a slight change in the peak position implies a great change in the magnitude of the cross section.

2.3 Implementation of Formalism

In principle a Bayesian evaluation can be performed either by Monte Carlo techniques or in case of normal distribution by Generalized least Square (GLS) method. In this contribution we restrict ourselves to multivariate normal distributions and make use of the modified GLS-technique [1]. Thus we obtain expressions for the Bayesian update of the mean model cross sections $\vec{\sigma}_1$, the mean model error $\vec{\epsilon}_{\mathrm{mod}}$ and the associated covariance matrices $\mathbf{A}_1$ and $\mathbf{K}_1$,

$$\vec{\sigma}_1 = \vec{\sigma}_0 + \mathbf{A}_0\mathbf{S}^T\left(\mathbf{S}\mathbf{A}_0\mathbf{S}^T + \mathbf{S}\mathbf{K}_0\mathbf{S}^T + \mathbf{B}\right)^{-1}\left(\vec{\sigma}_{\mathrm{exp}} - \mathbf{S}\vec{\sigma}_M\right), \quad (5)$$

$$\vec{\epsilon}_1 = \mathbf{K}_0\mathbf{S}^T\left(\mathbf{S}\mathbf{A}_0\mathbf{S}^T + \mathbf{S}\mathbf{K}_0\mathbf{S}^T + \mathbf{B}\right)^{-1}\left(\vec{\sigma}_{\mathrm{exp}} - \mathbf{S}\vec{\sigma}_M\right), \quad (6)$$

$$\mathbf{A}_1 = \mathbf{A}_0 - \mathbf{A}_0\mathbf{S}^T\left(\mathbf{S}\mathbf{A}_0\mathbf{S}^T + \mathbf{S}\mathbf{K}_0\mathbf{S}^T + \mathbf{B}\right)^{-1}\mathbf{S}\mathbf{A}_0, \quad (7)$$

$$\mathbf{K}_1 = \mathbf{K}_0 - \mathbf{K}_0\mathbf{S}^T\left(\mathbf{S}\mathbf{A}_0\mathbf{S}^T + \mathbf{S}\mathbf{K}_0\mathbf{S}^T + \mathbf{B}\right)^{-1}\mathbf{S}\mathbf{K}_0. \quad (8)$$

where the matrix $\mathbf{S}$ provides the transformation from the model grid to the experimental grid. For simplicity we assumed in Eqs. (5) to (8) that the model grid contains all energies for which experimental data are given.

Due to the ansatz (2) the evaluated cross section is not $\vec{\sigma}_1$ but $\vec{\sigma}_{\mathrm{true}} = \vec{\sigma}_1 + \vec{\epsilon}_1$ and the corresponding covariance matrix

$$\mathbf{U}_1 = \mathbf{U}_0 - \mathbf{C}_0\left(\mathbf{S}\mathbf{A}_0\mathbf{S}^T + \mathbf{S}\mathbf{K}_0\mathbf{S}^T + \mathbf{B}\right)^{-1}\mathbf{C}_0^T, \quad (9)$$

with $\mathbf{U}_0 = \mathbf{S}\mathbf{A}_0\mathbf{S}^T + \mathbf{K}_0$ and $\mathbf{C}_0 = \mathbf{A}_0\mathbf{S}^T + \mathbf{K}_0\mathbf{S}^T$.

3 Application

In order to demonstrate the applicability of this formalism in the resonance range we apply the procedure on a schematic set of experimental data with known model defects. The example and the obtained results are discussed in detail in the following subsections.

3.1 Generation of schematic example

We start from a schematic R-matrix example of neutron-${}^{16}_{8}$O scattering including all partial waves up to $J = 5/2$. In three partial waves, i.e. $J^\pi = 1/2^+, 3/2^+$ and $5/2^-$, we assume the existence of one pole leading to the single-channel R-matrix

$$R^{J\pi}(E) = \frac{\gamma_i^2}{E_i - E} \quad \text{for} \quad J^\pi = 1/2^+, 3/2^+ \text{ and } 5/2^+ \quad (10)$$

and vanishing in all other partial waves. The chosen pole energies and widths are given in Tab 1. Using the R-matrix method the cross section σ_{calc} was determined for the energies between 0.005 MeV and 2.550 MeV with an increment of 0.005 MeV equivalent to a model grid of N=510 mesh points.

Table 1. Comparison of the originally chosen pole parameters of the R-matrix calculation (left) and the parameter values obtained by fit of the experimental data distorted by model defects.

		chosen	pole values	fitted	pole values
J^π	i	E_i[MeV]	γ_i[MeV]	E_i[MeV]	γ_i[MeV]
$1/2^+$	1	0.70	0.20	0.685905	0.195017
$3/2^+$	2	1.20	0.40	1.150608	0.407358
$5/2^-$	3	2.00	0.30	1.911950	0.355688

In order to simulate the model defect we determine the mean value of the experimental data by applying a distortion function

$$\bar{\sigma}_{\text{exp}}(E) = \left[1 + \alpha_d \frac{d\sigma_{\text{calc}}}{dE}(E)\right] \sigma_{\text{calc}}(E) \tag{11}$$

with $\alpha_d = 7 \times 10^{-6}$ MeV/mbarn. Choosing a subset $\{E_1, E_2, \ldots, E_M\}$ of the N mesh points we generate a set of schematic experimental data with a statistical uncertainty $\epsilon_S = 0.03$ and a normalisation error of $\epsilon_N = 0.03$ by

$$\sigma_{\text{exp}}(E_j) = (1 + \epsilon_N)\, \bar{\sigma}_{\text{exp}}(E_j) \cdot \mathcal{N}(1, \epsilon_S^2), \tag{12}$$

where $\mathcal{N}(1, \epsilon_S^2)$ refers to a normal distribution with mean value 1 and the standard deviation ϵ_S. Thus we are able to construct the experimental vector $\vec{\sigma}_{\text{exp}}$ which is of dimension M and the $M \times M$ covariance matrix $\mathbf{B}$ of experimental cross sections,

$$\mathbf{B} = \begin{pmatrix} (\bar{\sigma}_1^{\text{exp}})^2[\epsilon_s^2 + \epsilon_N^2] & \bar{\sigma}_1^{\text{exp}}\bar{\sigma}_2^{\text{exp}}\epsilon_N^2 & \cdots & \bar{\sigma}_1^{\text{exp}}\bar{\sigma}_M^{\text{exp}}\epsilon_N^2 \\ \bar{\sigma}_2^{\text{exp}}\bar{\sigma}_1^{\text{exp}}\epsilon_N^2 & (\bar{\sigma}_2^{\text{exp}})^2[\epsilon_s^2 + \epsilon_N^2] & \cdots & \bar{\sigma}_2^{\text{exp}}\bar{\sigma}_M^{\text{exp}}\epsilon_N^2 \\ \vdots & \cdots & \ddots & \vdots \\ \bar{\sigma}_M^{\text{exp}}\bar{\sigma}_1^{\text{exp}}\epsilon_N^2 & \bar{\sigma}_L^{\text{exp}}\bar{\sigma}_2^{\text{exp}}\epsilon_N^2 & \cdots & (\bar{\sigma}_M^{\text{exp}})^2[\epsilon_S^2 + \epsilon_N^2] \end{pmatrix}, \tag{13}$$

where $\bar{\sigma}_j^{\text{exp}} = \bar{\sigma}_{\text{exp}}(E_j)$.

In Fig. 1 a comparison of σ_{calc} and σ_{exp} at M mesh points is given.

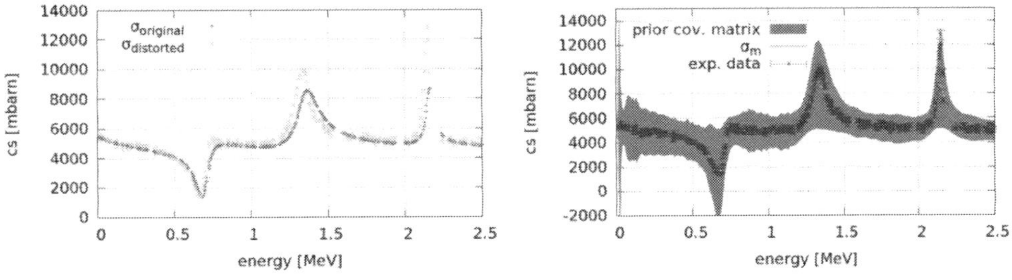

Figure 1. Left: Comparison of original and distorted data with $\alpha_d = 7 \times 10^{-6}$. Right: schematic experimental data and best parameter cross section $\sigma_M(E)$ with a-priori variance $\sqrt{A_0(E_j, E_j)}$.

3.2 Generation of prior

The prior consists of the a-priori mean value of the cross section and the associated covariance matrix $\mathbf{A}_0$. In order to obtain the mean value we make a fit of the schematic experimental data with the R-matrix method. The best set of resonance parameters obtained from the schematic experimental data is given in Tab. 1 and yields the cross section $\sigma_M(E)$ displayed in the right part of Fig. 1.

In the next step, we assume an uncertainty width of 50% for each of the 6 best resonance parameters and perform $n = 200$ calculations of the cross sections at all mesh points with different randomly chosen combinations of resonance parameters. Following [6] we are thus able to generate the *a*-priori covariance matrix $\mathbf{A}_0$ from the n Monte Carlo sweeps. Using $\sigma_M(E)$ and the parameter values $\delta_1 = 0.05$, $\lambda_1 = 0.5$ MeV, $\delta_2 = 1.50$, $\alpha = 0.2$ MeV and $\lambda_2 = 0.10$ MeV in Eq. (4) yields the *a*-priori covariance matrix $\mathbf{K}_0$ of the model defect ϵ_{mod}.

3.3 Bayesian update and results

In the previous subsection we described the generation of the schematic experimental data as well as the determination of the required *a*-priori information. Thus we are able to apply the Bayesian update procedure outlined in Subsect. 2.3. At this point it should be remarked that we use a surrogate model which means that we calculate our observables $\vec{\sigma}_M$ by the R-matrix method on the N energies of the evaluation mesh and use these observables as parameters of the model. Hence, every point on the mesh is a random variable, but cannot be arbitrarily varied because the properties of the model enter via the *a*-priori covariance matrix $\mathbf{A}_0$.

The only quantity we still need for the evaluation is the matrix $\mathbf{S}$ which transforms from the grid of the evaluation to the grid of energies where experimental data are given. The matrix is of dimension $M \times N$ and is particularly simple in our example, i.e. it has only non-vanishing elements ($S_{ij} = 1$) on energy points where both calculated and experimental data are available.

With this preparation the Bayesian evaluation of the set of schematic experimental data is straightforward and allows us to study the differences due to the inclusion of model defects. A comparison of the evaluation without and with inclusion of model defects is shown in Fig. 2. The evaluation without model defects reproduces the experimental data quite well in regions of smooth energy dependence. However, the evaluation becomes problematic at resonances which can be best identified in the third resonance at 2.14 MeV. The model can never reach the height of the experimental peak (Fig. 3). If we account for model defects the agreement of the evaluated resonance cross sections with the experimental data (Figs. 2 and 3) significantly improves.

The comparison between the built-in model defect σ_{exp} - σ_M and the evaluated mean model defect ϵ_{mod} in Fig. 4 shows a striking similarity in size and structure. This result is very promising because it indicates that the procedure is indeed capable to extract the proper size and energy dependent structure of model defects.

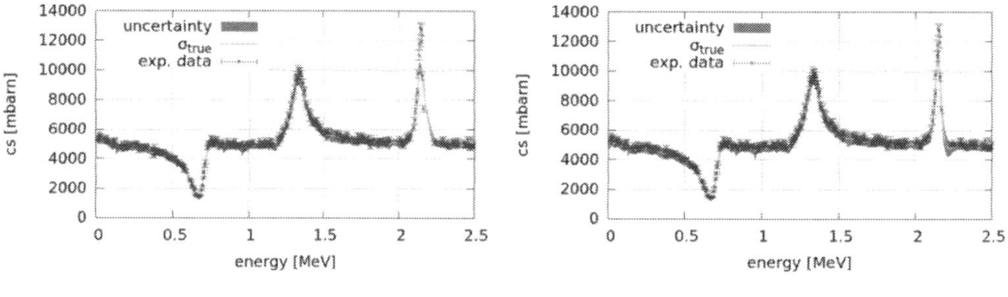

Figure 2. The true cross section σ_{true} evaluated without (left) and with (right) model defects.

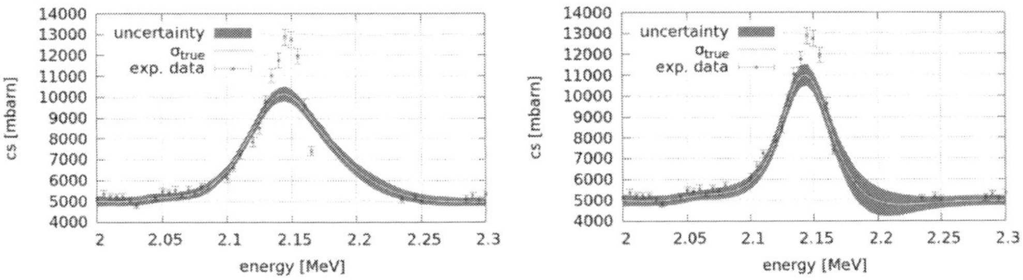

(a) Without model defects

(b) With model defects

Figure 3. Comparison of the third resonance, evaluated without (left) and with (right) model defects.

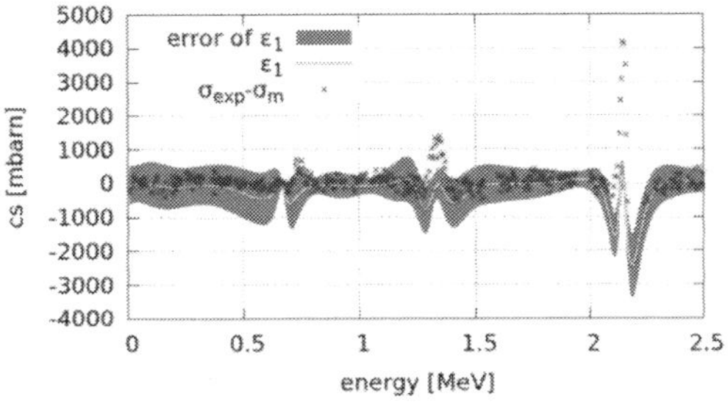

Figure 4. Comparison of the difference $\sigma_{\text{exp}} - \sigma_M$ with the calculated model error ϵ_1.

In Fig. 5 we consider the correlation matrices $\mathbf{A}_1$ updated without and with model defects. At first glance the correlations seem to be quite similar apart from the occurrence of anticorrelations in the presence of model defects. The size of the relative uncertainties differs

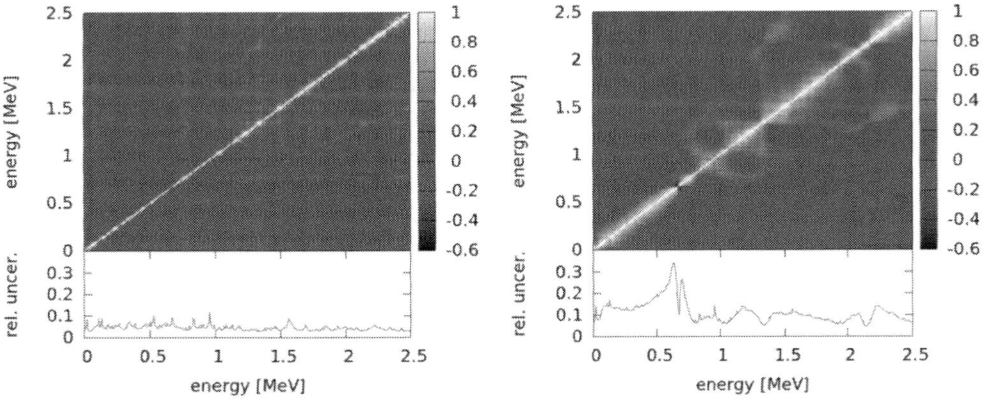

(a) Without model defects

(b) With model defects

Figure 5. Correlation associated to matrix $\mathbf{A}_1$ evaluated without (left) and with (right) model defects. In the inset below the relative uncertainty $\sqrt{\mathbf{A}_1(E_i, E_i)}/\sigma_{\text{true}}$ is given.

significantly and confirms the general finding that uncertainties tend to be underestimated if model defects are neglected in the evaluation.

In order to complete the comparison we show in Fig. 6 the correlation matrix $\mathbf{U}_1$ of the true cross section σ_{true} which is the final result which should enter in an evaluated data file. In addition the updated covariance matrix $\mathbf{K}_1$ of the model error ϵ_{mod} is also shown in Fig. 6.

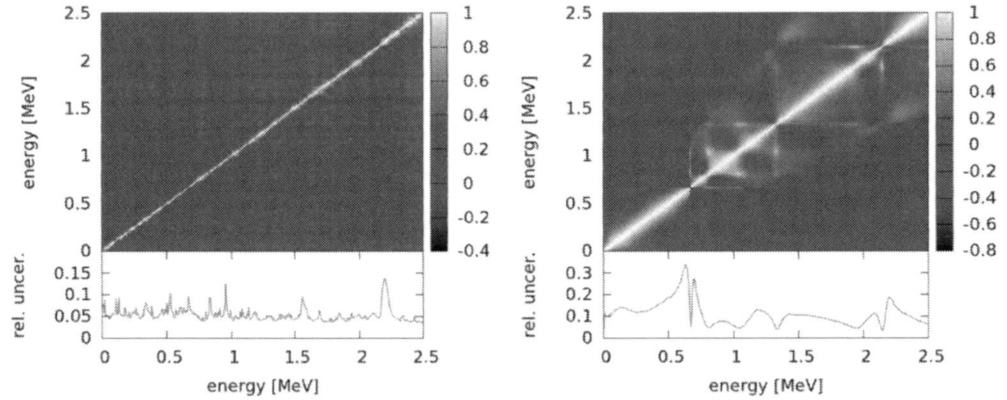

Figure 6. Correlation matrices associated to $\mathbf{U}_1$ (left) and $\mathbf{K}_1$ (right) with their relative uncertainties $\sqrt{\mathbf{U}_1(E_i,E_i)}/\sigma_{\text{true}}$ and $\sqrt{\mathbf{K}_1(E_i,E_i)}/\sigma_{\text{true}}$.

4 Summary and Outlook

We extended the recently developed formalism for model defects based on Gaussian processes [1, 5] to the resonance region. Key of the extension is the proposal of an alternative prior covariance matrix $\mathbf{K}_0$ of model defects, Eq. (4), which is primarily sensitive to the shape of resonances. This covariance matrix allows a Bayesian evaluation including model defects for the total energy range in a unified scheme simultaneously including the resonant and the smooth behavior of cross sections at low and intermediate projectile energies, respectively. The method also provides the magnitude of uncertainties as well as the associated covariance matrices including cross correlations between the different contributions.

References

[1] G. Schnabel, *Large Scale Bayesian Nuclear Data Evaluation with Consistent Model Defects*, PhD thesis, TU Wien (2015)
[2] P. Helgesson, H. Sjöstrand, Rev. Sci. Instr. **88**, 115114 (2017)
[3] H. Leeb, D. Neudecker, Th. Srdinko, Nucl. Data Sheets **109**, 2762 (2009)
[4] D. Neudecker, R. Capote, H. Leeb, Nucl. Instr. Meth. A **723**, 163 (2013)
[5] G. Schnabel, H. Leeb, EPJ Web of Conferences **111**, 09001 (2016)
[6] G. Schnabel, H. Leeb, Nucl. Instr. Method A **841**, 87 (2017)

EPJ Web of Conferences **211**, 07007 (2019) https://doi.org/10.1051/epjconf/201921107007
WONDER-2018

Monte Carlo integral adjustment of nuclear data libraries – experimental covariances and inconsistent data

Henrik Sjöstrand[1,*] and *Georg* Schnabel[1]

[1]Uppsala University, Department of Physics and Astronomy, Box 516, 751 20 Uppsala, Sweden

Abstract. Integral experiments can be used to adjust nuclear data libraries. Here a Bayesian Monte Carlo method based on assigning weights to the different random files is used. If the experiments are inconsistent within them-self or with the nuclear data it is shown that the adjustment procedure can lead to undesirable results. Therefore, a technique to treat inconsistent data is presented. The technique is based on the optimization of the marginal likelihood which is approximated by a sample of model calculations. The sources to the inconsistencies are discussed and the importance to consider correlation between the different experiments is emphasized. It is found that the technique can address inconsistencies in a desirable way.

1 Introduction

Integral experiments (benchmarks) can be used to adjust nuclear data (ND) libraries. The process is referred to as Integral adjustment (IA). The motivation is to use the full width of the experimental data available and hence increase the predictive power of the ND library. IA is usually done implicitly, when models, parameters, and experiments are chosen in order to assure proper performance of the ND library in respect to the integral data. Alternatively, the IA can be performed explicitly, where the ND-library is tuned to the integral data using fitting routines and mathematical models. This paper relates to such explicit IA, and from here on the use of IA will only refer to explicit IA.

IA is a well-established field [1, 2] and can be performed by deterministic adjustment techniques [1] or by Monte Carlo IA (MC-IA) [3, 4]; this paper concerns the later which sometimes is referred to as Bayesian Monte Carlo (BMC) [5]. However, since BMC can be used for both differential- and integral data and this particular paper addresses the use of integral data, we will use the term MC-IA

MC-IA has been explored by many authors, both for improving the best estimate (central value) of the ND file [6], but also to reduce the ND uncertainty [2 - 4]. In this paper, the MC-

* Corresponding author: henrik.sjostrand@physics.uu.se

IA method is extended by including a marginal likelihood optimization (MLO) technique to address inconsistencies between the experimental and calculated integral parameter (in this case k_{eff}). The paper also discusses the sources of these inconsistencies. Methods for treating inconsistent integral data has been proposed earlier, such as the adjustment margin (AM) method [1], and the $\Delta\chi^2$ -filtering [7]. The AM method does not consider correlations and is binary. The $\Delta\chi^2$ is also binary, and the choice of the rejection criteria appears to be rather arbitrary. In a binary method, an experiment is either kept or rejected. I.e., it is assumed that we have either full or no confidence it the experiment. We believe that this is a too course selection mechanism and that a continuous scale representing our trust in an experiment is more appropriate. The proposed method is statistical well-founded, and address the mentioned issues of the previous methods. The treatment of inconsistent data and handling of outliers is a wide research field, and the above-mentioned methods are examples of techniques that have been used in the IA community. Many other methods have been suggested for treatment of inconsistent data and Ref. [8] compares the performance of some different methods. In [8], the Bayesian procedure to adjust uncertainties performed best. The presented Bayesian procedure determines a factor to rescale the uncertainties of all data points. In contrast, our approach allows uncertainties to be adjusted individually. In the Bayesian setting, our MLO approach corresponds to the maximum a posteriori probability estimate of the uncertainties.

This paper is outlined in the following way, in chapter two the basics of MC-IA is explained as well as the underlying motivation for adding a calculation uncertainty when using IA. An example of an MC-IA with inconsistent data is also provided. Chapter three describes the MLO technique. Chapter 4 presents the results when using the MLO technique in combination with the MC-IA. Chapter 5 contains the conclusions.

2. Monte Carlo Integral adjustment

In IA, the ND is updated to reduce the deviation between the calculated and experimental value of the integral experiment. There are many different types of integral experiments and in this paper criticality experiments from the ICSBEP [9] is used. More specifically, the ICSBEP MCNP benchmarks of the integral experiments are used to calculate the k_{eff} of the benchmarks. This is subsequently compared to the reported benchmark values. A reported benchmark value is based on the experimental result of the underlying integral experiment and hence referred to as the experimental value.

MC-IA uses ND in the form of random files. Each ND random file is a full ENDF (but without a covariance file), which can be a possible realization of the underlying ND given the PDF of the ND. I.e., the set (e.g., 1000) of ND random-files is a way to provide the PDF of the ND. ND-random files can be obtained through the so-called Total Monte Carlo (TMC) method [10, 11]. These TMC random files are provided on the TENDL webpage or can be produced using the so-called T6 code package [10]. Alternatively, ND random files can be generated by sampling the covariance matrix of an ENDF (e.g., JEFF3.3). This sampling can be achieved by using tools like SANDY [12, 13], NUSS [14], or NUDUNA [15]. One advantage of using TMC random files is that these can contain higher order moments of the ND-PDF as well as more cross-correlations between different parts of the ND file.

Given a reasonably large set of random-files (at least 1000, but in many cases more files are needed for convergence), an MC-IA can be performed. Each random file is used in the benchmark calculations, e.g., calculating the k_{eff} for a set of benchmarks. The MC-IA is performed by assigning weights, w, to the different random files, reflecting the agreement between the calculated and the experimental value. The weight is set to,

$$w_i = \delta e^{-\frac{\chi_i^2}{2}}, \tag{1}$$

(see, e.g., [16 - 18]) where w_i is the weight for random file i, δ is an arbitrary normalization constant and χ_i^2 , is defined as

$$\chi_i^2 = \sum_b \frac{(C_{b,i} - E_b)^2}{\sigma_b^2}, \tag{2}$$

where b is the benchmark number, $C_{b,i}$ is the calculation result of benchmark b for random file i, E_b, is the experimental value of benchmark b and σ_b is the benchmark uncertainty. If correlations are also considered we instead us the generalized χ^2

$$\chi_i^2 = \left(\mathbf{C}_\mathrm{i} - \mathbf{E}_\mathrm{i}\right)^\mathrm{T} \mathbf{B}^{-1} \left(\mathbf{C}_\mathrm{i} - \mathbf{E}_\mathrm{i}\right), \tag{3}$$

where $\mathbf{C}_\mathrm{i}$ and $\mathbf{E}_\mathrm{i}$ are the vectors with the calculation and experimental results for all benchmarks. **B** is the covariance matrix for the benchmarks. **B** or σ_b should contain the reported experimental uncertainty $\sigma_{\mathrm{reported}}$, the statistical uncertainty of the MCNP calculations, σ_{stat}, and possibly other calculational uncertainties [4] (see Eq. 4 for details). In Refs. [19, 20] the same approach is used, but in [20] with a different definition for the weights.

When the weights for the random files have been calculated, a new PDF for the underlying ND can be determined [3]. In addition, a new weighted PDF for the calculated value of the benchmark or for any other macroscopic parameter for any system under investigation can be inferred. I.e., using the weights, a posterior best estimate and ND uncertainty can be deduced for, e.g., the benchmarks used in the IA. This is also how the weights are used in this work. An example of this is seen in Fig. 1 where 1000 TENDL 2014 random files for ^{235}U and ^{238}U are used to calculate k_{eff} for five benchmarks. These benchmarks and random files are used for all the results in the paper.

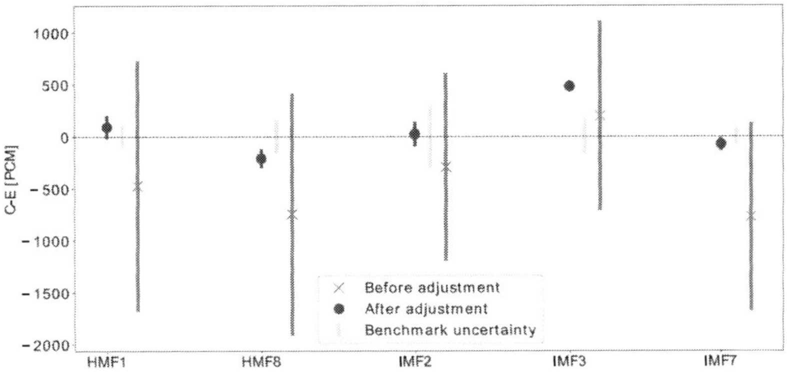

Fig. 1. (Color online) The figure shows the results for five criticality benchmarks. On the y-axis is the difference between the mean of $C_{b,i}$ (over i) and the experimental value in PCM. The red crosses are the results before the weighting of the random files. The blue dots are the result after the weighting of the random files using Eq. 1 The error-bars connected to the crosses and dots contain the ND uncertainty of the calculated k_{eff}. The bars without marks are the benchmark uncertainties.

As can be seen in Fig. 1, all except IMF3 obtain better agreement (in absolute values) after the IA, and the ND uncertainty is reduced for all the benchmarks. In fact, the ND uncertainty is reduced to such an extent that the calculation results are hardly consistent with the experimental results. A p-value of 6 % is obtained if both experimental and ND uncertainty is included. The p-value is defined as the probability that the difference between the experiments and the calculation would be at least as extreme as the obtained result, given the null hypothesis. In our case, the null hypothesis is that the difference between the experimental data and the model prediction is governed by a multivariate normal distribution with a zero mean-vector and a covariance matrix composed as the sum of the experimental covariance matrix and the model covariance matrix. The test statistic is the generalized chi-square value and "as least as extreme" has to be understood with respect to this test statistic. IMF3 cannot be adjusted to obtain a good agreement with the experimental value simultaneously as the other values were adjusted because of the strong calculation correlations between the benchmarks (see Fig. 2 left). E.g., if the fission cross-section in the fast energy-range is increased for ^{235}U, k_{eff} is increased for all the benchmarks.

In fact, by eye one can easily believe that the prior data (data before the adjustment) is in good agreement with the experimental data. However, if the p-value is calculated for the prior data including the full ND covariance a p-value of 11% is obtained. The reason for this low p-value is the strong correlations between the benchmarks. What we witness is that the calculation data disagrees with our set of experimental data. Differently expressed, given the calculational data there is a disagreement between the experiments. This disagreement can be due to:

- Model defects. I.e., the true ND is not within the PDF of the reported ND (in our case reported using random files).
- The adjusted ND (in this case ^{235}U and ^{238}U) contain errors that are not reported in the covariance file or represented by the random files. E.g., in many libraries, covariance information for angular distribution is missing.
- Unaccounted experimental uncertainties or covariances.
- Isotopes not considered. In this case, only ^{235}U and ^{238}U are adjusted. However, the actual experiments contain other isotopes that could be responsible for the disagreement. These isotopes can either be included in the adjustment procedure or should be marginalized, i.e., added to the benchmark uncertainty [4].

Whatever the reason, it is clear that the posterior uncertainties are too small given our set of experiments. This disagreement needs to be addressed in some way.

3 Addressing inconsistent experiments

As mentioned previously there are many reasons for the data to be inconsistent, both among the different experiments and between the experiments and the calculation. Many of these underlying causes can be treated by different approaches. E.g., model defects can be considered in new ND evaluations [21, 22] or better models can be used; ND evaluations can include more uncertainties (an example of this is the TENDL evaluations [10], which report uncertainties on most ND quantities); experiments can be re-evaluated to find unaccounted uncertainties; and more isotopes can be considered in the IA [4]. We encourage and are part of these efforts, but also acknowledge that at any given time an IA is performed, remaining inconsistencies need to be treated. In this section, we present the MLO technique for the treatment of such inconsistencies. In [23] a different version of the MLO approach based on the construction of a sensitivity matrix of the model by linearizing it was suggested. Here we do not linearize the model but build the MLO approach on top of a sample of model calculations.

We add individual extra uncertainties, $\sigma_{\mathrm{extra,b}}$, to each of the benchmarks to account for the inconsistencies. In addition, we add a fully correlated common uncertainty, $\sigma_{\mathrm{extra,c}}$, to all the benchmarks. I.e., for each benchmark the new benchmark uncertainty is calculated as

$$\sigma_b^2 = \sigma_{reported}^2 + \sigma_{stat}^2 + \sigma_{extra,b}^2 + \sigma_{extra,c}^2 . \qquad (4)$$

In this way, we try to model that there can be effects that are common for all the experiments and that there can be additional individual components in each experiment. The magnitudes of the σ_{extra} are found by using MLO. Details on the technique can be found in [23]. The likelihood for a σ_{extra} is described by,

$$L = \sum_i e^{-\frac{\chi_i^2}{2}} \Big/ \sqrt{(2\pi)^n |B|}, \tag{5}$$

where n is the number of experiments and $|B|$ is the determinant of the benchmark covariance matrix, with the components defined in Eq. 4. With increased σ_{extra}, χ^2 decreases and the numerator term increases. However, with increased σ_{extra}, the denominator increases as well. A derivation of Eq. 5 can be found in [24]. An example of the likelihood function in one dimension is shown in Fig. 2 (right), where σ_{extra} for HMF1 is varied and kept fixed for the other experiments.

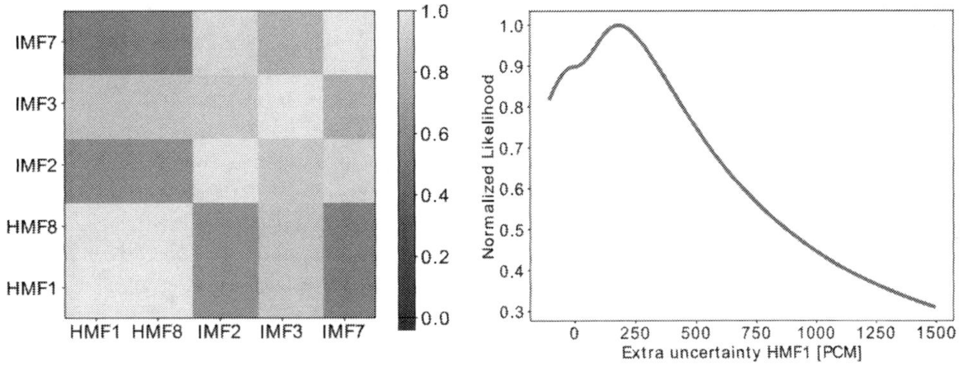

Fig. 2 (Left) The prior calculation correlation between the benchmarks. (Right) The normalized likelihood (y-axis) for the $\sigma_{\text{extra,b}}$ (x-axis) of HMF1.

As can be seen, the most likely σ_{extra} for HMF1 is close to 250 pcm in this example; this value should be used for σ_{extra}. In the full exercise, the six-dimensional space (with five individual $\sigma_{\text{extra,b}}$ and a $\sigma_{\text{extra,c}}$) was explored, and L was optimized (maximized). For the optimization, the minimum of the negative logarithm of L was found using the python scipy.optimize.minimize package with the 'SLSQP' method. For the calculation of the determinant, a Cholesky decomposition of B was performed using the python numpy.linalg.cholesky package.

Given no information from the data, we would generally assume that the σ_{extra} should be small since considerable work has been done in both evaluating the experiments and creating the ND files. Eq. 5. does not consider this prior belief. To promote small σ_{extra}, a new term is added to the likelihood,

$$L = \left(\sum_i e^{-\frac{\chi_i^2}{2}} \Big/ \sqrt{(2\pi)^n |B|} \right) e^{-\sum \beta \sigma_{extra}^2}, \tag{6}$$

where the summation is over both $\sigma_{\text{extra,b}}$ and $\sigma_{\text{extra,c}}$, β is a constant that allows us to shape our prior belief on the σ_{extra}. Here, β is set so that 500 pcm σ_{extra} has a prior probability (before inspecting the data) of 50 % compared to zero σ_{extra}. In the choice of β, expert judgment is introduced in the methodology. We use a Gaussian function in the added term in Eq. 6, but other functions have also been suggested [23]. It is not absolutely necessary to include the term. However, by not including it σ_b becomes independent of σ_{reported}. I.e., the information from the evaluation of the benchmarks uncertainties is not affecting the final results. This would be undesired, and hence we recommend using a prior term. In addition, β allows us to code expert knowledge into the likelihood. In our case β was set to the same value for all the experiments; however, β can be chosen to have individual values for different experiments to encode our belief in a particular experiment.

4 Results

The adjustment was tested including $\sigma_{extra,c}$; without $\sigma_{extra,c}$; using Eq. 5; and using Eq. 6. Fig. 3 illustrates the results using Eq. 6 and including $\sigma_{extra,c}$.

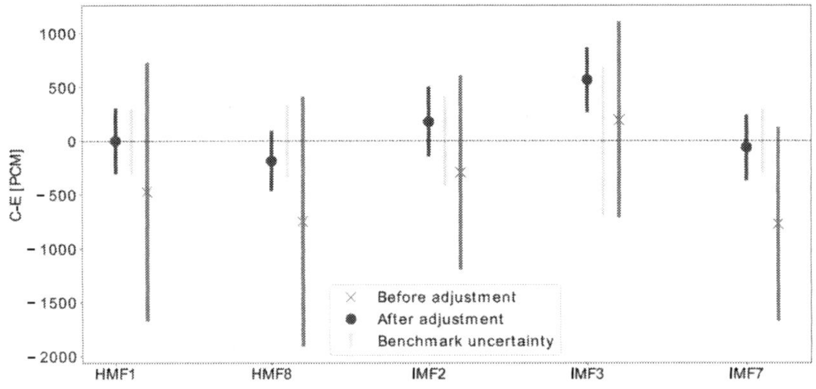

Fig. 3 Same as Fig. 2, but using the MLO technique to include σ_{extra}. The green bars without marker contain the benchmark uncertainty including σ_{extra}.

Fig. 3 now presents a consistent data set. If compared to the results without using the MLO (Fig. 2), one can note that the posterior ND uncertainty is dramatically increased (e.g., from 69 pcm to 253 pcm for HMF1) In this case, the total benchmark uncertainty is augmented for all benchmarks, mostly since a 209 pcm common uncertainty component is added to all the experiments. In fact, only IMF3 obtains a significant individual extra uncertainty added and the $\sigma_{b,IMF3}$ increases from 170 pcm to 468 pcm. The results become quite different if $\sigma_{extra,c}$ is assumed to be zero, as can be seen in Table 1. In the MLO $\sigma_{extra,b}$ were allowed to take negative values as long as the sum of $\sigma_{extra,b}$ and $\sigma_{extra,c}$ was larger than zero. The motivation for this was that part of the reported experimental uncertainties could stem from common causes in the different benchmarks. To assure to retain a minimum individual uncertainty the criterion $\sigma^2_{extra,b} > -0.5\sigma^2_{reported,b}$ was also included.

Table 1. Table of resulting σ_b. The second row contains $\sigma_{reported}$ (No MLO performed). In the third row $\sigma_{extra,c}$ is assumed to be zero. The third and fourth row uses Eq.5. The last row uses Eq. 6.

σ_b [PCM]	HMF1	HMF8	IMF2	IMF3	IMF7	$\sigma_{extra,c}$
No MLO	100	160	300	170	80	0
using Eq. 5	153	204	300	580	390	0
using Eq. 5	267	329	333	591	409	257
using Eq. 6	232	263	366	468	228	209

5 Conclusions and outlook

A new technique to address inconsistent integral experiments using an MC-MLO technique is proposed and applied to a set of benchmarks. The method can resolve the inconsistencies in the data, and the outcome of the results is desirable. To our knowledge, the MC version of the technique has not been proposed before in the context of ND-evaluation, and we believe it is the first time an MLO technique is proposed for IA.

The technique has many desirable features, e.g., being statistical well-founded; can be integrated with expert judgment in a transparent way; is not binary; takes into account both experimental and calculational correlations between the benchmarks; and results in reasonable posterior ND uncertainties. MLO is hence a promising technique to be used for integral adjustment in future library releases.

We suggest that the technique should be tested with a larger set of benchmarks, where also validation benchmarks are used to infer the predictive power of the method. In addition, the method should be compared quantitatively with other discrepant data techniques.

Acknowledgment

This work was supported by the Swedish research council and Swedish Nuclear Centre. A special thanks to Daniel Siefman for input in regards to established discrepant data methods. The benchmark suite used was prepared by Steven Van der Marck.

References

[1] M. Salvatores, et al., Nuclear Data Sheets 118, 38 (2014).
[2] C. De Saint Jean, P. Archier, E. Privas, G. Noguère, O. Litaize, and P. Leconte, Nucl. Data Sheets **123**, 178 (2015).
[3] D. A. Rochman, E. Bauge, A. Vasiliev, H. Ferroukhi, and G. Perret, EPJ Nucl. Sci. Technol. **4**, 7 (2018).
[4] E. Alhassan, H. Sjöstrand, P. Helgesson, M. Österlund, S. Pomp, A. J. Koning, and D. Rochman, Prog. Nucl. Energy **88**, 43 (2016).
[5] C. D. S. Jean, P. Archier, E. Privas, and G. Noguere, EPJ Web Conf. **146**, 02007 (2017).
[6] D. Rochman and A. J. Koning, Nucl. Sci. Eng. **169**, 68 (2011).
[7] M. L. Williams, B. L. Broadhead, M. A. Jessee, J. J. Wagschal, and R. Lefebvre, Oak Ridge Natl. Lab. ORNLTM-200539 **3**, (2009).
[8] S. I. Kafala, T. D. MacMahon, and P. W. Gray, Nucl. Instrum. Methods Phys. Res. Sect. Accel. Spectrometers Detect. Assoc. Equip. **339**, 151 (1994).
[9] *International Criticality Safety Benchmark Evaluation Project (ICSBEP).*
[10] A. J. Koning and D. Rochman, Nucl. Data Sheets **113**, 2841 (2012).
[11] D. Rochman, W. Zwermann, S. C. van der Marck, A. J. Koning, H. Sjöstrand, P. Helgesson, and B. Krzykacz-Hausmann, Nucl. Sci. Eng. **177**, 337 (2014).
[12] L. Fiorito, G. Žerovnik, A. Stankovskiy, G. Van den Eynde, and P. E. Labeau, Ann. Nucl. Energy **101**, 359 (2017).
[13] L. Fiorito, *Sandy* (https://luca-fiorito-11.github.io/sandy-docs/).
[14] T. Zhu, A. Vasiliev, H. Ferroukhi, and A. Pautz, Ann. Nucl. Energy **75**, 713 (2015).
[15] A. Hoefer, O. Buss, M. Hennebach, M. Schmid, and D. Porsch, Ann. Nucl. Energy **77**, 514 (2015).
[16] R. Capote, D. L. Smith, A. Trkov, and M. Meghzifene, J. ASTM Int. **9**, 1 (2012).
[17] P. Helgesson, H. Sjöstrand, A. J. Koning, J. Rydén, D. Rochman, E. Alhassan, and S. Pomp, Prog. Nucl. Energy **96**, 76 (2017).
[18] R. Capote and D. L. Smith, Nucl. Data Sheets **109**, 2768 (2008).
[19] C. D. S. Jean, P. Archier, E. Privas, and G. Noguere, EPJ Web Conf. **146**, 02007 (2017).
[20] E. Bauge and P. Dossantos-Uzarralde, J. Korean Phys. Soc. **59**, 1218 (2011).
[21] G. Schnabel and H. Leeb, EPJ Web Conf. **111**, 09001 (2016).
[22] P. Helgesson and H. Sjöstrand, Ann. Nucl. Energy **120**, 35 (2018).
[23] G. Schnabel, ArXiv180300960 Nucl-Ex Physicsnucl-Th Physicsphysics (2018).
[24] G. Schnabel, (http://urn.kb.se/resolve?urn=urn:nbn:se:uu:diva-381563).

Examples of Monte Carlo techniques applied for nuclear data uncertainty propagation

Oscar Cabellos[1,*] and *Luca* Fiorito[2]

[1]Department of Energy Engineering, Universidad Politécnica de Madrid, 28006 Madrid, Spain
[2]OECD/Nuclear Energy Agency, Boulogne-Billancourt, 92100, France

Abstract. The aim of this work is to review different Monte Carlo techniques used to propagate nuclear data uncertainties. Firstly, we introduced Monte Carlo technique applied for Uncertainty Quantification studies in safety calculations of large scale systems. As an example, the impact of nuclear data uncertainty of JEFF-3.3 ^{235}U, ^{238}U and ^{239}Pu is demonstrated for the main design parameters of a typical 3-loop PWR Westinghouse unit. Secondly, the Bayesian Monte Carlo technique for data adjustment is presented. An example for ^{235}U adjustment using criticality and shielding integral benchmarks shows the importance of performing joint adjustment based on different set of integral benchmarks.

1 Introduction

This paper focuses on the application of Monte Carlo (MC) methodologies for the propagation of nuclear data uncertainties. The first section of this paper presents MC technique to perform Uncertainty Quantification (UQ) analysis in large-scale systems which is increasingly necessary in reactor safety calculations of LWRs [1]. The second section is devoted to review current nuclear data adjustment methodologies based on a Bayesian approach in combination with a Monte Carlo technique.

2 Monte Carlo technique for Uncertainty Quantification

Two different methodologies have been developed for propagating nuclear data uncertainties to large scale systems. The first method is based on the MC approach, which uses random samples of nuclear data libraries and perform a separate reactor calculation for each random sample. The second one is based on the first order perturbation theory approach, which makes use of sensitivity coefficients and available covariance files.

The Sensitivity and UQ (S/U) is based on first order perturbation method. The UQ can be obtained through the sandwich rule, after sensitivity coefficients are computed. This technique allows an easy decomposition of the uncertainty in isotopic partial cross-section components. Different techniques have been developed for the determination of the sensitivity coefficients of the neutronics parameters to nuclear data such as the Standard Perturbation Theory (SPT) used for k_{eff} uncertainties, and Generalized Perturbation Theory

* Corresponding author: oscar.cabellos@upm.es

(GPT) used for power distribution uncertainty [2]. These S/U techniques show different weaknesses in uncertainty quantification studies. For instance, one can point out the following drawbacks: 1) a low efficiency for a large number of response functions because the number of adjoint functions increases linearly with the number of responses, 2) the applicability is only for small uncertainties because it is based on a linear-approach and large uncertainties may exceed the range of linearity of the model, and 3) it shows severe limitations as consequence of the non-linearity of multi-physics (neutronics, thermohydraulic, depletion, …) in reactor calculations.

The MC approach overcomes those limitations, but due to its stochastic nature it carries an inherent statistical uncertainty. To reduce the statistical uncertainty a large number of samples is often required, which implies a larger computational time. This can be seen as a significant disadvantage compared to first order approximation techniques. MC approach can also be classified by their nuclear data uncertainty inputs, thus "Total Monte Carlo" (TMC)[3] relies on model parameter covariances; NUDUNA[4] and SANDY[5] take as input the information provided by nuclear data evaluations or multigroup covariance applying them to continuous energy data, XSUSA[6] takes the form of multigroup covariance matrices applying only for multigroup data.

For the MC approach, the nuclear data are sampled at the beginning of the simulation, then the reactor core simulator that solves coupled multi-physics at different levels of approximation is executed. In this work, we took advantage of the SEANAP system [7] developed for and applied to 3-D PWR core analysis. Already in the past SEANAP has demonstrated a very good agreement with the broad sets of parameters and cycles analysed at the Spanish PWR units [8]. One of the major advantages of using SEANAP is its very low CPU time requirement as compared to other codes. A full core cycle and branching lattice calculations can be performed in only few minutes.

For this work, SANDY provides full MC sampling of the nuclear data files for transport and depletion calculations. Each nuclear data random file is used to perform a separate core calculation. With SANDY we samples reaction cross sections, fission neutron multiplicities and prompt fission neutron energy distributions using the evaluated nuclear data and covariances from JEFF-3.3. Then, the random files were processed with NJOY in WIMS format, which serve as input for the SEANAP code. The statistics of all SEANAP simulations finally yields the desired uncertainty quantification. In this work, sampling simulations for ^{235}U, ^{238}U and ^{239}Pu are carried out separately. Table 1 shows UQ analysis for the critical boron concentration in a 3-loop PWR Westinghouse unit along the burnup. The total uncertainty of boron concentration is around ~65 ppm. The main contributor to this uncertainty is ^{235}U-nubar, the contribution due to ^{238}U cross-section is constant during the cycle. For the ^{239}Pu, its contribution is larger than ^{235}U at end-of-cycle.

3 Bayesian Monte Carlo technique

In this section, we review the current nuclear data adjustment methodologies based on the Bayesian approach together with MC techniques. Hereafter, experimental data is referred to integral information, such as criticality integral benchmarks (e.g. k_{eff} and spectral indices) shielding/transmission benchmarks (e.g. neutron leakage), etc. It is well-recognized that integral benchmarks should be avoided for the adjustment of general-purpose libraries because this can lead to potential compensating effects. These compensating errors can be due to both the impact of other isotopes in the benchmark and errors in the calculation attributed to complicated multi-physics.

However, it is also recognized that such integral data may be used to perform tune or fine adjustment of specific nuclear data to improve the overall performance of an entire general-purpose library. Thus, "tuned" nuclear data adjustments should rely on high-fidelity

Table 1. UQ assessment for critical boron concentration in a PWR Westinghouse unit. SEANAP system is used to perform calculations. Calculation performed with a total of 2100 random files generated by SANDY tool (300 samples for each case). Measured and Calculated values are shown.

Power (%)	Burnup (GWd/tHM)	Boron Meas. (ppm)	Boron Calc. (ppm)	Uncertainties in ppm (Boron Concentration) due to JEFF-3.3 covariance data						
				Pu9 XS	Pu9 Nubar	Pu9 chi	U5 XS	U5 Nubar	U5 chi	U8 XS
75	0.031	1113	1085	18	15	9	27	46	10	24
100	0.134	985	1011	19	15	9	27	46	10	25
100	1.340	870	896	22	16	9	25	47	10	24
100	3.591	688	701	27	19	9	24	43	10	24
100	5.549	504	526	30	21	9	22	40	10	24
100	7.716	319	321	34	23	9	21	38	10	23
100	8.823	227	219	35	24	9	21	37	10	23
100	10.284	101	79	37	25	9	20	35	10	23
100	11.351	4	-29	39	26	9	20	34	10	23

experiments that can be used as simple (e.g. a single isotope), well-understood and so-called clean benchmarks. In this sense, one should avoid adjustments that include error sources unrelated to nuclear data into the evaluation procedure. Consequently, reactor core measurements (such as the experimental data presented in Table 1) can be used only to derive specific adjustments for nuclear data libraries for reactor applications.

Two distinct methods of nuclear data adjustment methodologies can be categorized into deterministic and stochastic methods.

3.1 Generalized Linear Least Squares (GLLS)

The GLLS is a deterministic method consisting on minimization, Eq. 1:

$$[E - C(\sigma)]^T V_E^{-1} [E - C(\sigma)] + [\sigma - \sigma_0]^T V_\sigma^{-1} [\sigma - \sigma_0] = \chi^2(\sigma) \qquad (1)$$

The GLLS is based on several assumptions [9]: 1) experimental and nuclear data are normally distributed, 2) linear approximations between all observables, and 3) model and experimental data are uncorrelated. Thus, GLLS can be seen as a Bayesian approach in the sense that experimental data are used to adjust prior values. Although probability density functions are not considered explicitly, the GLLS is used to derive the posterior means and covariances of nuclear data.

3.2 Monte Carlo nuclear data adjustment methodologies

MC nuclear data adjustment methodologies have been developed to avoid the need to linearize non-linear models, and to handle models which are not necessarily normally distributed [9], (e.g. model parameters (x) which lead through a model transformation

$\sigma=\mathcal{M}(x)$). These methodologies are able to include integral data, reducing uncertainties while performing better posterior nuclear data. The Bayesian MC techniques are based on a direct application of Bayes' Theorem, Eq.2:

$$p(\sigma|E) \propto p_0(\sigma|\sigma_C, V_C) \times L(y_E, V_E|\sigma) \quad (2)$$

The prior probability $p_0(\sigma|\sigma_C,V_C)$ and likelihood $L(y_E,V_E|\sigma)$ are independent probability density functions. The principle of maximum entropy will ensure that $p_0(\sigma|\sigma_C,V_C)$ and $L(y_E,V_E|\sigma)$ are known and can be approximated by multivariate normal distributions. In case the normality assumption is not acceptable, σ may be mapped onto an approximately normally distributed vector by an invertible transformation [4].

Under these conditions the Bayes' Theorem yields a multivariate normal posterior probability density function. This Bayesian approach can be defined as Multivariate Normal Bayesian Model (MNBM) and the resulting equations are widely known as the MOCABA equations [4,10]. A stochastic methodology can be used to randomly sample the "a priori" nuclear data. There are codes such as SANDY and NUDUNA able to provide samples of nuclear data. They work on normal (or log-normal) distributions because no further information on the distribution of the nuclear data are provided in evaluated files. In this work, SANDY code is used assuming normal distributions. The sampling can be also performed at the level of nuclear parameters in nuclear reaction codes such as TALYS [11,12]. In this case, it is unlikely that $p_0(\sigma|\sigma_C,V_C)$ will be normal, resulting in a loss of information about the prior function if it is approximated just by a normal distribution.

3.3 Bayesian Monte Carlo approach

The requirement of a well-known prior and likelihood probability density function is a serious limitation of the previous approach if non-informative prior distributions are known [9,11]. To overcome this limitation the Bayesian Monte Carlo (BMC) approach [9,11,12,13] takes advantage of TMC method to generate "a priori" random files which are not explicitly normal. The BMC technique will incorporate this integral "a priori" information through likelihood factors that are defined in Eq. 3:

$$L(y_E, V_E|\sigma) \sim e^{-\chi_k^2/2}, \text{ with } \chi_k^2 = [E - C(\sigma_k)]^T V_E^{-1} [E - C(\sigma_k)] \quad (3)$$

Based on this definition, one can calculate a "weight" for any k-sample set, as follows in Eq.4, which normality assumption is implicitly given by chi-squared:

$$\omega_k = e^{-\chi_k^2/2} \quad (4)$$

These weights are used to calculate "a posteriori" moments (see Eq.5).

$$\sigma_i' = \frac{\sum_{k=1}^{N} \omega_k \times \sigma_{i,k}}{\sum_{i=k}^{N} \omega_k}; \; V'_{\sigma_{ij}} = \frac{\sum_{k=1}^{N} \omega_k \times (\sigma_{i,k} - \sigma_i')^T (\sigma_{j,k} - \sigma_j')}{\sum_{k=1}^{N} \omega_k} \quad (5)$$

One can work with different definitions of weights. Simpler definitions of ω_k values can lead to very small weights. To overcome this limitation different renormalized ω_k-functions have been introduced. For instance, the Backward-Forward MC method (BFMC) [12,13] proposed a renormalization function to avoid such large dispersion of weights. Recently, other methods have been developed with BMC to compensate for the model deficiencies and inconsistent data. [14,15]

 https://doi.org/10.1051/epjconf/201921107008

3.4 Selection of integral Benchmarks

The selection of dedicated integral experiments determines the scope of the nuclear data adjustment. Nuclear data evaluations have used mainly criticality benchmarks to match C/E values (e.g. k_{eff} and spectral indices). However, the k_{eff} value is a global parameter that can be achieved through many possible combinations of nuclear data. It can lead to inherent compensating effects in the adjustment work [12,13]. In recent works, other sources of integral data are pointed out to extend the application and the prediction capabilities of future evaluations. Apart of the well-known spectral index measurements, one can considerer other measurements such as delayed neutron fraction and shielding/transmission leakage neutron spectra. [16]

As an example of BMC adjustment, two different integral benchmarks have been selected to see the impact of ^{235}U adjustment between 500 keV and 10 MeV: For criticality, the Godiva (HMF1) benchmark; for shielding, the LLNL-pulsed sphered for ^{235}U. A comparison of sensitivity profiles is performed to assess the importance of those benchmarks. Fig. 1 shows the experimental values for neutron leakage and their sensitivities to the continuum-inelastic scattering (MT91) cross-section of ^{235}U, sensitivities are calculated with MCSEN5 code [17]. For this analysis, MT91 is chosen because it is the main inelastic channel between 2 to 10 MeV.

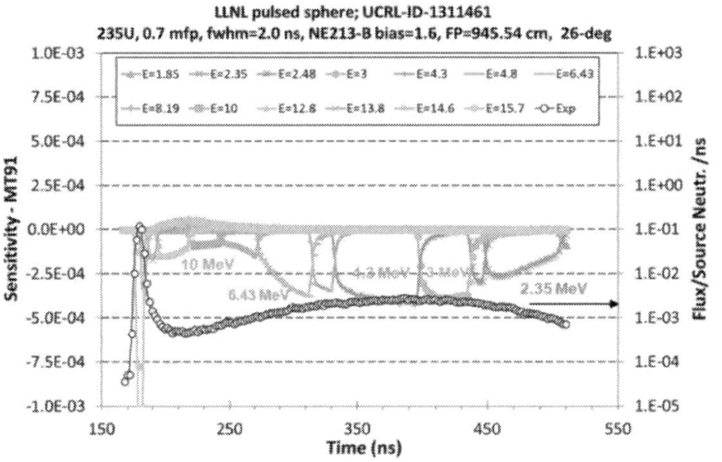

Fig. 1. Sensitivity coefficient for the ^{235}U continuum inelastic cross section (MT91) in the LLNL/235U pulsed sphere. Sensitivity calculation performed with MCSEN5 code.

Fig.2 shows the results of the BMC technique for ^{235}U adjustment. The ^{235}U random files are obtained from TENDL-2014, a total number of 5000 samples were used in this work. This number seems enough to reach an adequate convergence in BMC [18].The high sensitivity of the fission cross section in Godiva shows an increase in fission around +1.4% if only Godiva is used in the adjustment. LLNL-pulsed sphere goes in the opposite direction, with -1.0%. However, the high-sensitivity in MT91 for the pulsed sphere gives an increase up to +2.5%. These changes are dominant in the case of a jointly adjustment.

BMC technique also predicts "a posteriori" covariances. The Godiva experiment leads to a strong reduction of the uncertainty in fission cross-section of around 1.5%. The LLNL-pulsed sphere only changes the inelastic uncertainty with a reduction of the uncertainty in 0.6%. In addition, the Godiva experiment leads to a cross-correlation between fission and nubar as a consequence of the criticality constraint (see Fig.3-right). However, no cross-correlation is seen for the LLNL-pulsed sphere (see Fig.3 left).

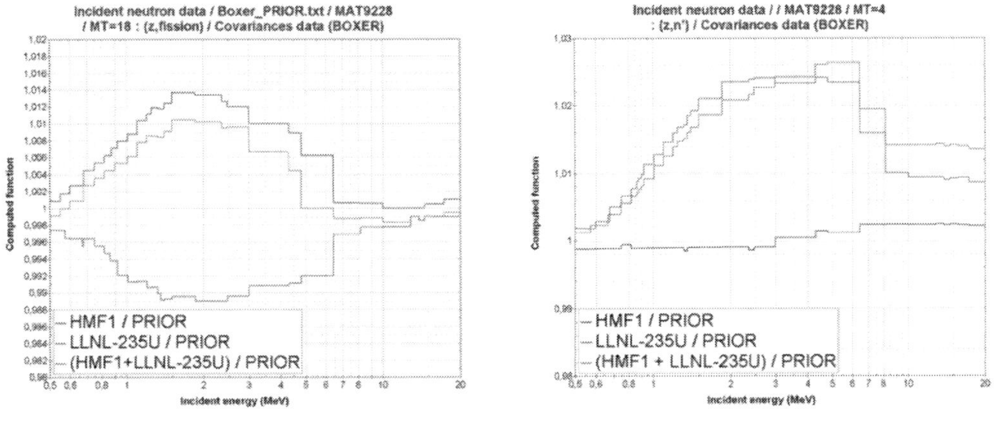

Fig. 2. Results for BMC adjustment, fission (left) and total inelastic (right) cross-sections. The figures show the predicted posterior values using experimental adjustments with Godiva(HMF1), LLNL-pulsed sphere or both benchmarks together divided by prior cross-sections.

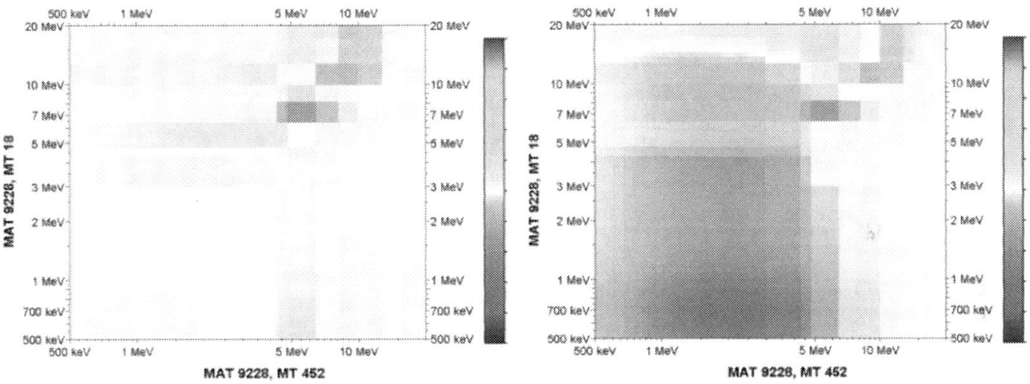

Fig. 3. Posteriori correlation matrix using LLNL-^{235}U pulsed sphere (left) and Godiva(right).

In Table 2, prior and posterior results for k_{eff} in criticality, and chi-squared in shielding case are shown. It can be concluded that a joint constraint is important to get a better and reliable adjustment of nuclear data and for avoiding compensation effects.

Table 2. Results of BMC adjustment for Godiva and LLNL/235U pulsed sphere Benchmarks. Calculations are performed with MCNP-6.1. The k_{eff} statistical uncertainty < 10pcm.

	HMF1/Godiva		**LLNL/235U -0.7mfp pulsed sphere**
Margin	**k_{eff} (mean)**	**Δk_{eff} (1σ) due to nuclear data unc.**	**Total χ^2**
Prior	0.99504	0.01119	3.52
A posteriori …only HMF1	0.99992	0.00100	3.69
A posteriori …only LLNL-^{235}U	0.99146	0.01069	3.13
A posteriori ...Both HMF1 & LLNL-^{235}U	0.99985	0.00099	3.37

4 Conclusion

Two different examples have been used to demonstrate the importance of Monte Carlo techniques in the procedure of evaluation of nuclear data. First, reactor core measurements have been used to perform an UQ assessment which can help to assess nuclear data trends and determine the main contributors and uncertainty targets for the evaluation procedure. The second example has shown the importance of selection of benchmarks in the Bayesian approach avoiding as much as possible compensations in the "a posteriori" evaluated data. Benchmarks should be selected focusing on an individual isotope, energy range, reaction type and different constraints.

References

1. W. Wieselquist, M. Williams, D. Wiarda, M. Pigni, U. Mertyurek, NUREG/CR-7249 (2018)
2. A. Santamarina, P. Blaise, N. Dos Santos, C. Vaglio, C. De Saint-Jean, PHYSOR 2014. Kyoto, Japan, September 28 – October 3, 2014
3. A.J. Koning, D. Rochman, Annals of Nuclear Energy **35** (2008) 2024–2030
4. A. Hoefer, O. Buss, M. Hennebach, M. Schmid and D. Porsch, Annals of Nuclear Energy **77** (2015) 514.
5. L. Fiorito, G. Zerovnik, A. Stankovskiy, G. Van den Eynde, P.E. Labeau, Annals of Nuclear Energy **101** (2017) 359–366.
6. W. Zwermann, L. Gallner, M. Klein, I. Pasichnyk, A. Pautz, and K. Velkov, EPJ Web of Conferences **42** (2013).
7. O. Cabellos, E. Castro, C. Ahnert, C. Holgado, Nuclear Engineering and Tech., **46** No.3 (2014)
8. C. Ahnert, J.M. Aragonés, Nucl. Technol., **69**, 350 (1985)
9. D.L. Smith, Proceed. of the 8th International Topical Meeting on Nuclear Applications and Utilization of Accelerators, Pocatello, July 29 - August 2, 2007, p. 736.
10. D.J. Siefman, M. Hursin, P. Grimmy, A. Pautz, Conference: M&C 2017 - International Conference on Mathematics & Computational Methods Applied to Nuclear Science & Engineering, Jeju, South Korea (2017)
11. A.J. Koning, Nuclear Data Sheets **123** (2015) 207–213
12. C. De Saint Jean, P. Archier, E. Privas, G. Noguere, B. Habert, P. Tamagno, Nuclear Data Sheets **148** (2018) 383–419
13. D. Rochman, E. Bauge, A. Vasiliev, H. Ferroukhi and G. Perret, EPJ-N, **4** (2018) 7.
14. G. Schnabel, H. Leeb, EPJWeb of Conferences **111**, 09001 (2016)
15. H. Sjöstrand, P. Helgesson, G. Schnabel, WONDER 2018, Aix-en-Provence, France. October 8-12 (2018)
16. I. Kodeli, L. Plevnik, Progress in Nuclear Energy **106** (2018) 215–230
17. R.L. Perel, J.J. Wagschal, Y. Yeivin, Nuc. Sci. and Eng., **124** (1), 197–209 (1996)
18. D.A. Rochman, E. Bauge, A. Vasiliev, H, Ferroukhi, G. Perret, EPJ Nuclear Sci. Technol. **4**, 7 (2018)

AUTHOR INDEX

AUTHOR INDEX

AUTHOR INDEX

AUTHOR INDEX